R. J. Wijngaarden
A. Kronberg
K. R. Westerterp

Industrial Catalysis

Optimizing Catalysts and Processes

WILEY-VCH

R. J. Wijngaarden
A. Kronberg
K. R. Westerterp

Industrial Catalysis

Optimizing Catalysts and Processes

WILEY-VCH

Weinheim · New York · Chichester · Brisbane · Singapore · Toronto

Dr. ir. Ruud J. Wijngaarden
Shell Chemical Company
Geismar Plant
P.O. Box 500
7594 Highway 75
Geismar, LA 70734
USA

Dr. A. Kronberg
Universitiy of Twente
Faculty of Chemical Technology
P.O. Box 217
NL-7500 AE Enschede
The Netherlands

Prof. dir. ir. K. R. Westerterp
Universitiy of Twente
Faculty of Chemical Technology
P.O. Box 217
NL-7500 AE Enschede
The Netherlands

Library of Congress Card No. applied for
A CIP catalogue record for this book is available form the British Library.

Die Deutsche Bibliothek - CIP-Einheitsaufnahme
Wijngaarden, Ruud J.: Industrial Catalysis – Optimizing Catalysts and Processes /
R. J. Wijngaarden ; A. Kronberg ; K. R. Westerterp. - Weinheim ; New York ;
Chichester ; Brisbane ; Singapore ; Toronto: WILEY-VCH, 1998
ISBN 3-527-28581-4

Printed on acid-free and chlorine-free paper.

Composition: OLD-Satz digital, D-69239 Neckarsteinach
Printing: Strauss Offsetdruck, D-69509 Mörlenbach
Bookbinding: Wilhelm Osswald & Co., D-67933 Neustadt

Printed in the Federal Republic of Germany.

Preface

Catalysis is the acceleration of a chemical reaction by a small quantity of a substance, which may take part in the reaction, but in the end is not changed by the reaction. This alien substance can be dissolved in the reaction mixture or one of its components, which is referred to as homogeneous catalysis. It can also form its own separate phase, yielding heterogeneous catalysis. Homogeneous and heterogeneous catalysis are equally important from an industrial point of view. Homogeneous catalysis will not be the subject of this book. We will focus on heterogeneous catalysis, where the catalyst is a porous material itself or is impregnated in such material.

In heterogeneous catalysis reactants have to be transported to the catalyst and (if the catalyst is a porous, solid particle) also through the pores of the particle to the active material. In this case all kinds of transport resistance's may play a role, which prevent the catalyst from being fully effective in its industrial application. Furthermore, because appreciable heat effects accompany most reactions, heat has to be removed from the particle or supplied to it in order to keep it in the appropriate temperature range (where the catalyst is really fully effective). Furthermore, heterogeneous catalysis is one of the most complex branches of chemical kinetics. Rarely do we know the compositions, properties or concentrations of the reaction intermediates that exist on the surfaces covered with the catalytically effective material. The chemical factors that govern reaction rates under these conditions are less well known than in homogeneous catalysis. Yet solid catalysts display specificities for particular reactions, and selectivity's for desired products, that in most practical cases cannot be equaled in other ways. Thus use of solid catalysts and the proper (mathematical) tools to describe their performance are essential.

The selection of a solid catalyst for a given reaction is often still empirical and based on prior experience or analogy. However, at the same time there are many aspects of this complex situation which are really well understood. For example we know how the true chemical kinetics, which are an intrinsic property of the catalyst, and all the many aspects of transport of material and heat around the catalytic particles interact. The chemist searching for new and better catalysts should always consider these physical factors, for they can be brought under control, and often in this way definite gains can usually be made both in activity and in selectivity.

Considerations based on the known physical phenomena can guide the choice of catalyst porosity and porous structure, catalyst size and shape and reactor type and size. These considerations apply both to laboratory as well as to large-scale operations. Many comprehensive reviews and good books on the problem of reactor design are available in the literature. The purpose of this book is to teach the reader the mathematical tools that are available for calculating interaction between the transport phenomena and true chemical kinetics, allowing optimization of catalyst performance. The discussed theories are elucidated with examples to provide training for application of the mathematics.

RJW, AEK, KRW

Contents

Symbols and Abbreviations

a	amount of gas adsorbed	mol
a	specific contact area between reaction and nonreaction phase per unit reactor volume	$m^2\,m^{-3}$
a_J	chemical activity of component J	$mol\,m^{-3}$
a_m	amount of gas adsorbed in monolayer	mol
$A, B, P, Q, \ldots$	component $A, B, P, Q, \ldots$	
An	modified Arrhenius number	
An_0	zeroth Aris number (becomes equivalent to $1/\eta^2$ for low values of η)	
An_1	first Aris number (becomes equivalent to $1\text{-}\eta^2$ for values of η close to one)	
A_p	external surface area catalyst pellet	m^2
c_p	heat capacity at constant pressure, per unit mass	$J\,kg^{-1}\,K^{-1}$
C	total molar concentration	$mol\,m^{-3}$
C_A	molar concentration of species A	$mol\,m^{-3}$
$\overline{C}_A$	concentration of component A in well-mixed fluid bulk	$mol\,m^{-3}$
$D_{AP}, D_{AD_i}, \ldots$	binary diffusivities for system $A\text{-}P$, $A\text{-}D_i, \ldots$	$m^2\,s^{-1}$
D^e_{ij}	$= (\varepsilon_p/\gamma_p)D_{ij}$ effective binary diffusivity for system $i\text{-}j$ $m^2\,s^{-1}$	
D_{Ak}	Knudsen diffusivity of component A	$m^2\,s^{-1}$
$D_{A,surf}$	surface diffusivity of component A	$m^2\,s^{-1}$
D_e	effective diffusion coefficient	$m^2\,s^{-1}$
d	molecular diameter	m
d_e	equivalent pore diameter	m
d_p	pellet diameter	m
E_a	energy of activation	$J\,mol^{-1}$
h	longitudinal coordinate	m
H	height of the catalyst pellet	m
$(-\Delta H_{ads})$	adsorption enthalpy	$J\,mol^{-1}$
$(-\Delta H)$	reaction enthalpy (exothermic reactions $(-\Delta H) > 0$; endothermic reactions $(-\Delta H) < 0$)	$J\,mol^{-1}$
J	mole flux through the cross-sectional area of the catalyst pellet (see Appendix E)	$mol\,m^{-2}s^{-1}$
J_i	mole flux of species i through the cross-sectional area of the catalyst pellet	$mol\,m^{-2}s^{-1}$
J^*	mole flux through the cross-sectional area of the catalyst pores (see Appendix E)	$mol\,m^{-2}s^{-1}$
k	reaction rate constant (dimensions depend on kinetic expression)	
k	reaction rate constant per unit outer surface of the pellet (dimensions depend on kinetic expression)	
k_g	mass transfer coefficient	$m\,s^{-1}$
K	adsorption constant for Langmuir-Hinshelwood kinetics	$m^3\,mol^{-1}$
Kn	$= \lambda/d_e$ Knudsen number	
Kn^*	Knudsen number, modified for multicomponent systems	
L	length (thickness) of a catalyst pellet	m
L	reactor length	m
m	reaction order for the component in excess in bimolecular reactions	

m	mass	kg
m	distribution coefficient	
M	molar mass	$kg\,mol^{-1}$
$\langle M \rangle$	average molar mass of the gas mixture	$kg\,mol^{-1}$
n	reaction order of reactant not in excess for bimolecular reactions	
n	number of moles	mol
N	conversion rate	$mol\,s^{-1}$
N	molar flux at pellet external surface	$mol\,m^{-2}\,s^{-1}$
N_A	Avogadro number (6.02×10^{23} molecules/mole)	
P	total pressure	Pa
P_A	partial pressure of species A	Pa
P_0	adsorbent saturation pressure	Pa
$\bar{P}$	mean pressure across the porous particle	Pa
r	radial coordinate	m
r_e	equivalent pore radius	m
R	gas constant	$J\,mol^{-1}\,K^{-1}$
R	conversion rate per cubic metre catalyst volume	$mol\,m^{-3}\,s^{-1}$
R	distance between centre and outer surface of the catalyst pellet for an infinite slab, infinite cylinder or sphere	m
R_{ads}	adsorption rate	$mol\,m^{-2}\,s^{-1}$
R_{des}	desorption rate	$mol\,m^{-2}\,s^{-1}$
R_i	inner radius of a ring-shaped catalyst pellet	m
R_u	outer radius of a ring-shaped catalyst pellet	m
R_R	$= \Phi_R/\Phi_v$ recycle ratio	
$\lvert R \rvert$	conversion rate per cubic metre catalyst volume	$mol\,m^{-3}\,s^{-1}$
$\tilde{R}$	reaction rate per unit pore surface	$mol\,m^{-2}\,s^{-1}$
R	reaction rate per unit external pellet surface	$mol\,m^{-2}\,s^{-1}$
$\langle R \rangle$	average reaction rate per unit pellet volume	$mol\,m^{-3}\,s^{-1}$
s_0	surface area occupied by a molecule of an adsorbed species	m^2
S	internal surface area per unit catalyst volume	$m^2 m^{-3}$
S_g	internal surface area per unit gram of catalyst	$cm^2 g^{-1}$
S_R	reactor cross-section	m^2
S_t	total internal surface area of the catalyst	m^2
Sc^*	Schmidt number, modified for multicomponent systems	
T	temperature	K
$\bar{v}$	mean molecular speed	$m\,s^{-1}$
v_i	velocity of species i	$m\,s^{-1}$
V	characteristic number for viscous flow inside catalyst pores	
V_A	$= \frac{r_P^2}{8\eta_g} \cdot \frac{P}{D_{Ak}} \cdot \frac{\nabla P}{\nabla(\kappa_A P)}$ (Equation A.105)	
V_g	pore volume per unit gram of catalyst	$cm^3\,g^{-1}$
V_p	volume of the catalyst pellet	m^3
W	width of a catalyst pellet	m
x	rectangular coordinate	m
x_A	mole fraction of species A	
X_0	$= V_p/A_p$ characteristic, shape-generalized dimension	m
y	rectangular coordinate	m
Y_p	yield	
z	rectangular coordinate	m

Greek symbols

α — dimensionless adiabatic temperature rise inside a catalyst pellet

$$\alpha = \frac{T_m - T_s}{T_s} = \frac{(-\Delta H) D_{e,A} C_{A,s}}{\gamma_P T_s} \quad \text{(Equations 6.15 and 7.1)}$$

β — measure for the influence of the component in excess for bimolecular reactions

$$\beta = \nu_B \frac{D_{e,A}}{D_{e,B}} \frac{C_{A,s}}{C_{B,s}} \quad \text{(Equation 7.30)}$$

γ — measure for the dependence of the effective diffusion coefficient on the gas composition and pressure

$$\gamma = \frac{\sqrt{\nu_P} + 1 - \psi}{1 + Kn^* + \psi\left(\sqrt{\nu_P} - 1\right)\kappa_{A,s}} \cdot \left(\sqrt{\nu_P} - 1\right)\kappa_{A,s} \quad \text{(Equation 7.78)}$$

γ_J — activity coefficient or fugacity coefficient of component J

γ_P — tortuosity of the catalyst pores

$\gamma(\rho,\omega)$ — dimensionless concentration

$$\gamma(\rho,\omega) = \frac{C_A(\rho,\omega)}{C_{A,s}} \quad \text{(Equation A.37)}$$

Γ — geometry factor

$\delta_{\beta,0}$, $\delta_{\beta,1}$ — relative error in η and $1\text{-}\eta$ respectively, introduced if for a bimolecular reaction the concentration of the component in excess is taken as constant

$\delta_{\gamma,0}$, $\delta_{\gamma,1}$ — relative error in η and $1\text{-}\eta$ respectively, introduced if the dependence of the effective diffusion coefficient on the gas composition and pressure is neglected

$\delta_{\xi,0}$, $\delta_{\xi,1}$ — relative error in η and $1\text{-}\eta$, respectively, introduced if intraparticle temperature gradients are neglected

$\delta^{2\text{-}3}$ — relative error, introduced if instead of the three-parameter model the two-parameter model is used (see Equation 7.7)

$\delta^{2\text{-}3}_{max}$ — maximum value of $\delta^{2\text{-}3}$; i.e. the value of $\delta^{2\text{-}3}$ for $C_A = 0$

Δ — relative deviation between η and the approximation η for $\eta < ½$ and $1\text{-}\eta$ for $\eta > ½$

Δ_{max} — maximum value of Δ for the interval η e [0,1]

$\Delta\sigma$ — relative deviation between σ and the approximation $\bar{\sigma}$

ε — $= E_a/RT_s$ dimensionless energy of activation or Arrhenius number

ε_p — porosity of the catalyst pellet

η — effectiveness factor

$\tilde{\eta}$ — approximation for η

$\tilde{\tilde{\eta}}$ — approximation for $\tilde{\eta}$

ζ — relative degree of conversion

η_1	effectiveness factor for first order kinetics and arbitrary catalyst geometries	
η_g	dynamic viscosity of the gas mixture	Pa s
η_m	modified effectiveness factor for ring-shaped pellets	
θ	degree of occupation of the surface	
θ	dimensionless temperature	
θ_m	$=\sqrt{An_0}$ the generalized modified Thiele modulus	
ι	$= R_i/R_u$ dimensionless inner radius of a ring-shaped catalyst pellet	
λ	$= H/R_u$ dimensionless height of a ring-shaped catalyst pellet	
λ	molecular mean free path	m
λ_p	heat conductivity of the catalyst pellet	$W\,m^{-1}K^{-1}$
μ	viscosity	Pa s
μ	first moment of the response curve	
ν_B	stoichiometric coefficient, moles of component B per ν_A moles of key reactant A	
ν_g	kinematic viscosity of the gas mixture	$m^2\,s^{-1}$
$\Delta\nu$	net yield of molecules per molecule of key reactant converted (for bimolecular reactions $\Delta\nu = \nu_P + \nu_Q - \nu_B - 1$)	
ξ	$= \alpha\varepsilon$, measure for the importance of intraparticle temperature gradients	
ξ	parameter according to Karanth, Koh and Hughes	–

$$\xi = \frac{1}{\nu_B}\frac{D_{e,B}}{D_{e,A}}\frac{C_{B,s}}{C_{A,s}} - 1 = \frac{1-\beta}{\beta} \quad \text{(Equation 6.20)}$$

ρ	$= r/R_u$ dimensionless radial coordinate	
ρ_g	density of the gas mixture	$kg\,m^{-3}$
ρ	density of the reaction mixture	$kg\,m^{-3}$
σ	$= (P-P_s)/P_s$ dimensionless intraparticle pressure	
$\bar{\sigma}$	approximation for σ	
σ_P	selectivity	
σ'_P	differential selectivity or selectivity of the catalyst	
σ^2	second central moment of the response curve	s^2
τ_L	average residence time	s
ϕ	$= H/(R_u-R_i)$ measure for the geometry of a ring-shaped catalyst pellet	
ϕ_1	shape-generalized Thiele modulus for first-order kinetics of Aris	

$$\phi_1 = X_o\sqrt{\frac{k}{D_e}} \quad \text{(Equation 6.2)}$$

ϕ_{hc} Thiele modulus for a hollow cylinder

$$\phi_{hc} = R_u\sqrt{\frac{k}{D_e}} \quad \text{(Equation A.40)}$$

ϕ_p modified Thiele modulus for a plate or an infinite slab

$$\phi_P = \frac{1}{\eta} \quad \text{for low } \eta \text{ (Equation 6.6)}$$

ϕ_s modified Thiele modulus for a sphere

$$\phi_s = \frac{3}{\eta} \quad \text{for low } \eta \text{ (Equation 6.5)}$$

ϕ_T Thiele modulus according to Thiele

$$\phi_T = R\sqrt{\frac{R(C_{A,s})}{D_{e,A}C_{A,s}}} \quad \text{(Equation 6.1)}$$

Φ_v	volume flow rate	$m^3\,s^{-1}$
Φ_R	recycle volume flow rate	$m^3\,s^{-1}$
χ	defined by Equation 7.37	

$$\chi = \frac{\eta^{\sim}|_{\beta} - \eta^{\sim}|_{\beta=0}}{\eta^{\sim}|_{\beta=1} - \eta^{\sim}|_{\beta=0}}$$

ψ defined by Equation 7.75

$$\psi = \frac{(Kn^*)^2}{(Kn^*)^2 + \{(\nu_P - 1)\kappa_{A,s} + \sqrt{\nu_P} + 2\}Kn^* + 1/2.(\sqrt{\nu_P} + 1)}$$

ψ defined by Equation 7.99

$$\psi = \nu_P\sqrt{\frac{M_P}{M_A}} + \nu_Q\sqrt{\frac{M_Q}{M_A}} - \nu_B\sqrt{\frac{M_B}{M_A}} - 1$$

ω $= h/H$, dimensionless longitudinal coordinate

ω mass fraction

Subscript

$A, B, P, Q, \ldots$	for component $A, B, P, Q, \ldots$
g	gas
l	liquid
m	at the centre of the catalyst pellet
p	for the catalyst pellet
s	solid
s	at the outer surface of the catalyst pellet
β	modified for bimolecular reactions
0	initial or inlet condition
$+$	forward reaction
$-$	reverse reaction

Superscripts

H	in the longitudinal direction
R	in the radial direction
*	modified for non-diluted gases
+	modified for anisotropy

Mathematical functions and symbols

$erf\,(x)$ error function

$$erf(x)=\frac{2}{\sqrt{\pi}}\int_0^x e^{-y^2}dy$$

$F(\phi,K)$ elliptic integral of the first kind

$$F(\phi,k) = \int_0^\phi \frac{1}{\sqrt{1-k^2\sin^2\theta}}\,d\theta$$

$I_0(x), I_1(x)$ modified Bessel functions of the first kind of zeroth and first order, respectively

$$I_n(x)=(x/2)^n\sum_{i=0}^{\infty}\frac{(x/2)^{2i}}{i!(n+i)!}$$

$K_0(x), K_1(x)$ modified Weber functions (Bessel functions of the second kind of zeroth and first order, respectively

$$K_n(x)=1/2.\,(x/2)^n\cdot\sum_{i=0}^{n-1}\left(\frac{(n-i-1)!}{i!}\cdot\left(-x^2/4\right)^i\right)+$$
$$(-1)^{n+1}\cdot In\,(x/2)\cdot I_n(x)+$$
$$(-1)^n\cdot 1/2.\,(x/2)^n\sum_{i=0}^{\infty}\{(\psi(i+1)+\psi(n+i+1))\}\cdot$$
$$\frac{(x/2)^{2i}}{i!\cdot(n+1)!}\}$$

with the Psi or digamma function

$$\psi(m) = -\gamma+\sum_{i=1}^{m}\frac{1}{i}$$

and γ Euler's constant

$$\gamma = \lim_{m\to\infty}\left\{\sum_{i=1}^{m}\frac{1}{i} - In(m)\right\} \approx 0.5772156649$$

$\approx$	is approximately equal to
$\sim$	is equivalent to
$\rightarrow$	approaches
$\downarrow$	approaches from above
$\uparrow$	approaches from below
$\forall$	universal kwantor
e	is an element of
∇	gradient of a parameter field, or divergence of a gradient field
∇^2	Laplacian operator

1 Introduction

1.1 Introduction

Catalysis is the acceleration of a chemical reaction by a small quantity of a substance, which may take part in the reaction, but in the end is not changed by the reaction. This alien substance can either be mixed with or dissolved in the reaction mixture or one of its components and forms one single phase with it (we speak of **homogeneous catalysis** in this case), or it can form its own separate phase. In the last case we speak of **heterogeneous catalysis**. Both methods of homogeneous as well as heterogeneous catalysis are equally important from an industrial point of view. Homogeneous catalysis is not the subject of this book; we focus on heterogeneous catalysis, where the catalyst is a porous material itself or is impregnated in such material.

In heterogeneous catalysis the reactants first have to be transported to the catalyst and – if the catalyst is a porous, solid particle – also through the pores of the particle to the active material, that truly enhances the reaction rate. In this case all kinds of transport resistance's may play a role, which prevent the catalyst from being fully effective in its industrial application. Reactants have to transported to the catalyst particle and the products away from it. Furthermore, because appreciable heat effects accompany most reactions, heat has to be removed from the particle or supplied to it in order to keep it in the appropriate temperature range (where the catalyst is really fully effective). Heterogeneous catalysis, in which gases and liquids are contacted with solids so the reactions are accelerated, is one of the most complex branches of chemical kinetics. Rarely do we know the compositions, properties or concentrations of the reaction intermediates that exist on the surfaces covered with the catalytically effective material. The chemical factors that govern reaction rates under these conditions are less well known than in homogeneous catalysis. Yet solid catalysts display specificities for particular reactions, and selectivities for desired products that, in most practical cases, cannot be equaled in any other way. Thus use of solid catalysts is essential.

The selection of a solid catalyst for a given reaction is to a large extent still empirical and based on prior experience or analogy. However, there are now many aspects of this complex situation that are quite well understood. For example we know how the true chemical kinetics, which are an intrinsic property of the catalyst, and all the many aspects of transport of material and heat around the catalytic particles, interact. In other words, the physical characteristics around the catalyst system and their effects on catalyst performance are well known today. The chemist searching for new and better catalysts should always consider these physical factors, for they can be brought under control, and often in this way definite gains can usually be made both in activity and in selectivity. Further, this knowledge enables us to avoid some common errors of interpretation of apparent reaction rates, as observed in experiments.

Considerations based on the known physical phenomena can guide the choice of catalyst porosity and porous structure, catalyst size and shape and reactor type and size. These considerations apply both to the laboratory as well as to large-scale operations. Many comprehensive reviews and good books on the problem of reactor design are available in the literature. The basic theory for porous catalysts is summarized in this book and simple rules are set forth to aid in making optimum choices to obtain fully effective catalyst particles, which give the best performance from an economic point of view.

1.2 Catalysis in an Industrial Reactor

Consider a long, slender cylinder filled with catalyst pellets and a flow of reactants entering at one end. The reactant stream may be a gas, a liquid or a mixture of them. The flow occurs in the interstices between the catalyst granules. The rate of flow, relative to the catalyst particle, is the prime factor, together with temperature and catalyst properties, in determining the amount of conversion. The velocity and turbulence of the flow determine how rapidly molecules are carried from the fluid phase to the exterior surfaces of catalyst pellets. Rapid transfer from fluid to solid outer surface is obtained with highly turbulent flow, which means a highly irregular flow pattern with momentary velocities strongly deviating from the main flow direction. High turbulence is obtained with high flow velocities, large particles and low viscosities. The amount of turbulence also has a strong influence on the rate of heat transfer between the catalyst pellet and the fluid, and also to the wall of the cylinder.

Flow conditions may cause a more or less broad distribution of residence times for individual fluid packets or molecules. Effects of variable residence times have to be taken into account in the design and operation of large industrial reactors: with adequate precautions the chemical engineer can prevent the undesirable effects of a residence time distribution, or utilize them.

Once a molecule has arrived at the exterior boundary of a catalyst pellet it has a chance to react. Most of the active surface, however, will be inside the pellet and will only be reached after diffusion for an appreciable distance. Usually we need a much larger active surface than is available on the exterior surface of a pellet. For practical particle sizes of porous catalysts, the interior surface greatly exceeds the exterior. For example, with 3 mm diameter pellets of an interior surface area of 100 $m^2 g^{-1}$, the interior total surface area is 100 000 times larger than the exterior surface of the pellets. There are, however, some catalysts so active that reaction on the outer surface of a nonporous catalyst of a convenient size suffices for obtaining useful rates. The platinum gauze used for the ammonia oxidation is an example. Such nonporous catalysts are ideal if they can be used, for in this case we are free from the usually harmful diffusion effects.

When the reaction mixture diffuses into a porous catalyst, simultaneous reaction and diffusion have to be considered when obtaining an expression for the overall conversion rate. The pore system is normally some kind of complex maze and must be approximated to allow mathematical description of mass transport inside the particles, such as a system of uniform pores. It is important to note that catalyst pellets can be of many structural types, which all have an influence on the degree of diffusion limitation of the overall conversion rates. An extreme type is the gel catalyst with tiny pores, which have quite low effective diffusion coefficients: silica, alumina and zeolites are examples of this structure. Another extreme type of pellet is an aggregate of quite large particles cemented together at points of contact. The channels in the ultimate particles may be as big as 100 μm in diameter. A third type is an aggregate of aggregates, where there are large pores leading into the interior and smaller pores leading to the major portion of the active surface. Such a structure is close to ideal from the standpoint of low diffusion resistance and high internal surface area.

If diffusion is fast relative to the rate of reaction, then all the interior surfaces of a catalyst pellet are bathed in fluid of the same composition. This composition will depend on the position of the pellet in the reactor; that is, on the distance from the inlet. But if the diffusion rate is of the same order of magnitude as the reaction rate, or slower, the concentration of the reactants and product will vary within the pores: we then speak of a diffusion limitation of the transport of reactants, generally resulting in a lower conversion rate. Diffusion limitation can have quite profound effects. Some of these are listed below:

- the apparent activity of the catalyst is generally lowered;
- the apparent order of the reaction may be changed;
- the selectivity may be altered markedly;
- the temperature gradient within a pellet may become large;
- inner portions of pellets may be more or less rapidly deactivated than outer portions, or vice versa.

All these effects are significant in our interpretation of a catalytic experiment or the behavior of an industrial reactor. If we do not recognize the role of diffusion, we may be badly misled in our interpretations. However, with the aid of knowledge about diffusion, we can sometimes improve our results over what they otherwise would have been.

Temperature differences between catalyst and fluid will exist even without a diffusion limitation. For an exothermic reaction the entire catalyst pellet must be somewhat hotter than the surrounding, flowing fluid, as only this temperature difference serves as the driving force for removal of the heat of reaction. This excess temperature is usually only some tenths of a degree to a few degrees centigrade. When reaction rates are extremely high compared to the heat removal capabilities, the temperature difference between pellet and fluid may become very large and can be around the local **adiabatic temperature rise** of the reaction. This adiabatic temperature rise ΔT_{ad} is an extremely important property of a reaction and is given by

$$\Delta T_{ad} = (-\Delta H)_A C_A / \rho c_p \tag{1.1}$$

Where $(-\Delta H)_A$ is the heat of reaction evolved, C_A the inlet concentration of the reactant in the fluid and ρc_p the sensible heat per unit of volume of the reaction mixture. The value of ΔT_{ad} should always be determined first, before studying a reaction. When it is low, heat effects are relative unimportant, for high values heat removal is a must in catalyst utilization.

1.3 Catalytic Reactors

Industrial catalytic reactors exhibit a great variety of shapes, types and sizes. It is not our aim here to discuss all possibilities; a survey of the most important reactor types is given by Ullmann [1]. In general, heterogeneous catalytic reactors can be divided in two categories, depending on the size of the catalyst particles, large and small.

1.3.1 Large Particle Catalyst

Large particle catalysts can be kept stationary, so that packed in a bed they can be kept in the reactor and the reaction mixture passes through the bed of particles. Of course, the aim is to keep the catalyst charge as long as possible in the reactor, say for many years. This method avoids all the trouble of eliminating the catalyst from the product stream coming out of the reactor. Whatever the shape of the particles, their size in industrial reactors is usually larger than 2 mm. In Figure 1.1 some examples are given of reactors with catalyst beds. In Figure 1.1(a) the common adiabatic packed bed reactor is shown. This reactor is used for single-phase reaction mixtures, either gases or liquids with moderate heat effects. Also, systems with multiple beds with cooling in between the beds are frequently used, such as for NH_3 and CH_3OH production. For reactions with high heat effects the cooled tubular reactor is used (Figure 1.1(b)). Here the reactor consists of a large number of parallel tubes, which are cooled with a coolant flowing around the outside of the tubes. To maintain the catalyst bed in the tubes as isothermal as possible, only a small number of particles across a tube diameter is permitted, say between 3 and 20 particles. Depending on the required reactor capacity the number of tubes varies between from 30 to 30 000. Adiabatic packed bed and cooled tubular reactors can be used both for gaseous and liquid reactor feeds.

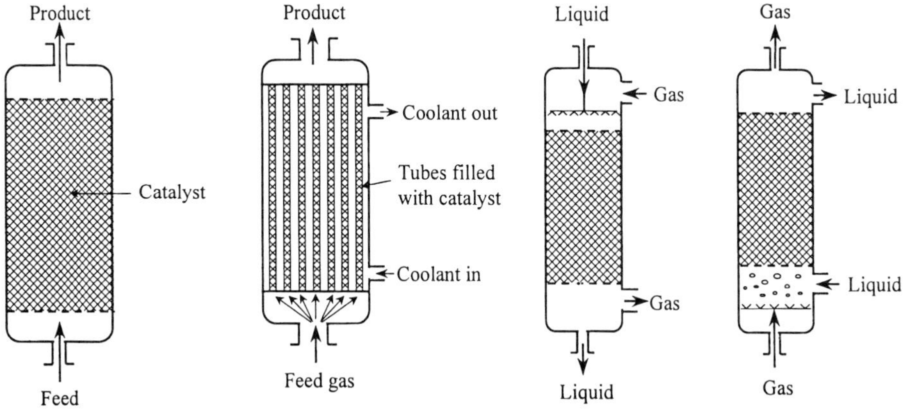

Figure 1.1 Examples of reactors with fixed catalyst beds: (a) adiabatic packed bed; (b) cooled tubular reactor; (c) cocurrent trickle bed reactor; (d) packed bubble column.

When a mixture of gas and liquid is to be fed to a packed bed reactor then, depending on the required residence times in the reactor, two reactor types are commonly used. In Figure 1.1(c) the trickle flow reactor is shown, in which the gas and the liquid streams are fed cocurrently at the top of the reactor. The liquid wets the catalyst particles and slowly trickles to the bottom through the bed. The gas dissolves in the liquid and is transported to the catalyst surface, where reaction takes place with reactants coming from the liquid phase. A counter current flow of gas and liquid can also be applied. Residence times for the liquid phase in industrial trickle flow reactors can be as high as 10-15 min. If much larger residence times are required for the liquid phase, the packed bed bubble column reactor in Figure 1.1(d) is often used. Here gas and liquid are both fed to the bottom of the reactor. The reactor is filled with liquid through which the gas bubbles slowly upwards. In these reactors liquid residence times of the order of hours can easily be achieved; the catalyst is fully wetted with the liquid phase.

1.3.2 Small Particle Catalyst

For small catalyst particles completely different reactor types are used. The catalyst is now suspended in the flowing reaction mixture and has to be separated at the reactor exit or is carried along with the fluid. Particle sizes are now from 10 μm up to 1 mm. In Figure 1.2 some of the common reactor types are shown. Figure 1.2(a) shows the fluid bed reactor, where the gaseous feed keeps the small catalyst particle in suspension. Catalyst carried over in the exit stream is separated, for example, in cyclones. For even shorter contact times, riser reactors are used in which the solid catalyst is transported in the gas stream. Fluid bed reactors are also used for feed mixtures of a liquid and a gas.

In cases when the feed stream is a liquid, which requires rather long residence times, the suspension bubble column or an agitated tank reactor is used (Figures 1.2(b) and 1.2(c)). Here, in the reactor exit, quite elaborate filtering systems are required to remove the catalyst from the liquid stream. In these reactors a gas generally is supplied, because these suspension reactors are mostly used for hydrogenations and oxidations.

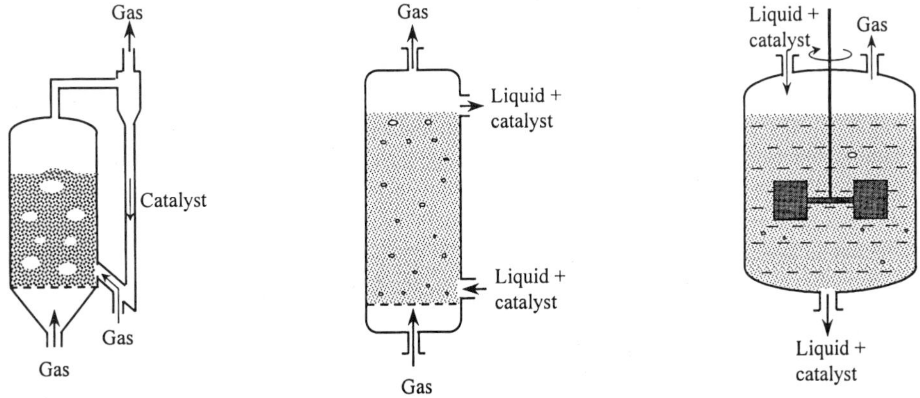

Figure 1.2 Common reactor types with moving catalyst beds: (a) fluid-bed reactor; (b) bubble column with suspended catalyst; (c) sparged stirred tank with suspended catalyst.

1.4 Characteristics of Reactor Performance

For the quantitative description of a reactor and the individual catalyst particles in it, an expression must be available for the chemical production rate [2]. The production R_A rate of component A, is defined as the number of moles of A produced per unit time and unit volume. If A is consumed in a chemical reaction, R_A is negative; if A is produced then R_A is positive. The production rate in mass units is found by multiplying R_A by the molar mass M_A.

For convenience, the concept of the chemical conversion rate $|R|$ is often used. $|R|$ is always positive when a reaction proceeds in the direction of the arrow in the reaction equation. The conversion rate is expressed in moles of **key reactant** consumed or produced per unit time and unit volume. When a reaction proceeds according to

$$\nu_A \mathrm{A} + \nu_B \mathrm{B} \rightarrow \nu_P \mathrm{P} + \nu_Q \mathrm{Q}$$

and A is chosen as key reactant, then

$$|R| = -R_A = -R_B \frac{\nu_A}{\nu_B} = R_P \frac{\nu_A}{\nu_P} = R_Q \frac{\nu_A}{\nu_Q} \tag{1.2}$$

The coefficients ν_A, ν_B, ν_P, and ν_Q are called the stoichiometric coefficients. Often, and also in this text, the value of ν_A for the key reactant is taken equal to one, so that ν_A/ν_B(etc.) can be taken as $1/\nu_B$ (etc.).

In this text, the conversion rate is used in relevant equations to avoid difficulties in applying the correct sign to the reaction rate in material balances. Note that the chemical conversion rate is not identical to the chemical reaction rate. The chemical reaction rate only reflects the chemical kinetics of the system, that is, the conversion rate measured under such conditions that it is not influenced by physical transport (diffusion and convective mass transfer) of reactants toward the reaction site or of product away from it. The reaction rate generally depends only on the composition of the reaction mixture, its temperature and pressure, and the properties of the catalyst. The conversion rate, in addition, can be influenced by the conditions of flow, mixing, and mass and heat transfer in the reaction system. For homogeneous reactions that proceed slowly with respect to potential physical transport, the conversion rate approximates the reaction rate. In contrast, for homogeneous reactions in poorly mixed fluids and for relatively rapid heterogeneous reactions, physical transport phenomena may reduce the conversion rate. In this case, the conversion rate is lower than the reaction rate.

For a single reaction as given above, the mass fractions of the reactants, ω_A and ω_B, decrease and those of the products, ω_P and ω_Q, increase as the reaction proceeds from left to right. For a closed system (no material added or withdrawn),

$$\omega_A + \omega_B + \omega_P + \omega_Q = \text{constant} = \omega_{tot}$$

For such a system the **relative degree of conversion** ζ can be introduced, which is a measure of the extent to which the reaction has proceeded. It can be defined quite generally as the fraction of the amount of a reactant, fed prior to and during reaction that has been converted. For a closed system, ζ_J can be expressed in terms of mass fractions as

$$\zeta_J = \frac{\omega_{J,0} - \omega_J}{\omega_{J,0}} \tag{1.3}$$

with $J = A, B$, respectively. Since, by definition, the total mass within a closed system m is constant, the relative degree of conversion can also be expressed in terms of the number of moles:

$$\zeta_J = \frac{\frac{\omega_{J,0} m}{M_J} - \frac{\omega_J m}{M_J}}{\frac{\omega_{J,0} m}{M_J}} = \frac{n_{J,0} - n_J}{n_{J,0}} \tag{1.4}$$

Here M_J is the molar mass of either A or B. If the total volume of the reaction mixture does not depend on the relative degree of conversion, the latter can be expressed relatively easily in terms of concentrations:

$$\zeta_J = \frac{\frac{n_{J,0}}{V} - \frac{n_J}{V}}{\frac{n_{J,0}}{V}} = \frac{C_{J,0} - C_J}{C_{J,0}}$$

where V is the volume of the reaction mixture. If this volume changes during the reaction, the equation becomes

$$\zeta_J = \frac{C_{J,0}\frac{m}{\rho_0} - C_{J,0}\frac{m}{\rho}}{C_{J,0}\frac{m}{\rho_0}} = \frac{C_{J,0} - \frac{\rho_0}{\rho} C_J}{C_{J,0}} \tag{1.5}$$

with ρ_0 being the density of the reaction mixture for zero conversion and ρ the density for a conversion ζ_J.

The chemical reaction rate is usually dependent on the molar concentrations of the reactants and not on their mass fractions, because it depends on the chance of collision of molecules. However, here the definition of ζ in terms of mass fractions is preferred, because it can readily be incorporated into mass balances. A definition in terms of moles or molar concentrations might invite the use of mole balances instead of mass balances. Since, contrary to conservation of mass, there is no such thing as conservation of moles (because one molecule might divide into several molecules, or several might condense into one), the use of mole balances is strongly dissuaded. More information concerning the definition of conversion can be found elsewhere [2].

In industrial chemical operations, a reactant A or reactants A and B very often react to form not only desired products, but also undesirable ones. Therefore, in chemical reactor technology the concepts of selectivity and yield are often used.

The **selectivity** σ_P is the ratio between the amount of desired product P obtained and the amount of key reactant A converted; both quantities are usually expressed such that σ_P ranges between 0 (no P formed) and 1 (all A converted to P). This definition leads to

$$\sigma_P = \frac{(\omega_P - \omega_{P,0})M_A}{(\omega_{A,0} - \omega_A)M_P\nu_P} \tag{1.6}$$

For constant density of the reaction mixture, this can be written as

$$\sigma_P = \frac{(C_P - C_{P,0})}{(C_{A,0} - C_A)\nu_P} \tag{1.7}$$

The yield η_P is the ratio between the amount of desired product P obtained and the amount that could be obtained if all of key reactant A were converted to P with 100% selectivity. Therefore, Y_P can be calculated from

$$Y_P = \sigma_P \zeta_A \tag{1.8}$$

Like σ_P, the yield Y_P usually varies between 0 and 1. It is high when both the selectivity and the relative degree of conversion are high; it is low when either of them is low.

If in the product separation section of the reactor, the key reactant A can be removed and recycled to the reactor, selectivity is the key factor in economical operation of the plant. However, if A cannot be recycled, then yield is the key factor.

References

1. Westerterp, K.R., Wijngaarden, R.J., Principles of Chemical Reaction Engineering and Plant Design, *Ullmann's Encyclopedia of Industrial Chemistry* (1992) Vol. B4 (5th edn.). Weinheim:VCH.
2. Westerterp, K.R., Van Swaaij, W.P.M., Beenackers, A.A.C.M. (1987) *Chemical Reactor Design and Operation*. Chichester: Wiley.

2 Kinetics

2.1 General

Elementary reactions are individual reaction steps that are caused by collisions of molecules. The collision can occur in a more or less homogeneous reaction medium or at the reaction sites on a catalyst surface. Only three elementary kinetic processes exist: mono-, bi-, and trimolecular processes. Of these, trimolecular processes are rarely found, because the chance of three molecules colliding at the same time is very small. Each elementary reaction consists of an activation of the reactants, followed by a transition state and decomposition of the latter into reaction products:

$$A(+B)\left[\underset{Deactivation}{\overset{Activation}{\leftrightarrow}}\begin{Bmatrix}Energized\\molecule\end{Bmatrix}\rightarrow\begin{Bmatrix}Transition\\state\\molecule\end{Bmatrix}\right]\rightarrow \text{products}$$

Energized and transition state molecules are unstable, and as such cannot be isolated.

For elementary reactions, the influence of the temperature and composition of the reaction mixture can be represented separately. The influence of temperature is accounted for by the reaction rate constant k, referred to as a constant because it does not depend on the composition of the reaction mixture. The temperature dependence of k is

$$k = k_{\infty} \exp\left\{-\frac{E_a}{RT}\right\} \tag{2.1}$$

Where E_a is called the activation energy of the reaction. Equation 2.1 is illustrated in Figure 2.1, where the reaction rate constant is plotted as a function of temperature (both

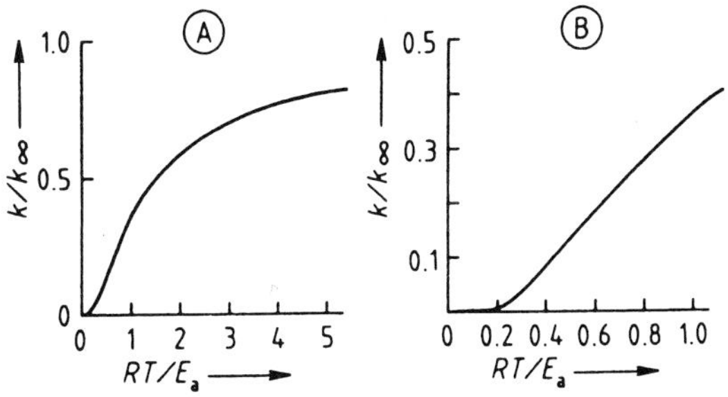

Figure 2.1 A) Temperature dependence of the reaction rate constant (note that as E_a/R ranges from 5000 K to 35 000 K; in practice only the region $RT/E_a < 1$ is of interest). B) Enlargement of the region $RT/E_a < 1$.

quantities in dimensionless form). The term E_a/R is sometimes referred to as the activation temperature. For most reactions the activation energy lies in the range 40-300 kJ mol^{-1}; hence the activation temperature E_a/R ranges from 5000 K to 35 000 K.

Therefore, for industrial purposes, only the range $RT/E_a < 1$ is of interest. This region is shown enlarged in Figure 2.1(b). As can be seen from Figure 2.1(a), although at extreme temperature k approaches k_∞, in practice it is several orders of magnitude lower.

The temperature increase required for doubling a reaction rate constant (ΔT_{double}) as a function of the temperature of the reaction mixture and activation energy is given in Figure 2.2. The higher the activation energy, the more temperature sensitive are the reactions. To double the reaction rate at T = 100 °C, for example, a temperature increase of 3 °C (E_a = 300 kJ mol^{-1}) to 20 °C (E_a = 50 kJ mol^{-1}) is required. Also note that in the higher temperature range much larger temperature changes are necessary to double the reaction rate at constant activation energy.

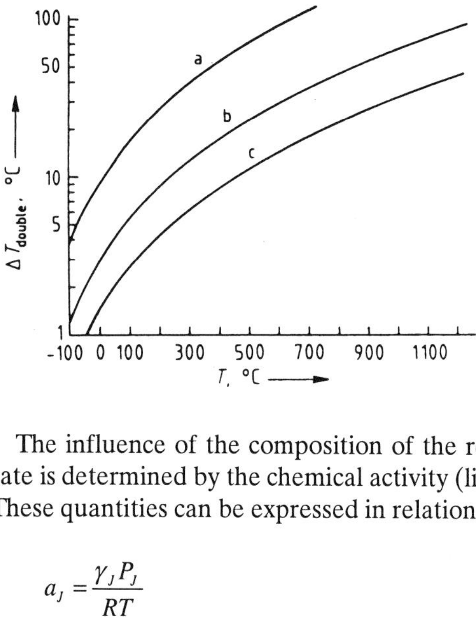

Figure 2.2 Temperature increase required to double the reaction rate constant (ΔT_{double}) for different values of the activation energy: (a) E_a = 50 kJ mol^{-1}; (b) E_a = 150 kJ mol^{-1}; (c) E_A = 300 kJ mol^{-1}.

The influence of the composition of the reaction mixture on the chemical reaction rate is determined by the chemical activity (liquids) or fugacity (gases) of the reactants. These quantities can be expressed in relation to each other as

$$a_J = \frac{\gamma_J P_J}{RT}$$

where a_J is the chemical activity, γ_J the activity coefficient or fugacity coefficient, and P_J the partial pressure of component J. The product $\gamma_J P_J$ is called the fugacity. For ideal reaction mixtures, $\gamma_J = 1$ and the reaction rate is determined by the molar concentration or partial pressure of the reactants.

For elementary reactions and ideal reaction mixtures, the reaction rate is proportional to the concentration of each of the reactants, since the number of molecular collisions per unit time is proportional to it. For example, for a bimolecular elementary reaction:

$$A + B \rightarrow \text{products}$$

$$|R| = -\frac{dC_A}{dt} = -\frac{dC_B}{dt} = kC_AC_B$$

If A is chosen as key reactant, $|R|$ is expressed in moles of A converted per cubic meter of reaction mixture per second and k has the dimension of cubic meters of reaction mixture per mole of B per second.

The notation given here is convenient for reactions carried out in liquids. For gases a more convenient notation is

$$|R| = k' P_A P_B$$

with

$$k' = \frac{k}{(RT)^2}$$

and

$$|R| = -\frac{1}{RT}\frac{dP_A}{dt} = -\frac{1}{\nu_B RT}\frac{dP_B}{dt} = k' P_A P_B$$

For nonideal elementary reaction mixtures, the chemical activities and fugacities, respectively, must replace the concentrations and partial pressures in the above equations.

Most reactions are overall reactions. The exact mechanism behind them is often unknown, which means that kinetic data cannot be fitted to the exact rate equation. The chemical reaction rate can often be approximated in some range of concentrations by a power law equation:

$$|R| = kC_A^n \tag{2.2}$$

If Equation 2.2 is used, the reaction is said to be of the order n with respect to A. The orders do not have to be whole numbers, as explained in Section 2.2, depending on the region covered by the kinetic data (temperature, concentration $C_{A,0}$). For multimolecular overall reactions power laws can also be used. For example, for the bimolecular overall reaction

$$A + \nu_B B \rightarrow \text{products}$$

the following equation can be written:

$$|R| = k\ C_A^n\ C_B^m \tag{2.3}$$

The reaction is of order n with respect to A and of order m with respect to B; the total order of the reaction is $(n+m)$. Again the orders n and m can be any numbers, not necessarily positive; negative values can also be found.

Hence, different sets of kinetic data, measured for different temperature or concentration ranges, yield different values for n and m. Therefore, if kinetic data or rate equations are obtained (e.g. from the literature), one should always ascertain that the region to which the kinetic data will be applied has indeed been explored experimentally. Extrapolation of kinetic data is generally impossible or very dangerous at least.

The principles discussed here can be used for any kinetic scheme in general. This is now illustrated with some overall reactions often encountered in the practice of catalysis.

It was assumed above, that the reactions described are irreversible. In practice, reactions are often reversible, i.e. the products can be converted into the reactants again:

$$A+B \underset{k_-}{\overset{k_+}{\rightleftarrows}} P+Q$$

If both reactions are elementary and the reaction mixture is ideal, the chemical reaction rate is given by

$$|R| = k_+ C_A C_B - k_- C_P C_Q \qquad (2.4)$$

The activation energies of the forward reaction E_+ and the reverse reaction E_- are related as

$$E_+ = E_- - \Delta H_r$$

where $(-\Delta H_r)$ is the reaction enthalpy. (For exothermic reactions $(-\Delta H_r) > 0$; for endothermic reactions $(-\Delta H_r) < 0$.)

Because of this, the temperature dependence of the reaction rates for exothermic and endothermic equilibria are different (Figure 2.3). Initially, the rate increases with increasing temperature for both types of reaction. For **exothermic reactions** where $E_+ < E_-$, the value of k_- increases faster than the value of k_+. Above a certain temperature this can have the net result that $|R|$ in Equation 2.4 decreases with increasing temperature. Thus for exothermic reversible reactions, $|R|$ can, at constant reaction mixture composition, pass through a maximum as a function of temperature; it will become zero for the temperature at which the reaction mixture is at equilibrium (i.e. $k_+C_AC_B = k_-C_PC_Q$) and negative beyond this value. For **endothermic reactions**, k_+ increases faster with increasing temperature than k_-; hence, $|R|$ continues to increase with increasing temperature, and the equilibrium shifts more and more to the right-hand side.

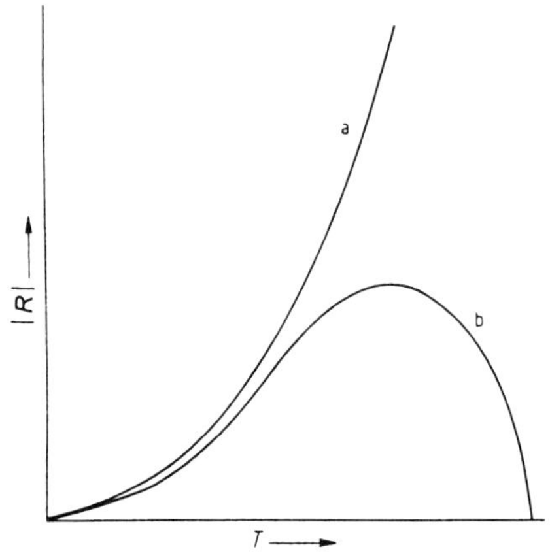

Figure 2.3 Dependence of reaction rate on temperature for a reversible reaction: (a) endothermic reaction; (b) exothermic reaction.

The temperature at which the maximum rate is found for exothermic reactions can be calculated from

$$T_{max} = \frac{(-\Delta H_r)}{R}\left[\ln\left(\frac{C_P C_Q\, k_- E_-}{C_A C_B\, k_+ E_+}\right)\right]^{-1}$$

Hence, this temperature depends on the composition of the reaction mixture. Therefore, for exothermic equilibrium reactions the temperature should be lowered as more reactant is converted, to keep the rate high. Thus this type of reaction is often carried out in multiple stages (e.g. multiple reactors or multiple zones within a reactor), each with its own temperature, adapted to the reaction mixture composition.

If chemical equilibrium is reached the reaction rate becomes zero and the reaction mixture composition must fulfil the equilibrium condition

$$\frac{C_P C_Q}{C_A C_B} = K$$

Where K is the equilibrium constant. Since $|R| = 0$ at equilibrium, K can be calculated as:

$$K = k_+ / k_-$$

2.2 Heterogeneous Catalytic Reactions

Heterogeneous catalytic reactions are reactions that take place on the surface of a solid catalyst. The following phenomena are significant in determining the microkinetics of these reactions:

- adsorption of the reactants;
- reaction of the adsorbed reactants (e.g. with components in the gas phase or other adsorbed species);
- desorption of the products.

The reaction itself can be monomolecular or bimolecular. Both situations are discussed below.

A general mechanism for heterogeneously catalyzed **monomolecular reactions** is

$$A_{(gas)} \underset{k_{-a}}{\overset{k_{+a}}{\rightleftharpoons}} A_{(ads)} \qquad \text{adsorption}$$

$$\xrightarrow{k_r} P_{(ads)} \qquad \text{reaction}$$

$$\xrightarrow{k_d} P_{(gas)} \qquad \text{desorption}$$

A kinetic expression for the above reaction can be obtained by assuming a pseudo-steady state for the adsorbed species $A_{(ads)}$ and $P_{(ads)}$. The measure of the concentrations of these species is the **degree of occupation** of the surface, θ_A and θ_P, respectively. θ_A and θ_P are defined such that they range from zero (no active surface area covered at all) to unity (all active surface area covered). We further assume a pseudosteady state for the degrees of occupation of the catalytically active surface area, so that θ_A and θ_P do not change in time, so $d\theta_A/dt = d\theta_P/dt = 0$. If the concentration of **active sites** in the catalyst pellet is C^*, then the free sites, on which A can be adsorbed, are given by $C^*(1 - \theta_A - \theta_B)$ and the rate of adsorption by

$$R_{ads\,A} = k^*_{+a} P_A C^* (1 - \theta_A - \theta_P)$$

$$= k_{+a} P_A (1 - \theta_A - \theta_P) \qquad k_{+a} = k^*_{+a} C^*$$

The rates of desorption are $R_{des\,A} = -k_{-a}\theta_A$ $R_{desP} = -k_d\theta_P$ and the rate of conversion of A to chemisorbed P is $R_P = +k_r\theta_A$. Pseudosteady state assumption to the rate equations of the adsorbed species gives:

$$\frac{d\theta_A}{dt} \approx k_a P_A (1 - \theta_A - \theta_P) - k_{-a}\theta_A - k_r\theta_A = 0 \tag{2.5}$$

$$\frac{d\theta_P}{dt} \approx k_r\theta_A - k_d\theta_P = 0 \tag{2.6}$$

The overall chemical reaction rate is obtained from

$$|R| = -\frac{dC_A}{dt} = \frac{dC_P}{dt} = k_d\theta_P \tag{2.7}$$

Combining Equations 2.5 – 2.7 substituting expressions for θ_A and θ_P, gives

$$|R| = \frac{k_r \frac{k_a}{k_{-a}} P_A}{\left(1 + \frac{k_r}{k_{-a}}\right) + \left(1 + \frac{k_r}{k_d}\right)\frac{k_a}{k_{-a}} P_A} \tag{2.8}$$

If the adsorption and desorption rates in the above scheme are much higher than the rate of the surface reaction (i.e. $k_a, k_{-a}, k_d >> k_r$), Equation 2.8 can be approximated by

$$|R| \approx k_r \frac{\frac{k_a}{k_{-a}} P_A}{1 + \frac{k_a}{k_{-a}} P_A} = k_r \frac{b_A P_A}{1 + b_A P_A} \tag{2.9}$$

Which is generally written as

$$|R| = k\frac{K_A P_A}{1+K_A P_A} \tag{2.10}$$

where k and K_A are lumped parameters, which can be obtained by comparing Equations 2.9 and 2.10. The above expression is referred to as **Langmuir-Hinshelwood kinetics**. For low partial pressures of A the rate is first order in A, whereas for high partial pressures it becomes zero order because $K_A P_A >> 1$. For K_A,

$$K_A = b_A = \frac{k_a}{k_{-a}} = b_{A,\infty} \exp\left[\frac{(-\Delta H_{ads})}{RT}\right]$$

where $(-\Delta H_{ads})$ is the **adsorption enthalpy** of A on the active catalyst surface. Since the adsorption of a given component is an exothermic process, the term $-\Delta H_{ads}$ is positive. An Arrhenius plot (i.e. the logarithm of the reaction rate versus the reciprocal absolute temperature) for this situation is given in Figure 2.4. For low temperatures, b_A is large (θ_A close to unity) and the rate is zero order:

$$|R| \approx k_r = k_{r,\infty} \exp\left(-\frac{E_a}{RT}\right)$$

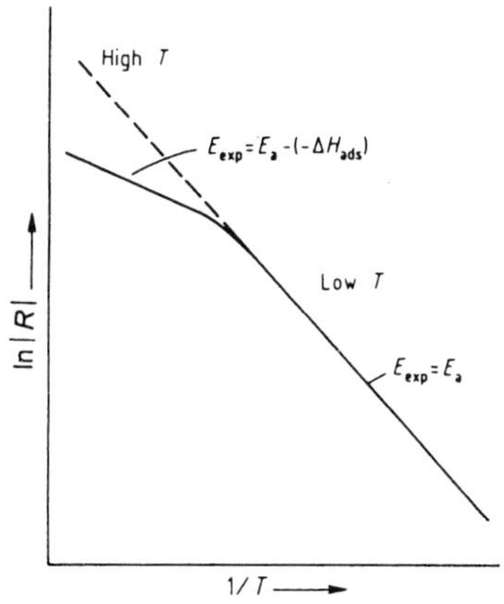

Figure 2.4 Arrhenius curve for monomolecular Langmuir-Hinshelwood kinetics.

Hence, for low temperatures the energy of activation equals E_a and is relatively large. As the temperature increases, the value of b_A (or θ_A) decreases and the reaction becomes first order:

$$|R| \approx k_r b_A P_A = k_{r,\infty} b_{A,\infty} \exp\left[\frac{E_a - (-\Delta H_{ads})}{RT}\right] P_A$$

So, in this regime, a smaller, apparent activation energy is found that is equal to $E_a - (-\Delta H_{ads})$.

For **bimolecular, heterogeneously catalyzed reactions**, two mechanisms can be distinguished:

- Eley-Rideal mechanism. One of the reactants is adsorbed on the catalyst surface, the other reacts from the gas phase:

$$A_{(gas)} \underset{k_{-a}}{\overset{k_{+a}}{\rightleftharpoons}} A_{(ads)}$$

$$A_{(ads)} + B_{(gas)} \underset{k_{-r}}{\overset{k_{+r}}{\rightleftharpoons}} P_{(ads)}$$

$$P_{(ads)} \xrightarrow{k_d} P_{(gas)}$$

- Langmuir-Hinshelwood mechanism. Both reactants are adsorbed on the catalyst surface:

$$A_{(gas)} \underset{k_{-a}}{\overset{k_{+a}}{\rightleftharpoons}} A_{(ads)}$$

$$B_{(gas)} \underset{k_{-b}}{\overset{k_{+b}}{\rightleftharpoons}} B_{(ads)}$$

$$A_{(ads)} + B_{ads} \underset{k_{-r}}{\overset{k_{+r}}{\rightleftharpoons}} P_{(ads)}$$

$$P_{(ads)} \overset{k_d}{\rightarrow} P_{gas}$$

Both mechanisms are illustrated in Figure 2.5. The pseudosteady state assumption for the adsorbed species in the Eley-Rideal mechanism yields

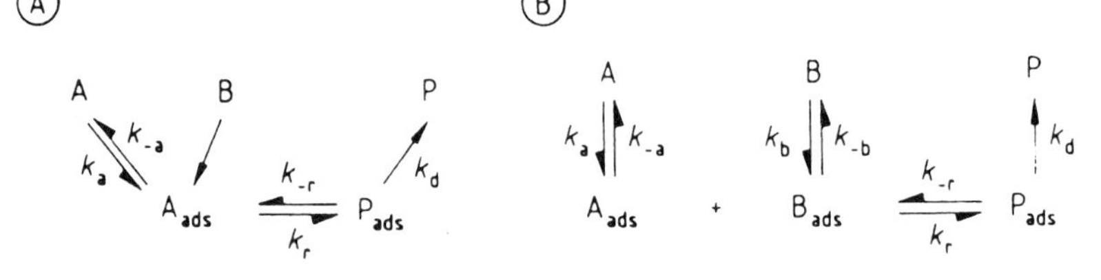

Figure 2.5 Elementary steps involved in bimolecular, heterogeneously catalyzed reactions: A) according to Eley-Rideal; B) according to Langmuir-Hinshelwood.

$$\frac{d\theta_A}{dt} \approx k_a P_A (1-\theta_A-\theta_P) - k_{-a}\theta_A - k_r P_B \theta_A + k_{-r}\theta_P = 0$$

$$\frac{d\theta_P}{dt} \approx k_r P_B \theta_A - k_{-r}\theta_P - k_d \theta_P = 0$$

The overall chemical reaction rate is now obtained from

$$|R| = k_r P_B \theta_A - k_{-r}\theta_P$$

Combining the above equations gives

$$|R| = \frac{k_r \frac{k_a}{k_{-a}} P_A P_B}{1 + \frac{k_{-r}}{k_d} + \frac{k_a}{k_{-a}}\left(1 + \frac{k_{-r}}{k_d}\right) P_A + \frac{k_r}{k_{-a}} P_B + \frac{k_a k_r}{k_{-a} k_d} P_A P_B} \tag{2.11}$$

From this formula, several limiting regimes can again be obtained. For example, if the adsorption equilibrium rates are still much higher than the rate of the surface reaction (i.e. k_a, k_{-a}, $k_d >> k_r$, k_{-r}) Equation 2.11 can be approximated by

$$|R| = k_r \frac{b_A P_A}{1 + b_A P_A} P_B$$

with

$$b_A = \frac{k_a}{k_{-a}} \tag{2.12}$$

From the Langmuir-Hinshelwood mechanism a formula similar to Equation 2.11 can be derived. This formula, however, is rather complex and bulky. Therefore, only the limit is discussed here, where adsorption and desorption rates are again much higher than the rate of the surface reaction (i.e. k_a, k_{-a}, k_b, k_{-b}, $k_d >> k_r$, k_{-r}). For this situation,

$$|R| = k_r \frac{b_A b_B P_A P_B}{(1 + b_A P_A + b_B P_B)^2}$$

with b_A defined by Equation 2.12 and

$$b_B = \frac{k_b}{k_{-b}}$$

If the partial pressure of one of the components A is increased, the reaction rate does not necessarily have to increase: for a certain value of P_A the maximum rate is obtained; above this pressure the rate decreases with increasing P_A. The explanation is that for high P_A values, adsorption of A is so strong that it hampers adsorption of reactant B and thus retards the surface reaction. Consequently, negative reaction orders can also be found: for both components the reaction order is between -1 and 1.

Examples of systems exhibiting Langmuir-Hinshelwood type behavior are:

- the oxidation of CO on platinum and palladium (order for CO is -1, and for O_2 between 0 and 1);
- the hydrogenation of C_2H_4 on nickel (order for C_2H_4 is between -1 and 0, and for H_2 between 0 and 1);
- the reaction $N_2O + H_2 \rightarrow N_2 + H_2O$ on platinum and gold (order for N_2O is between 0 and 1, and for H_2 -1).

2.3 Catalyst Performance

The two most important performance properties of a catalyst are deactivation behavior and its selectivity.

2.3.1 Deactivation

Deactivation is caused mostly by fouling (e.g. coke formation) or by poisoning. In most organic reactions a decomposition may occur, which leads to fine carbonaceous deposits, which generally are called coke. Especially when these coke deposits occur at the inlet of the pores of a porous catalyst, deactivation may occur rather rapidly [1]. Deactivation also may occur due to extremely small amounts of contaminants in the feed to the reactor. Sulfur is a well-known poison, as are traces of many metals. The deactivation behavior of a catalyst has to be tested in long-duration experiments with the feed, under plant conditions. Usually, for cases of slow deactivation, reactor conditions are gradually made more severe (e.g. by increasing the reactor temperature) up to a level, where a practical operating limit has been reached. The catalyst then has to be changed for a fresh batch.

To reactivate a coked catalyst, the coke can usually be removed by burning off the deposits at a controlled temperature with a mixture of air and an inert diluent, such as nitrogen or steam. The temperature level at which the coke deposits ignite has to be determined experimentally. The allowable O_2 content in the air-diluent mixture can be calculated [2].

2.3.2 Selectivity

The selectivity of a catalyst plays a role as soon as multiple reactions occur (Figure 2.6). We distinguish the following main types of multiple reactions:

- parallel reactions;
- consecutive reactions;
- combination reactions, consisting of a combination of parallel and consecutive reactions.

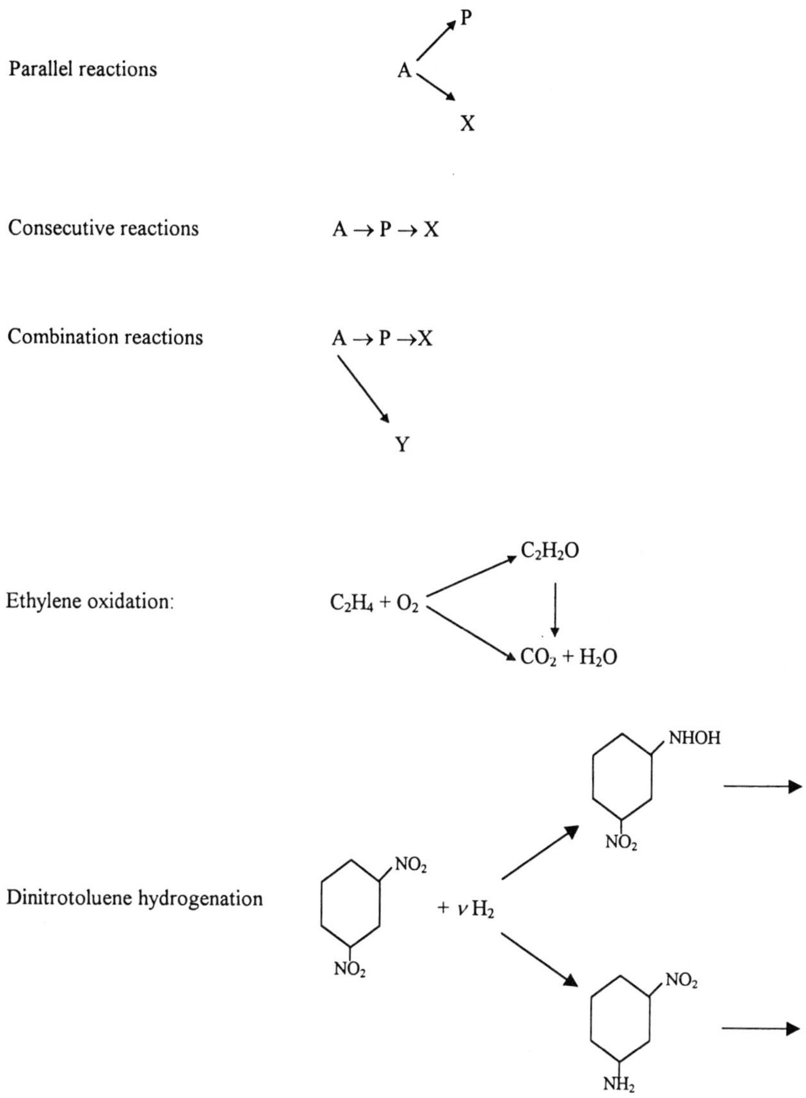

Figure 2.6 Examples of multiple reactions (ν is the stoichiometric coefficient).

In Figure 2.6 A stands for the reactants, P for the desired products and X and Y for the undesired products. In the same figure examples are given of industrial reactions, which in later chapters will be studied in further detail. Depending on the orders of the relevant reactions and the choice of reaction conditions, by careful selection of reactor type and concentration level the selectivity can be controlled favorably. These reactor properties were defined in Section 1.4.

The selectivity can also be influenced by a proper catalyst choice. In a reactor the overall selectivity or the yield of the reactor is the relevant economic factor. As conditions often change over the reactor length, the selectivity achieved at the individual catalyst pellets may also change accordingly. The selectivity of a catalyst pellet therefore is called the local or the differential selectivity, to distinguish it from the integral selectivity of the reactor. In general, the differential selectivity is given by

$$\sigma'_P = \frac{|R_P|}{|R_A|\nu_P}$$

Hereinafter the differential selectivity is called the selectivity of the catalyst.

In Chapter 9, Example 9.1 describes the evaluation of the kinetics of the hydrogenation of 2,4-dinitrotoluene (DNT) in methanol [3]. This reaction proceeds via three intermediate products, eventually giving 2,4-diaminotoluene DAT and is catalyzed by Pd either on active carbon or on an alumina carrier. The reactions occur according to a Langmuir-Hinshelwood mechanism on two active sites – one site for the hydrogen and one for the aromatic components. From the discussion of the example it can be learnt how:

- At low temperatures the catalyst reaches complete occupancy with hydrogen with increasing partial pressures of H_2. Beyond a certain pressure, further increase has no additional effect, because the catalyst is fully saturated with hydrogen. At higher temperatures an occupancy of around 100% ceases in the pressure range studied.
- As a function of temperature the initial rates do not exhibit an Arrhenius behavior: a plot of the initial rate versus $1/T$ is not a straight line. This is caused by the combined and distinct influences of the temperature on the kinetic and the chemisorption constants. The initial selectivity decreases with increasing temperature.
- DNT chemisorbs very strongly, much more than the consecutive intermediates. Over the whole range of DNT concentrations the occupancy with DNT is almost complete up till DNT conversions above 90%. This implies that almost all the DNT must be converted before the conversion of 2-nitro-4-hydroxylaminotoluene and of 2-amino-4-nitrotoluene can start. This is called **molecular queuing** – the intermediates queue to be adsorbed on the active sites until the DNT has almost completely disappeared.

2.4 Kinetics in Practice

The previous sections discussed the normal, theoretically based rate equations. The experimental determination of kinetics is discussed in more detail in Chapter 5. The following points are some of the possible difficulties found in practice:

- Kinetic rate equations may change per batch of catalyst manufactured and may vary between different lots.
- Kinetic rate expressions may vary in accuracy and in parameter values.
- More than one rate equation may fit the measured kinetics equally well.
- Kinetic rate data may depend on the test equipment used and the method of temperature measurement.
- The same active material deposited on different carriers may give different reaction rates at equal experimental conditions.
- In one and the same catalyst bed scatter may occur if the individual samples contain only a small number of catalyst particles.
- The reaction rates may depend on the method of activation of the catalyst and on the operating regime in which it is used.
- The procedure used to fit rate expressions to experimental data has an influence on the parameter values obtained.

This is only a short list of possible pitfalls of the use of kinetic rate expressions in practice. Generally it is wise practice to test a catalyst experimentally before using available rate expressions (Chapter 5). The following gives a few examples to give weight to the above remarks.

Kinetics measured may differ per batch of catalyst as manufactured by the supplier. It is evident that these differences will never be too large, at least for the experienced manufacturers (not more than a factor of 2-3 under identical test conditions) and that these small changes can be compensated for by changing the operating conditions, such as the reaction temperature. In the same batch of catalyst the individual catalyst pellets will exhibit a stochastic distribution of their properties. Therefore, for catalyst testing and rate determinations, a catalyst sample has to be taken sufficiently large that it is representative of the average of the entire batch. This often has to be determined empirically.

The kinetics measured and the data obtained, when fitted to a rate equation (both the accuracy as well as the parameters values in the rate equation) may depend on the experimental method of data acquisition.

In Chapter 9 an example is given, where four different rate equations taken from literature for the synthesis of methanol according to $CO + 2\ H_2 \leftrightharpoons CH_3OH$ are analyzed (Example 9.2). Two equations were based on data obtained in an integral reactor filled with catalyst and equipped with a sheath with two travelling thermocouples, the other two in an internally mixed catalytic reactor of the Berty type, which can be considered as approaching isothermal conditions. Along a sheath considerable heat conduction may take place, so the actual temperatures may be considerably higher than the measured ones. In the example the degrees of occupation of the catalyst, according to the rate equations as they were fitted to the data, are also given.

Plots of data can be informative, for example parity plots and relative residual plots. Figure 2.7 shows parity plots, in which measured data are plotted against the data as cal-

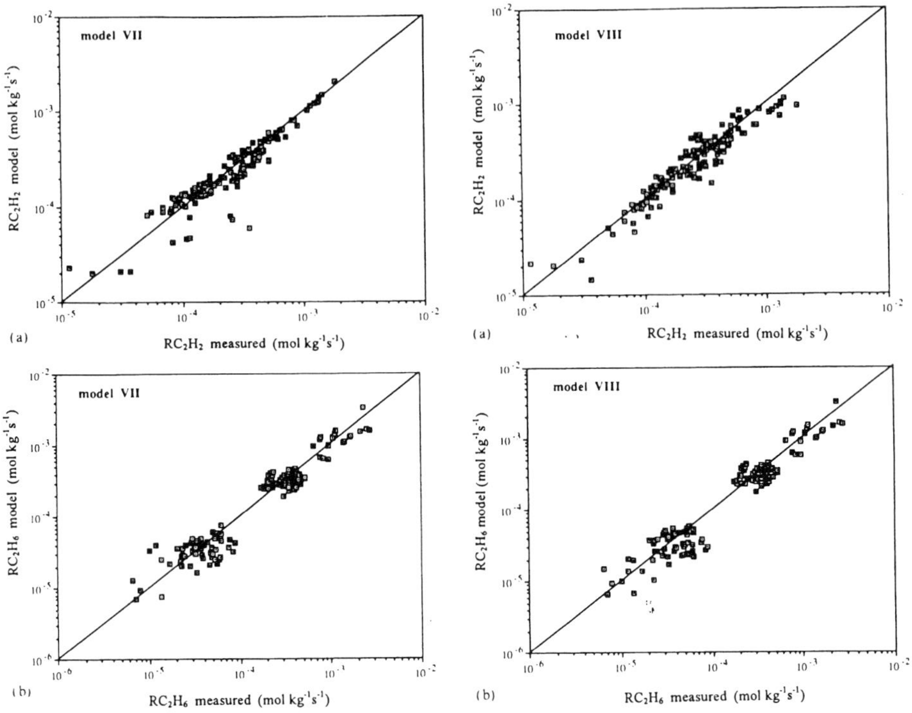

Figure 2.7 Parity plots of measured and correlated data for the hydrogenation of (a) acetylene and (b) ethane [4]. In a parity plot the data as calculated with the correlation, and as measured experimentally, coincide with the diagonal if they are exactly equal.
Reprinted from Chemical Engineering and Processing, **32**, A.N.R. Bos et. al., A kinetic study of the hydrogenation of ethyne and ethene on a commercial Pd/Al_2O_3 catalyst, 53-63, 1993, with kind permission from Elsevier Science S.A., P.O. Box 564, 1001 Lausanne, Switzerland.

culated by the rate equation fitted to the data. The data are taken from the determination of rate equations for the hydrogenation of acetylene and ethylene in mixtures of these two components [4]. They presented three correlation equations, which fitted the measured data almost equally well. The parity plots are given for two of their equation sets.

In Figure 2.8 plots are given of the relative residuals of two correlation equations, which were discarded. We observe that the residuals start to deviate of the average systematically for higher acetylene and ethylene conversions.

For the same catalytically active material but with different catalyst carriers, different reaction rates and rate equations can be expected. Consider the hydrogenation of 2,4 DNT as discussed in Section 9.2 for 5% Pd on an active carbon catalyst with an average particle size of 30 μm [3]. These experiments were later repeated but with a Pd on an alumina catalyst [5]. This catalyst consisted of 4 × 4 mm pellets, crushed to sizes of lower than 40 μm in order to avoid pore diffusion limitations. In Figure 2.9 the measured conversion rates are given as a function of the averaged catalyst particle diameter, showing that above a diameter of 80 μm the rate measured diminishes. For small particles they determined the rate equations under conditions where there were no pore diffusion lim-

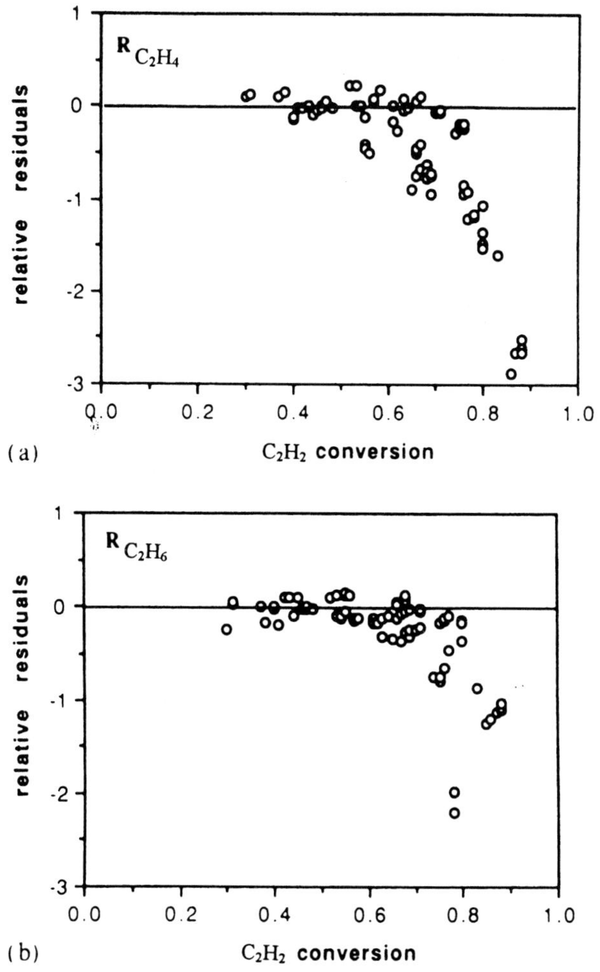

Figure 2.8 Plot of the residuals for the hydrogenation of (a) acetylene and (b) ethane. At high conversions the residuals deviate from an equal distribution around zero.
Reprinted from Chemical Engineering and Processing, **32**, A.N.R. Bos et. al., A kinetic study of the hydrogenation of ethyne and ethene on a commercial Pd/Al_2O_3 catalyst, 53-63, 1993, with kind permission from Elsevier Science S.A., P.O. Box 564, 1001 Lausanne, Switzerland.

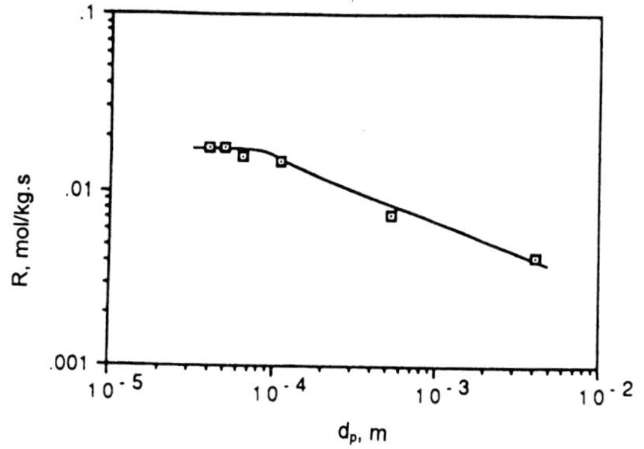

Figure 2.9 Kinetics of the hydrogenation of 2,4-dinitrotoluene. Initial consumption rates as a function of the averaged catalyst particle diameter at 345 K, P_{H_2} = 0.5 MPa, C_{A0} = 196 mol m^{-3}, m_{cat} = 20 g for a Pd on alumina catalyst.
Reprinted from Chemical Engineering Science, **47**, E.J. Molga and K.R. Westerterp, Kinetics of the hydrogenation of 2,4-dinitrotoluene over a palladium on alumina catalyst,1733-1749, 1992, with kind permission from Elsevier Science Ltd, The Boulevard, Langford Lane, Kidlington OX5 1GB, UK.

itations. In Table 2.1 the parameter values for the best fit to the same rate equations are given, which show differences in activation and chemisorption energy values. The more acid alumina carrier, as compared to the neutral active carbon carrier, causes this.

Table 2.1 Evaluation of the two best fitting rate equations for synthesis of methanol

		Kuczysnki et al. [6]	Bos et al. [4]
Rate equation 1:		$\frac{k(a_{CO}a_H^2 - a_M / K_{eq})}{(1 + Aa_{CO} + Ba_{H_2})^3}$	
k	:	$2.68 \times 10^9 \exp(-18400/T)$	$50.2 \times 10^6 \exp(-15950/T)$
A	:	0.069	$65.6 \times \exp(-3428/T)$
B	:	$61.9 \times 10^9 \exp(6610/T)$	$149.5 \times \exp(-4643/T)$
Rate equation 2:		$\frac{k(a_{CO}a_H^2 - a_M / K_{eq})}{1 + Ca_{CO} + Da_{CO}a_{H_2}^2}$	
k	:	$18.0 \times \exp(-9032/T)$	$160 \times \exp(-10343/T)$
C	:	$297 \times 10^{-6} \exp(3540/T)$	$45.7 \times 10^{-18} \exp(17200/T)$
D	:	259×10^{-6}	$0.85 \times \exp(-4690/T)$

(The indices CO, H_2 and M refer to CO, H_2 and methanol respectively; a is the fugacity in Pa.)

From the above it is clear that kinetic rate equations are afflicted with uncertainties, which can be caused by inaccurate data and the fitting procedures used, and also by the experimental methods and the differences in catalyst material and its quality. It is not possible to conclude from the fitted rate equations, what the reaction mechanism of the catalysis is.

References

1. Bartholomew, C.H. (1984) *Chem. Eng.*, Nov. **12**, 96.
2. Westerterp, K.R, Fontein, H.J., Van Beckum, F.P.H. (1988), *Chem. Eng. Technol.*, **11**, 367.
3. Janssen, H.J., Kruithof, A.J., Steghuis, G.J. Westerterp, K.R. (1990), *Ind. Eng. Chem. Res.*, **29**, 754,1822.
4. Bos, A.N.R., Borman, P.C., Kuczynski, M., Westerterp, K.R., (1989) *Chem. Eng. Sci.*, **44**, 2435.
5. Molga, E.J., Westerterp, K.R. (1992) *Chem. Eng. Sci.*, **47**, 1733.
6. Kuczynski, M., Browne, W.I., Fontein H.J., Westerterp, K.R. (1987) *Chem. Eng. Process.*, **21**, 179.

3 Production and Physical Characteristics of Solid Catalysts

3.1 Introduction

In the evaluation of industrial catalysts the total conversion rate per unit reactor volume is of importance. Without mass and heat transfer limitations this rate is proportional to the active surface area exposed to the fluid per unit volume or weight of the catalyst. A high surface area is achieved with small particles or even powders. However, there are practical limitations to the size of particles that can be used, such as:

- the pressure drop in packed bed reactors (small particles may produce an excessive pressure drop through the bed);
- the difficulties when handling powders in slurry reactors (removal and separation by settling or filtration);
- entrainment in fluid bed reactors.

The usual way to obtain a large amount of catalytic surface area is to use a porous material with small pores. This can be understood assuming the pores are cylindrical and of the same radius. The relationship between the pore surface per unit particle volume S, the porosity ε_p, and equivalent pore radius r_e, is given by

$$S = \frac{2\varepsilon_p}{r_e}$$

Typically, commercial catalysts have a void fraction (porosity) ε_p of about 0.5. Therefore an approximate relationship between the internal surface area per unit particle volume and average pore size r_e is:

$$S \approx \frac{1}{r_e}$$

Thus, the area per unit volume is much larger when the particle contains small pores. Pores found in industrial catalysts are divided in three groups:

- macropores (diameter > 50 nm);
- mesopores (diameter between 2 nm and 50 nm);
- micropores (diameter < 2nm).

For mesopores the above estimate gives specific surface areas of $(40-1000)\times10^6$ m^2m^{-3}. The ranges of the macro-, meso- and micropores correspond approximately to the limits of the available physical methods for pore size measurements. Usually, the specific surface area is presented per unit gram of catalyst $S_g = S/\rho_a$, where ρ_a is the apparent density of the particle. For most catalysts ρ_a is roughly 1-10 g cm^{-3}, corresponding to $(1-10)\times10^6$ $m^2\,m^{-3}$. Typical values of the amount for internal surface area, in conventionally used units, range from 10 $m^2\,g^{-1}$ to 250 $m^2\,g^{-1}$. Ranges of internal surface area S_g, pore volume V_g and pore size for some selected industrial catalysts are shown in Table 3.1.

Table 3.1 Values of internal surface area, pore volume, and average pore radius for typical catalysts (from Perry [1])

Catalyst	S_g ($m^2\,g^{-1}$)	V_g ($cm^3\,g^{-1}$)	$r_e = 2\,V_g/S_g$ (nm)
Activated carbons	500-1500	0.6-0.8	1-2
Silica gels	200-700	0.4	1.5-10
Silica-Alumina cracking catalyst ≈ 10-20% Al_2O_3	200-700	0.2-0.7	1.5-15
Silica-Alumina (steam deactivated)	67	0.519	15
Silica-magnesia microspheres: Nalco, 20% MgO	630	0.451	1.4
Activated clays	150-225	0.4-0.52	10
Activated alumina (Alorico)	175	0.388	4.5
CoMo on alumina	168-251	0.26-0.33	2-4
Fe synthetic NH_3 catalyst	4-13	0.12	20-100

The enormous internal surface areas present in typical high-area catalysts can be visualized by recognizing that a quantity of such catalyst, easily held in one's hand, has a total surface area greater than a football field [2]. Highly impressive is also the total length of the pores [3]. If we imagine that all pores are aligned and have the same radius r, the length of this cylinder in a gram of the catalyst can be found from

$$L_g = \frac{S_g}{\pi r_e^2}$$

Taking $S_g = 1000$ $m^2\,g^{-1}$, and $r_p = 1$ nm (values typical for activated carbon) gives

$$L_g = \frac{10^3}{2\pi 10^{-9}}\ \text{m} \approx 160\times10^6\ \text{km}$$

Note that the average distance between the Earth and the Sun is 149.6×10^6 km.

A high surface area, however, is not always an advantage for catalytic reactions. In certain reactions, the fine pores impede intraparticle mass and heat transfer, which may result in lowering of the apparent catalyst activity, an unfavorable product distribution and/or in sintering of the catalyst. That is why surface areas higher than about 1000 $m^2\,g^{-1}$ are impractical.

The real pore structure of catalyst particles is very complicated. It is sometimes represented as an ensemble of cylindrical pores of equal radius or as an ensemble of small equal spherical particles and free space in between them. This, however, gives only a very rough description of the real pore geometry. A more detailed characterization of the pore geometry is given by the pore size distribution function $f(r)$. This function usually is defined by such a way that $f(r)dr$ represents the volume dV_g of pores per unit weight of the catalyst which have a radius between r and $r + dr$. Because of the large range of sizes commonly met in porous catalysts it is suitable to use log(r) as an argument of the distribution function rather than r. A typical pore size distribution is shown in Figure 3.1(a). In these figures

$$\frac{dV_g}{d\log(r)} = \frac{dV_g}{dr}\frac{d\log(r)}{dr} = \frac{f(r)}{r\ln(10)} = \frac{f(r)}{2.303r}$$

The pore size distribution and average pore sizes depend on the manufacturing procedure. Many other distribution curves are possible. An important case found in many commercial catalysts is shown in Figure 3.1(b). Such a porous structure is termed bimodal or bidisperse. It would be obtained, for example, if the particle is made of smaller porous particles pressed together. In this case the second peak represents the coarser pore structure formed by the passageways around the compacted particles, whereas the first peak represents the fine pore structure within each of the particles of the original powder. For some industrial catalysts a bidisperse structure is deliberate, because it affords a considerable amount of the internal surface area (small pores) and it facilitates mass transfer between the external surface and the center of the particle through the large pores.

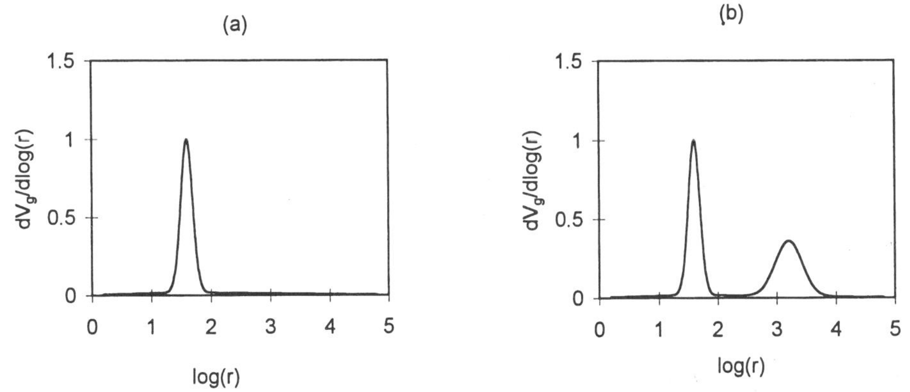

Figure 3.1 Pore-size distribution in catalysts (r in nm): (a) monodisperse; (b) bidisperse.

Manufactured as a powder, the catalyst is formed into particles, whose shape and size are determined by the end use. Common types, with a brief description of each, are given in Table 3.2.

The first three types (pellets, extrudates and granules) are primarily used in packed bed operations. Usually two factors (the diffusion resistance within the porous structure and the pressure drop over the bed) determine the size and shape of the particles. In packed bed reactors, cooled or heated through the tube wall, radial heat transfer and heat transfer from the wall to the bed becomes important too. For rapid, highly exothermic and endothermic reactions (oxidation and hydrogenation reactions, such as the ox-

Table 3.2 Common catalyst particles

Type	Method of manufacture	Shape	Size	Use
Pellets	made in high-pressure press	cylindrical, very uniform, rings	2-10 mm	packed tubular reactors
Extrudates	squeezed through holes	irregular lengths of different cross-section	> 1 mm	packed tubular reactors,
Granules	made by aging liquid drops	spherical	1-20 mm	Packed tubular reactors, moving beds
Granules	made by fusing and crushing or particle granulation	irregular	8-14 to 2-4 mesh, >2 mm	packed tubular reactors
Flakes	powder encapsulated in wax	irregular		liquid phase reactors
Powders	spray-dried hydrogels		20-300 μm 75-200 μm	fluidized reactors slurry reactors

idation of methanol and ethylene and the conversion of methane to synthesis gas) these factors are often decisive.

In fluidized bed reactors a broad particle size distribution is desired for good fluidization characteristics. Therefore, the catalyst is usually supplied as a powder or the powder is produced in the normal course of operations by the gradual attrition of the catalyst in a highly turbulent environment [4]. The particles generally range from about 20-300 μm in diameter, the mean being about 50-70 μm. Particles smaller than 20 μm are entrained from the reactor; particles larger than 300 μm fluidize poorly. There is no problem here with intraparticle diffusion.

Reactions in the liquid phase require small particles, or even powder, because the diffusion rate in liquids is smaller by several orders of magnitude, compared to gaseous diffusion rates. For slurry reactors small particles are also desirable, to keep them in suspension. Unfortunately, fine powders are difficult to remove by settling or filtration.

3.2 Catalyst Manufacture

3.2.1 Materials and Methods

A wide variety of materials are used in the preparation of industrial catalysts. These materials are divided in three major constituents: the active catalytic agent, the support and the promoters:

- *Active catalyst agent.* This is the constituent primarily accountable for the catalytic function and it includes metals, semiconductors and insulators. The type of electrical conductivity (mainly for convenience) classifies the active components. Both electrical conductivity and catalytic properties depend on the atomic, electronic configurations, although there is no relationship between them.

- *Support or carrier.* Materials frequently used as catalyst supports are porous solids with high total surface areas (external and internal), that provide a high surface area for the active component. The support also gives to the catalyst its shape and its mechanical strength and in some instances it influences catalytic activity.
- *Promoter.* Compounds added to enhance physical or chemical functions of the catalyst are termed promoters. Although the promoters are added in relatively small amounts their choice is often decisive for the catalyst properties. Promoters can be compared with spices in cooking [5]. They may be incorporated into the catalyst during any step in the chemical processing of the catalyst constituents. In some cases, promoters are added during the course of reaction.

For many catalysts, the major component is the active material. Examples of such **unsupported catalysts** are the aluminosilicates and zeolites used for cracking petroleum fractions. One of the most widely used unsupported metal catalysts is the precious metal gauze as used, for example, in the oxidation of ammonia to nitric oxide in nitric acid plants. A very fast rate is needed to obtain the necessary selectivity to nitric oxide, so a low metal surface area and a short contact time are used. These gauze's are woven from fine wires (0.075 mm in diameter) of platinum alloy, usually platinum-rhodium. Several layers of these gauze's, which may be up to 3 m in diameter, are used. The methanol oxidation to formaldehyde is another process in which an unsupported metal catalyst is used, but here metallic silver is used in the form of a bed of granules.

In numerous other catalysts, the active material is a minor component, which is deposited on a more or less inert porous support. Widely used supports include:

- alumina
- silica gel
- activated carbon
- zeolites
- silicon carbide
- titania
- magnesia
- various silicates.

Most catalysts, where a metal is an active component, are supported catalysts, because a prime requirement here is the use of a large metal surface area. Examples of supported catalysts are activated-carbon-supported Pt and Pd, and Ni on alumina.

Industrial catalysts are manufactured by a variety of methods involving one or more processing steps, such as:

- precipitation
- washing
- drying
- calcination
- impregnation
- leaching
- thermal fusion forming.

The process used for catalyst manufacture depends on many factors, such as the chemistry of the catalyst component and its possible precursors, the concentrations of the many components required, the physical strength required and the reaction conditions of the catalyst in use. Most industrial catalysts are made either by precipitation, when the active phase and support are produced and precipitated together, or by the impregnation of an active phase on a preformed support.

The basic manufacturing methods (as well as some other ways in which industrial catalysts are made) are outlined below. Several books and reviews give more details on this topic. A book by Le Page et al. (Chapter 5) [6] and the *Catalyst Handbook* edited by Twigg [7] give detailed information on equipment and general procedures used in catalyst manufacture. Comprehensive descriptions are available, from different points of view, in the books of Richardson [5] and Satterfield [2].

3.2.2 Precipitated Catalysts

Precipitation is usually understood as obtaining a solid from a liquid solution. In the production of precipitated catalysts, the first step is the mixing of two or more solutions or suspensions of materials, causing the precipitation of an amorphous or crystalline precipitate or gel. The wet solid is converted to the finished catalyst by filtration, washing, drying, forming, calcination and activation. Adjusting production conditions can vary cristallinity, particle size, porosity, and composition of the precipitate or gel.

Precipitated catalysts are generally prepared by rapid mixing of concentrated solutions of metal salts. The product precipitates in a finely divided form with high surface areas. Mixed hydroxides or carbonates are normally prepared by precipitation. Typically, the following reaction is used:

$$2Ni(NO_3)_2 + 2NaOH + Na_2CO_3 \rightarrow Ni_2(OH)_2CO_3 + 4NaNO_3$$

After filtration and washing the precipitate is dried. The final crystallite size in the precipitated catalyst typically is in the range 3-15 nm, while overall surface areas can be in the range of 50 – 200 $m^2\,g^{-1}$ or more.

The main challenges, when employing a precipitation process for catalyst manufacture, are the intimate mixing of the catalyst components and the formation of very small particles with a high surface area.

If a carrier is to be incorporated in the final catalyst, the original precipitation is usually carried out in the presence of a suspension of the finely divided support or, alternatively, a compound or suspension, which will eventually be converted to the support, may initially be present in solution. Thus, a soluble aluminum salt may be converted to aluminum hydroxide during precipitation and ultimately to alumina. Alternatively, a supported nickel catalyst could be prepared from a solution of nickel nitrate, containing a suspension of alumina, by precipitation of a nickel hydroxide with ammonia.

The final size and shape of the catalyst particles are determined in the shape formation process, which may also affect the pore size and pore size distribution. Larger pores can be introduced into a catalyst by incorporating in the mixture 5 to 15 % wood, flour, cellulose, starch, or other materials that can subsequently be burned out. As a result, bidisperse catalyst particles are obtained.

After the catalyst particles are dried and given shape, the catalyst precursor is **activated**; that is, it is converted into its active form through physical and chemical changes. This typically involves heating to cause calcination or decomposition, followed by reduction if a metallic catalyst is being prepared.

Some advantages of precipitation are:

- It generally provides for a more uniform mixing on a molecular scale of the various catalyst ingredients.
- The distribution of the active species through the final catalyst particle is uniform.
- It is usually easier to achieve high concentrations of the catalytically active phase.
- Better control over the pore size and pore size distribution is possible.

There are also some disadvantages of this method compared to the impregnation method, as discussed in the next section.

3.2.3 Impregnated Catalysts

One of the widely used preparation methods for catalysts is impregnation of the porous support with a solution of the active component. In the absence of specific interactions between the support and the components of the impregnating solutions, the impregnation process steps are quite simple. A solution is made up containing the component to be put on the catalyst. In the next stage either the support is dipped into this solution or the solution is sprayed onto the support. In both cases the take-up of the solution is governed by the porous structure of the support, so that the amount of active component incorporated in the catalyst depends on the solution concentration and its penetration into the support. After absorption of the solution in the pore system of the support, a drying stage is used to remove water. This has to be done so that the impregnated component remains in the pore system and does not migrate to the exterior surface of the support. If this stage is done correctly, the support has crystallites of the impregnated component, typically a metal nitrate, internally on the pore surface. Most impregnated catalysts are calcined in air after drying, thus converting the soluble salt into the insoluble oxide. The carrier is then dried and the catalyst is activated as in the case of a precipitated catalyst.

To ensure a thorough impregnation, the air in the pores of the support is removed by evacuation or is replaced by steam or a soluble gas (CO_2, NH_3).

The degree of uniformity varies with the adsorptive properties of the carrier and the method of manufacture. Thus, the use of an alcoholic solution may lead to a substantially different concentration distribution than with an aqueous solution. Often-repeated alternating steps of impregnation then calcination gives a better dispersion and a higher activity at a lower loading than a single impregnation.

Impregnation with a multicomponent solution may result in an uneven distribution of components, because of chromatographic effects. The more strongly adsorbed component will be deposited preferentially on the surface, whereas the weakly adsorbed ingredient will penetrate into the interior of the support. Such an uneven distribution can often be corrected by leaving the impregnated catalyst in a moist state for some time to enable equilibration. Particles can also be impregnated evenly by charging the support first with NH_4^+ ions or by converting the ions to be introduced to electronically neutral or weakly adsorbed complexes.

Drying of the impregnated supports may change the distribution of active components, unless these components are strongly adsorbed to the support. From dilute solutions the active components precipitate and crystallize when the saturation limit is exceeded. Hence, a uniform distribution of the active material can be obtained only if all the liquid evaporates spontaneously. However, evaporation starts at the surface of the catalyst particle and continues preferentially in the large pores, in which the vapor pressure is higher. Capillary forces replace liquid lost from the small pores by evaporation with liquid from the large pores. If the drying process is slow and the active component soluble, the active material is preferentially concentrated in the small pores in the center of the catalyst particle, where it is not readily accessible to the reactants. Therefore, the use of saturated solutions for impregnation of the catalyst or drying it as fast as possible, are advantageous.

The great virtue the of the impregnation process is the separation of the production of the active phase and of the support phase, which is clearly not possible in the case of the manufacture of precipitated catalysts.

Impregnated catalysts have several advantages compared to precipitated catalysts. Their pore structure and surface area are determined by the support that is used. Because supports are commercially available in almost every desired range of surface area, porosity, shape, size and mechanical strength, the impregnation method permits production of supported catalysts appropriate for the mass transport and reaction conditions prevailing in a given reactor. Impregnated catalysts are also more economical than precipitated catalysts, because they require smaller amounts of the usually costly active component.

Examples illustrating the advantages of the impregnation process are the noble metal catalysts. For diffusion-controlled reactions on noble-metal catalysts, it is desirable to have the active component on or near the surface of the catalyst pellet. This **eggshell-type layer** can be formed by a short impregnation of a strongly adsorbing support, followed by immediate drying; or by spraying the proper amount of the impregnation solution on the support particles agitated in a heated rotating drum. Impregnation is the preferred process in preparing supported noble metal catalysts, for which it is usually economically desirable to spread out the metal as finely divided as possible. The noble metal is usually present in the order of 1 wt % or less. This allows for the best use of a very expensive ingredient. With complex reactions, selectivity is usually diminished in the presence of significant intraparticle concentration gradients through a porous catalyst (a low effectiveness factor). Confining the active catalyst to a thin outside layer provides a method of eliminating this problem while retaining a catalyst particle of a size easy to work with. For catalysts exposed to attrition, deeper impregnation is desirable.

The disadvantage of impregnation lies in the limited amount of material that can be incorporated in a support by impregnation.

3.2.4 Skeletal Catalysts

Skeletal catalysts consist of a metal skeleton remaining after the less noble components of an alloy has been removed by leaching with base or acid. The skeleton metals belong almost exclusively to Groups IB and VIIB of the periodic table (Fe, Co, Ni, Cu, and Ag), whereas Al, Zn, Si, and Mg are the most commonly used alloy components. The alloys are prepared by fusion of the components in the proper proportion. Raney pioneered the development of skeleton catalysts. A widely used Ni and Co catalyst, which is highly active for hydrogenation reactions, still bear his name. The catalyst is prepared from a nickel-aluminum alloy by leaching out almost all the aluminum with a caustic solution to leave behind a porous nickel catalyst. Leaching of the 50-50 alloy with a 20 – 30 % solution of sodium hydroxide gives a highly active Ni catalyst containing 90 – 97 % Ni, 4 – 8 % metallic Al, 0.3 – 0.5 % aluminum oxide, and 1 – 2 % hydrogen dissolved in the skeleton.

3.2.5 Fused and Molten Catalysts

Industrial ammonia synthesis catalysts are produced by fusion of magnetite, Fe_3O_4, containing small amounts of Al_2O_3, K_2O, and CaO. On reduction, Al_2O_3 and CaO occupy separate regions and serve as structural promoters by preventing the sintering of Fe. The reduced Fe is covered by an adsorption layer of K and O atoms, which acts as an electronic promoter facilitating the chemisorption of N_2 during the ammonia synthesis.

Fused V_2O_5 in a microspherical form and as a coating on alumina spheres has been used in the selective oxidation of naphthalene to phthalic anhydride.

3.2.6 Calcination

Calcination is a heat treatment carried out in an oxidizing atmosphere at a temperature slightly above the projected operating temperature of the catalyst. The object of calcination is to stabilize the physical, chemical, and catalytic properties of the catalyst. During calcination several reactions and processes can occur:

- Thermally unstable compounds (carbonates, nitrates, hydroxides, and organic salts) decompose under gas evolution and are usually converted to oxides.
- Decomposition products can form new compounds by solid-state reactions.
- Amorphous regions can become crystalline.
- Various crystalline modifications can undergo reversible conversions.
- The pore structure and mechanical strength of precipitated catalysts can change.

3.2.7 Reduction

Metal catalysts are often prepared by the reduction of precursors that contain the active component as an oxide or chloride. In the reduction with H_2 or other reducing gases at elevated temperatures, the metal nuclei usually form during an induction period. As soon as many nuclei have accumulated, the reduction proceeds rapidly with a considerable heat evolution. Starting with an H_2 stream diluted with N_2 or using milder reducing agents (e.g. alcohol vapors) is advantageous to avoid sintering of the metal during the initial stages of reduction.

3.2.8 Shape Formation of the Catalyst Particles

Precipitated catalysts and supports for impregnation need to be formed into suitably sized particles for use in the reactor. The size and shape of catalyst particles depend on the nature of the reaction, the reactants and the reactor used. Usually, pore diffusion effects in the catalyst particle, pressure drop across the packed bed reactor, radial heat transfer in the wall cooled or heated packed bed reactors, and the multiphase flow behavior in the fluidized bed and slurry reactors are considered when choosing the particle size and shape.

Catalysts are formed by a variety of methods depending on the rheology of the materials. The products of different processes have been compared in general terms in Table 3.2. The choice of the method depends on the size, shape and density of the catalyst particle required, on the strength required and on the properties of the starting material. The three main processes used in catalyst manufacture to make conveniently sized particles from powders are pelletizing, extrusion and granulation.

Microspherical particles are obtained by spray drying a solution or slurry. This method is used to produce all fluid cracking catalysts. The production of spherical gel-type catalyst of uniform size involves the injection of droplets of a liquid into oil, just after the addition of the precipitation agent. For example, silica gel spheres are formed from a mixture of nitric acid and an aqueous solution of sodium silicate. The droplets descend in a column of oil as beads and are removed at the bottom for an aging step and subsequent careful washing. Spherical catalysts can also be prepared on a tilted rotating pan into which the catalyst powder containing a binder is fed. The angle of the tilt and the speed of rotation regulate the size of the spheres formed. Other methods for forming spherical particles include tumbling short, extruded cylinders in a rotating drum and also briquetting. In the latter technique, powder and binders are fed between two rotating rolls, the surfaces of which are provided with matching hollowed out hemispheres.

A common form of catalyst is the cylindrical pellet, produced either by extrusion or by tableting (pelletizing). Extrusion produces low-density high-porosity pellets containing many macropores. Extruded pellets below 1 mm diameter are obtainable from pastes of fine powders or gels, by using either ring-roll or auger-type extruders. In the ring-roll method, the mix is fed to a rotating cylinder drilled with numerous holes of a given size. Inside the cylinder one or more compression rolls press the mix through the holes, while on the outside of the cylinder a knife cuts and removes the extrudate. In the Auger extruder a screw presses a perforated die through the mix and a rotating knife cuts off the extrudate to form pellets.

Cross-sectional shapes with a higher external surface per unit particle volume than cylinders (e.g. a clover leaf) can give even larger external specific areas on extruded catalyst; these have been used in some hydrodesulfurization catalysts.

Tableting produces strong pellets suitable for charging tall reactors requiring particles of a high mechanical strength. Tableting consists of compressing a certain volume of usually dry powder in a die between two moving punches, one of which also serves to eject the formed pellet. To ensure uniform filling of the dies in the tableting machine, the feed must be a dry, fine material with good flow characteristics. Irregular shaped particles, dusty powders, and crystalline materials are hard to tablet. Tablet sizes below 3 mm are not practical because of the high costs and the difficulties connected with filling small dies uniformly. Tableted catalysts are unsuitable for reactions in which diffusion limitations affect selectivity, because the surface layers of the tablets may be too strongly compressed and almost exclusively contain micropores. Pelletizing is the preferred forming technique for a catalyst with dimensions of a few millimeters, when catalytic reactions are not or only a bit pore-diffusion limited.

3.3 Physical Characterization of Catalysts

The chemical composition is not the only factor determining the activity of catalysts. In many cases, the physical characteristics of the catalysts, such as the surface area, particle porosity, pore size, and pore size distribution influence their activity and selectivity for a specific reaction significantly. The importance of the catalyst pore structure becomes obvious when one considers the fact that it determines the transport of reactant and products from the outer catalyst surface to the catalytic surface inside the particle.

Several experimental methods are available to characterize catalyst pore structure. Some of them, useful in quantifying mass transfer of reactant and product inside the porous particle, will be only briefly discussed here. More details concerning methods for the physical characterization of porous substances are given by various authors [5,8,9].

3.3.1 Void Fraction

The porosity of a catalyst or support can be determined simply by measuring the particle density and solid (skeletal) density or the particle and pore volumes. Particle density ρ_p is defined as the mass of catalyst per unit volume of particle, whereas the solid density ρ_s as the mass per unit volume of solid catalyst. The particle volume V_p is determined by the use of a liquid that does not penetrate in the interior pores of the particle. The measurement involves the determination by picnometry of the volume of liquid displaced by the porous sample. Mercury is usually used as the liquid; it does not penetrate in pores smaller than 1.2 μm at atmospheric pressure. The particle weight and volume give its density ρ_p. The solid density can usually be found from tables in handbooks: only in rare cases is an experimental determination required. The same devices as for the determination of the particle density can be used to measure the pore volume V_{pore}, but instead of mercury a different liquid that more readily penetrates the pores is used, such as benzene. More accurate results are obtained if helium is used as a filling medium [10]. The porosity of the particle can be calculated as:

$$\varepsilon_p = \frac{V_{pore}}{V_p} = \frac{\rho_s - \rho_p}{\rho_s}$$

3.3.2 Surface Area

3.3.2.1 Physical Adsorption Methods

The most common method of measuring surface areas involves the principles of physical adsorption by van der Waals' electrostatic forces. It is similar in character to the condensation of vapor molecules onto a liquid of the same composition. The surface area is determined by measuring the amount of gas adsorbed in a monolayer. The total surface area S_t is obtained from the product of two quantities: the number of molecules needed to cover the sample with a monolayer of adsorbate $a_m N_A$ and the amount of surface occupied by a molecule of a particular adsorbed species s_0:

$$S_t = a_m N_A s_0$$

Here, a_m is the number of moles of gas adsorbed in the monolayer and N_A is the Avogadro number (6.02×10^{23} molecules per mole). Values of s_0 are available for different gases [11].

The values of a_m can be found from the measurement of the amount of adsorbed gas and an **adsorption isotherm** (the relationship between the partial pressure of the adsorbate and the amount adsorbed at equilibrium at constant temperature). A simple adsorption isotherm is given by the Langmuir equation

$$\frac{a}{a_m} = \frac{bP / P_0}{1 + bP / P_0} \tag{3.1}$$

where a is the amount of gas adsorbed at pressure P, P_0 is the adsorbent saturation pressure at the experimental temperature and b the adsorption coefficient (independent of pressure).

The Langmuir equation for the adsorption isotherm is not suitable for physical adsorption because it only applies to monolayer adsorption. In practical work the semi-empirical equation of Brunauer, Emmet and Teller (BET equation) is used:

$$\frac{P / P_0}{a(1 - P / P_0)} = \frac{1}{a_m c} + \frac{(c-1)}{c a_m} \frac{P}{P_0} \tag{3.2}$$

where c is a dimensionless constant, dependent on the heats of adsorption and evaporation of the adsorbent. Equation 3.2 extends the Langmuir adsorption isotherm to multilayer adsorption; it takes into account the generation of additional adsorbed layers on some parts of the surface, before a complete monolayer has formed. The difference between two isotherms is shown in Figure 3.2.

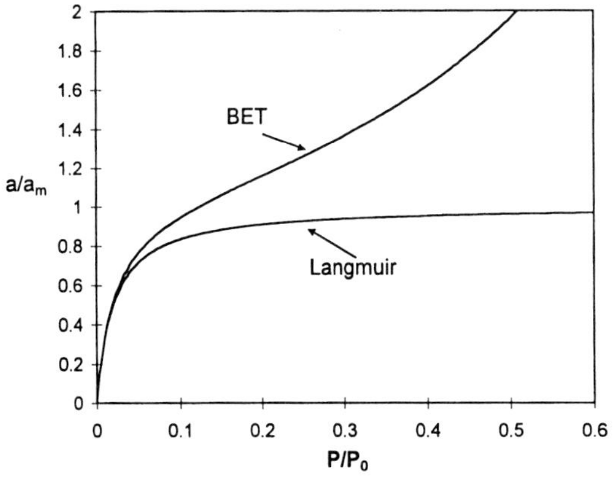

Figure 3.2 Physical adsorption isotherms: (a) Langmuir ($b = 50$); (b) BET ($c = 50$).

The BET equation contains two unknown constants a_m and c. Hence measurements of a at two pressures are enough to find a_m from Equation 3.2 and, consequently, S_t. Sometimes, for adsorbed gases $c >> 1$ and the BET equation simplifies to

$$a = \frac{a_m}{1 - P/P_0}$$

provided that $P/P_0 >> 1/c$. The last equation requires measurement only at one point. Usually $c \approx 1$, since the heat of adsorption and evaporation generally are almost equal. In that case, the more general BET equation must be used.

For many systems the BET equation holds in the range $P/P_0 = 0.05\text{-}0.30$. Any condensable inert vapor can be used in the BET method, but for reliable measurements the molecules should be small and approximately spherical. P/P_0 values of 0.05- 0.30 are conveniently attainable. The choice is usually nitrogen, in which case measurements are executed at cryogenic temperatures, using liquid nitrogen as coolant (the boiling point of nitrogen at atmospheric pressure is -195.8 °C). The effective cross-sectional area of an adsorbed molecule of nitrogen is usually taken to be 0.162 nm^2. The partial pressures of nitrogen gas are in the range 10 – 100 kPa in order to obtain values of P/P_0 in the range 0.05 – 0.30.

The amount of adsorbed gas can be measured either gravimetrically or volumetrically, and a variety of types of apparatus have been designed and used. A continuous procedure may also be used. In this method a nonadsorptive carrier gas, such as helium, is passed through the sample at a known rate with varying concentrations of nitrogen introduced in stages. The amount adsorbed can be determined by conventional gas chromatography.

The use of a single fixed value for the effective cross-sectional area of an adsorbed gas molecule s_0 assumes it is unaffected by the nature of the solid. For most adsorbates on most solids the effective value of s_0 varies slightly from solid to solid. Errors due to different values of s_0 can be avoided by comparing the experimental adsorption isotherm with those for materials, for which the surface has been determined by different, independent methods [12].

3.3.2.2 Gas Permeability Methods

There are methods for measuring the surface area using the flow of gases through a porous medium, by finding a pressure drop across the particle. The equation relating the mole gas flow rate J and the pressure drop ΔP can be written as [13]

$$J = \frac{K}{RT}\frac{\Delta P}{L} \tag{3.3}$$

where L is the sample length and K the permeability coefficient of the porous medium. For gas flow in porous media the permeability coefficient is given by

$$K = \frac{B_0 \overline{P}}{\mu} + \frac{4}{3} K_0 \overline{v} \tag{3.4}$$

where $\overline{P}$ is the mean pressure across the particle, μ is the gas viscosity, $\overline{v} = (8RT/\pi M)^{1/2}$ is the average speed of the gas molecules and B_0 and K_0 are the permeability coefficients that depend only on geometrical characteristics of the pores. The first term in Equation 3.4 gives the permeability of the porous material in the case of purely viscous flow, i.e. when the mean free path of the gas molecules is considerably greater than the mean pore diameter. The second term in Equation 3.4 accounts for the gas velocity at the wall of the pores not being equal to zero. This phenomenon is known as the gas slip and it can not be neglected in many practical situations of small pore sizes. The second term becomes predominant when the length of the mean free path is much larger than the pore size.

For broad classes of porous systems made up of discrete particles the following relationships are recommended for B_0 and K_0 [13]:

$$B_0 = \frac{1}{5S^2} \frac{\varepsilon_p^3}{(1-\varepsilon_p)^2} \tag{3.5}$$

$$K_0 = \frac{0.45}{S} \frac{\varepsilon_p^2}{(1-\varepsilon_p)} \tag{3.6}$$

Substituting Equations 3.5 and 3.6 into 3.3 gives the relationship between the permeability K and the specific surface area:

$$K = \frac{1}{5S^2} \frac{\varepsilon_p^3}{(1-\varepsilon_p)^2} \frac{\overline{P}}{\mu} + \frac{0.6}{S} \frac{\varepsilon_p^2}{(1-\varepsilon_p)} \overline{v} \tag{3.7}$$

Once the permeability K has been found from Equation 3.3 using the experimentally determined pressure drop and gas flow rate, the specific surface area S can be calculated from Equation 3.7. If the pore size is large compared to the mean free path the second term in Equation 3.7 can be neglected, and

$$S = \frac{\varepsilon_p^{3/2}}{1-\varepsilon_p} \sqrt{\frac{\overline{P}}{5K\mu}}$$

In the opposite limiting case of small pores, or Knudsen diffusion, the specific surface area can be found as

$$S = \frac{0.6}{K} \frac{\varepsilon_p^2}{(1-\varepsilon_p)} \overline{v}$$

The accuracy of the permeability method depends on the available relationship between the permeability parameters B_0 and K_0 and the structural parameters of porous media. When the pore texture is not sufficiently random and uniform, the accuracy of Equations 3.5 and 3.6 (and consequently of the method) becomes poor. This is a serious disadvantage of the permeability method, compared to the adsorption method, which does not depend on the pore texture. Nevertheless, permeability measurements are indicative of the porous structure and are useful for the determination of the parameters of the transport models. Various other experiments on flow and diffusion are also indicative of the texture of the porous particles. Some are discussed in Chapter 5.

3.3.3 Pore-size Distribution

3.3.3.1 Capillary Condensation

A curved liquid-vapor interface gives rise to a change in the saturation vapor pressure. This phenomenon is governed by the Kelvin equation

$$\ln\left(\frac{P}{P_s}\right) = -\frac{1}{2}\left(\frac{1}{r_1}+\frac{1}{r_2}\right)\frac{2\sigma V_{mol}}{RT} = -\left(\frac{1}{r_1}+\frac{1}{r_2}\right)\frac{\sigma V_{mol}}{RT} \tag{3.8}$$

where P is the saturation vapor pressure at the meniscus (pressure at which condensation occurs), P_s is the normal saturation vapor pressure (saturation pressure for a flat surface), $1/2\ (1/r_1 + 1/r_2)$is the meniscus curvature with the principal radii r_1 and r_2 (the radius is positive for a concave surface), V_{mol} is the molar volume of the liquid adsorbate and σ its surface tension. For a spherical meniscus $r_1 = r_2 = r$ and Equation 3.8 transforms to

$$\ln\left(\frac{P}{P_s}\right) = -\frac{2\sigma V_{mol}}{rRT} \tag{3.9}$$

For a cylindrical meniscus $r_1 = r$, $r_2 \rightarrow \infty$ and

$$\ln\left(\frac{P}{P_s}\right) = -\frac{\sigma V_{mol}}{rRT} \tag{3.10}$$

In a circular capillary of radius r partly filled with a liquid $r_1 = r_2 = r/\cos(\theta)$, where θ is the contact angle between adsorbate and solid, the Kelvin equation gives

$$\ln\left(\frac{P}{P_s}\right) = -\frac{2\sigma V_{mol}\cos(\theta)}{rRT}$$

The phenomenon of capillary condensation provides a method for measuring pore-size distribution. Nitrogen vapor at the temperature of liquid nitrogen for which $\cos(\theta) = 1$ is universally used. To determine the pore size distribution, the variation in the amount of nitrogen inside the porous particle is measured when the pressure is slightly increased or decreased. This variation is divided into two parts: one part is due to true adsorption and the other to capillary condensation. The variation due to adsorption is known from adsorption experiments with nonporous substances of known surface area, so the variation due to condensation can be calculated. The volume of this amount of nitrogen is equal to the volume of pores with the size as determined by the Kelvin equation. Once a certain model has been selected for the complicated pore geometry, the size of the pores can be calculated. Usually it is assumed that an array of cylindrical capillaries of uniform but different radii, and randomly oriented represents the porous medium. So the Kelvin equation in the form of Equation 3.9 is used. Since condensation is combined with adsorption, the thickness of the adsorption layer should also be taken into account; this correction is especially important for small pores. So the radius in Equation 3.9 is the difference between the actual pore radius and the thickness of the adsorbed layer. The method of analysis assumes the properties of the condensed phase in the capillaries are the same as those of the bulk liquid.

The capillary condensation method is widely used for the investigation of mesopores. The size of the largest pores, that can be measured, is limited by the rapid change of the meniscus radius with pressure as the relative pressure P/P_0 nears unity (because of massive condensation of the adsorbent around the boiling point). This holds for radii of about 100 nm. The smallest pore sizes that can be determined by this method are about 1.5 nm.

A somewhat different relationship between the amount of vapor sorbed and the pressure is often experimentally obtained upon decreasing rather than increasing the pressure. This is explained by the different curvature of the meniscus during adsorption and desorption. For example, consider a cylindrical pore. The adsorption is determined by the curvature of the liquid film on the pore surface. If the film thickness is much lower than the pore radius r, the curvature of the surface is $1/(2r)$ and the process is described by Equation 3.10. During desorption the liquid covers the entire cross-section of the pore and has a spherical meniscus of curvature $1/r$, and Equation 3.9 holds. Comparison of Equations 3.9 and 3.10 shows that a higher pressure is required to fill pores.

Since the difference in the meniscus curvature depends on the pore shape, the hysteresis loop, which is obtained by comparing adsorption and desorption isotherms, provides some information about the shape and interconnection of the pores.

3.3.3.2 Mercury Porosimetry

Most surfaces are not wetted by mercury, so an external pressure is required to force it into a capillary. The pressure difference that exists across a curved gas-liquid interface due to the surface tension is given by the Laplace equation

$$\Delta P = -\sigma\left(\frac{1}{r_1}+\frac{1}{r_2}\right) \tag{3.11}$$

where r_1 and r_2 are the principal radii of curvature and σ is the surface tension. For a cylindrical pore of radius r, partly filled with a liquid, we have $r_1 = r_2 = r/\cos(\theta)$ and Equation 3.11 then reduces to

$$\Delta P = -\frac{2\sigma\cos(\theta)}{r} \tag{3.12}$$

where θ is the contact angle with the surface and $\sigma = 0.471$ N m^{-1} for an air-mercury interface at 298 K. The contact angle between mercury and a wide variety of materials varies between 135° and 150° [14]. Equation 3.10 predicts, for $\theta = 140°$, that all cylindrical pores with radii larger than 7600 nm will be filled at atmospheric pressure. Pores with radii larger than 2.5 nm will be filled at a pressure above 300 MPa.

The increase in the amount of mercury forced into the porous material upon a pressure increase corresponds to filling of pores of a size given by Equation 3.10. The smallest pore sizes detected depend upon the pressure to which mercury can be subjected in a particular apparatus. High-pressure porosimeters have been built to measure pores down to 2.5 nm. At a pressure of 0.02 MPa, pores as large as 40 000 nm can be measured.

Because the shape of the pores is not exactly cylindrical, as assumed in the derivation of Equation 3.12, the calculated pore size and pore size distribution can deviate appreciably from the actual values shown by electron microscopy. If a pore has a triangular, rectangular or more complicated cross-section, as found in porous materials composed of small particles, not all the cross-sectional area is filled with mercury: due to surface tension mercury does not fill the corners or narrow parts of the cross-section. A cross-section of the solid, composed of a collection of nonporous uniform spheres during filling the void space with mercury is shown in Figure 3.3. Mayer and Stowe [15] have developed the mathematical relationships that describe the penetration of mercury (or other fluids) into the void spaces of such spheres.

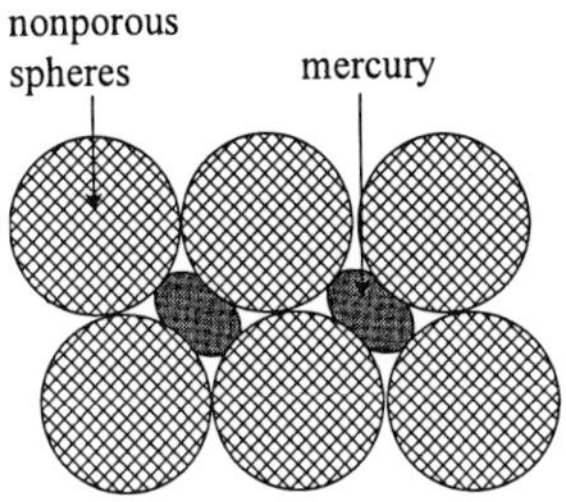

Figure 3.3 The filling of noncircular pores with mercury.

The porosimetry method permits the study of porous structures in a wider pore-size range than does the method of capillary condensation. The two methods do not always agree but porosimetry is better for sizes of 10 nm and higher, whereas the capillary condensation method is more suitable for sizes below 10 nm. A more direct overview of the pore structure of catalysts is obtained by using the stereoscan electron microscope [16].

3.4 Mass and Heat Transfer in Porous Catalysts

3.4.1 Introduction

When a chemical reaction takes place inside a porous catalyst, all reactants and products must find their way to and from the internal catalytically active surfaces. This transport can strongly influence the apparent reaction rates and selectivity, hence various mass transport models have been developed. Some of the best known and used models and the basic principles behind these models are described in this chapter. Since only little systematic information is available about transport in liquids, the discussion is mainly concerned gaseous mixtures.

Transport of molecules of any gaseous species in pores can be hindered by encounters with the wall of the pores and is influenced by collisions between molecules (also of other species).

The relative frequency of the intermolecular collisions and collisions between molecules and the pore wall can be characterized by the ratio of the length of the mean free path λ and the equivalent pore diameter $d_e = 2r_e$. This ratio determines the flow regime and is called the Knudsen number Kn:

$$Kn = \frac{\lambda}{d_e}$$

The mean free path λ is the average distance traveled by a molecule between successive collisions [17]. For a single gas and assuming the molecules are rigid spheres of diameter d, λ is given by

$$\lambda = \frac{1}{\sqrt{2}\pi d^2 n}$$

where $n = P/(kT)$ is the number of molecules per unit volume and k is the Boltzmann constant (1.38066×10^{-23} J K^{-1}). If the kinematic viscosity ν is known, the mean free path can be found as

$$\lambda = 3\nu\left(\frac{\pi M}{8RT}\right)^{1/2}$$

where M is the molecular weight, R the universal gas constant and T the absolute temperature.

Typical molecular diameters and mean free paths at atmospheric pressure and a temperature of 15 °C are given in Table 3.3 for different gases.

The mean free path is inversely proportional to the pressure and varies from 30-100 nm at 0.1 MPa to 0.3-1 nm at 10 MPa for normal gases. As previously described, the average size of the pores of commercial catalysts varies over a wide range (1- 200 nm). Therefore, typical values of the Knudsen number are 10^{-4} – 10^{2}. This rough estimate shows that different flow regimes of different complexity occur in practice.

The description of multicomponent gas transport in porous media at arbitrary Knudsen numbers is very complicated, especially in the transition region of $Kn \approx 0.1 - 10$. This problem has never been solved rigorously. The situation remains extremely complex, even if the porous medium is regarded as composed of long, straight, circular channels of equal diameter. In these circumstances a consideration of simplified and limiting situations, which are better understood, is very important. Moreover, the approximate relationship between mass fluxes of the species and concentrations can be found by an appropriate combination of the corresponding relationships for simpler situations.

The three types of simplified conditions generally considered for gas transport are:

- *Free-molecule or Knudsen flow* ($Kn >> 1$). The molecules are transported within the porous structure without intermolecular collisions; however, the molecules do collide with the pore walls. Collisions between molecules can be ignored compared to collisions of molecules with the walls of the porous medium or tube.

Table 3.3 Molecular diameter and mean free path in different pure gases at 0.1 MPa and 15 °C (from Kennard [18])

Gas	Diameter (nm)	Mean free path (nm)
H_2	0.274	117.7
He	0.218	186.2
CH_4	0.414	51.6
NH_3	0.443	45.1
H_2O	0.460	41.8
N_2	0.375	62.8
C_2H_4	0.495	36.1
C_2H_6	0.530	31.5
O_2	0.361	67.9
Ar	0.364	66.6
CO_2	0.459	41.9
Air	0.372	64.0

- *Continuum diffusion* ($Kn << 1$). The different species of a mixture move relative to each other under the influence of concentration gradients (ordinary or concentration diffusion), temperature gradients (thermal diffusion) or external forces (forced diffusion). Here molecule-wall collisions are neglected.
- *Viscous flow of a pure fluid* ($Kn << 1$). The gas acts as a continuum fluid driven by a pressure gradient, and both molecule-molecule collisions and molecule-wall collisions are important. This is sometimes called convective or bulk flow.

These situations, as well as an additional mode of transport known as a surface diffusion, are discussed below. For simplicity, only one-dimensional, steady-state transport is considered.

3.4.2 Ordinary Diffusion in Multicomponent Gases

3.4.2.1 Introduction

When considering ordinary diffusion only, the momentum conservation equation for the species i in a multicomponent system can be written as [17]

$$-\frac{dP_i}{dz}=\sum_{j=1}^{n}\frac{Px_ix_j(u_i-u_j)}{D_{ij}} \tag{3.13}$$

where P is the total pressure, $P_i = x_iP$ is the partial pressure of species i, x_i and x_j are the mole fractions of species i and j and u_i and u_j are the mean velocities of their molecules in the z direction; the coefficient D_{ij} is known as the binary diffusivity for gas pair i and j.

Equations 3.13 are known as the Stefan-Maxwell equations and are valid when the total pressure and temperature gradients as well as external forces can be neglected. They have the physical meaning that the rate of momentum transfer between two species is proportional to their concentrations and to the difference in their velocities. The molar average velocities of the species v_i and v_j are defined in a such way that the molar fluxes of the various species are

$$J_i = C_i v_i = C x_i v_i$$

where $C = P/RT$ is the total molar concentration. With these relationships the Stefan-Maxwell equations can be written in terms of molar fluxes J_i:

$$-\frac{dx_i}{dz} = \sum_{j=1}^{n} \frac{x_j J_i - x_i J_j}{CD_{ij}} \tag{3.14}$$

In the case of a two-component mixture of A and B Equation 3.14 for component A gives

$$-\frac{dx_A}{dz} = \frac{(1-x_A)J_A - x_A J_B}{CD_{AB}}$$

The last equation can be rewritten as

$$J_A = x_A(J_A + J_B) - CD_{AB}\frac{dx_A}{dz}$$

This arrangement shows the diffusion flux J_A is the resultant of two terms: the first is the molar flux of A resulting from the bulk motion of the fluid, and the second is the molar flux of A resulting from the diffusion superimposed on the bulk [17].

The ordinary diffusion equations have been presented for the case of a gas in absence of porous medium. However, in a porous medium, whose pores are all wide compared to the mean free path and provided the total pressure gradient is negligible, it is assumed that the fluxes will still satisfy the relationships of Stefan-Maxwell, since intermolecular collisions still dominate over molecule-wall collisions [19]. In the case of diffusion in porous media, the binary diffusivities are usually replaced by effective diffusion coefficients, to yield

$$-\frac{dx_i}{dz} = \sum_{j=1}^{n} \frac{x_j J_i - x_i J_j}{CD_{ij}^e} \tag{3.15}$$

Here,

$$D_{ij}^e = B_1 D_{ij} \tag{3.16}$$

where B_1 is a dimensionless parameter determined by the geometry and size of the pores. For practical purposes it is suitable to define the fluxes in Equation 3.15 per unit total area of the particles rather than per unit free area in the particle. The presence of the solid phase limits the cross-sectional area for flow of species into the medium. Thus, B_1 should be proportional to the particle void fraction or porosity ε. It should also take into account that the molecules cannot diffuse in straight lines, but have to follow the indirect paths of the pores. As a consequence, molecules have to diffuse over a longer path when going from one point to the other. This is accounted for by the tortuosity γ_P. Thus it is common to present B_1 in the form

$$B_1 = \varepsilon_p / \gamma_p$$

As a result, D^e_{ij} is typically a factor 10-100 less than D_{ij}, because usually $0.01 < \varepsilon_p/\gamma_p < 0.1$. The diffusion coefficients D_{ij} and D^e_{ij} are referred to as bulk diffusion coefficients. Their values and methods for their calculation and experimental determination will be discussed later.

It should be noted that the fluxes J_i and J_j in Equation 3.15 have a different meaning than in Equation 3.14. In Equation 3.15 they cannot be interpreted as quantities having meaningful values at a point. Rather, they have to be interpreted as quantities averaged over a small neighborhood of the point in question – a neighborhood that is small with respect to the dimension of the catalyst particle but large with respect to the dimensions of the passages inside the porous particle.

3.4.2.2 Free-molecule or Knudsen Flow

When the size of the pores is much smaller than the molecular mean free paths in the gas mixture, collisions of gas molecules with the pore wall are more frequent than inter-molecular collisions. This type of gas transport is also known as Knudsen diffusion. In this case each molecule acts independently of all the others and each component in a mixture behaves as though it were present alone. The movement of each molecule can be conveniently pictured as a random walk between the walls of pores. This leads to the following expression for the molar flux of species A:

$$J_{Ak} = -D_{Ak} \frac{dC_A}{dz} \tag{3.17}$$

Where the diffusion coefficient, known as the Knudsen diffusivity D_{Ak}, is proportional to the product of the average speed of molecules and an appropriate mean step size λ_k, the following holds true:

$$D_{Ak} \sim \lambda_k \, \bar{v}_A = \lambda_k \left(\frac{8RT}{\pi M_A} \right)^{1/2}$$

The average speed of a molecule in a given gas depends only on the temperature. At normal temperature and for gases of molecular mass $\approx$ 30 g mol^{-1} this is of the order of 500 m s^{-1}. Light hydrogen molecules move with an average speed of 2000 m s^{-1} at the same temperature. We may expect that, for a porous medium with a monodisperse porous structure, λ_k is of the order of the mean pore diameter:

$$\lambda_k \approx d_e = \frac{4V_p}{S_p}$$

Thus,

$$D_{Ak} = B_2 \frac{V_p}{S_p} \left(\frac{8RT}{\pi M_A} \right)^{1/2}$$

where the proportionality coefficient B_2 is determined by pore geometry and the law for molecule scattering on a surface. Generally, it should be determined experimentally. This coefficient can be calculated for pores that can be represented by capillaries, i.e. which are very long relative to their diameter. For such model pores we obtain

$$D_{Ak} = \frac{2\varepsilon_p r_e}{3\gamma_p} \left(\frac{8RT}{\pi M_A} \right)^{1/2} \tag{3.18}$$

The void fraction ε_p and tortuosity γ_p are introduced in Equation 3.18 for the same reason as in Equation 3.16. In deriving this equation it has been assumed that the pore size is still large with respect to molecular dimensions.

3.4.2.3 Viscous Flow

The basic law governing viscous flow of pure fluid through a porous medium with pores much larger than the mean free path is that of Darcy [13]. This law states that the rate of flow is directly proportional to the pressure gradient causing the flow. In terms of mole flow rate Darcy's law can be written as

$$J = -\frac{B_0 C}{\mu} \frac{dP}{dz} \tag{3.19}$$

in which μ is the fluid viscosity and B_0 the permeability of the porous medium. Permeability is a geometrical characteristic of the porous structure and has the unit of m^2. The molar flux in this equation has the same meaning as the fluxes in Equations 3.15 and 3.17. It is the superficial flux (molar rate of flow through a unit cross-sectional area of the solid plus fluid) averaged over a small regional space – small with respect to macroscopic dimensions in the flow system but large with respect to the pore size.

For a porous medium it is usually easier to determine B_0 experimentally than to calculate it from geometrical considerations, because of the complicated structure of real pores [13]. As a rough approximation the permeability can be calculated (see Equation 3.5) as

$$B_0 = \frac{1}{5S^2} \frac{\varepsilon_p^3}{(1-\varepsilon_p)^2}$$

where S is the pore surface area per unit particle volume. In the case of a uniform gas mixture instead of a pure gas, it is assumed a mixture behaves the same as a single gas. This means fluid flow does not cause a mixture to separate into its components. The flux of each species separately is then given by

$$J_i = x_i J$$

3.4.2.4 Surface Diffusion

In adsorbing systems, molecules adsorbed on the surface of pores may still have a considerable mobility. Migration of molecules in the adsorbed phase is referred to as surface diffusion. Its contribution to the total flux depends on the particular adsorbate-adsorbent pair, system temperature, the thickness of the adsorbed layer, etc. Although the mobility of molecules in the adsorbed layer is generally much lower than in the gas phase, the concentrations in the layers are much higher. Thus the contribution of surface diffusion can be significant.

Governed by the nature of the surface species, diffusion may be categorized as diffusion of physically adsorbed molecules and of chemisorbed species, and self-diffusion. The latter refers to the diffusion of atoms, ions, and clusters on the surface of their own crystal lattices and has been studied mostly for metals. All three categories of surface diffusion are of importance in catalysis. A review concerning all categories is available [20].

The use of the surface concentration gradient as the driving force for surface diffusion is the most popular approach with which to describe the mobility of adsorbed molecules. When each adsorbed species is assumed to be at adsorption equilibrium and transported along the surface independently of the other species, the molar flux of species i due to surface diffusion can be written as

$$J_{i,surf} = -D_{i,surf} \frac{dC_i}{dz}$$

where $D_{i,surf}$ the surface diffusion coefficient of component i. This has the same dimensions as ordinary and Knudsen diffusivities, $m^2 s^{-1}$.

One of the most intriguing aspects of surface diffusion is the strong dependence of the diffusivity on sorbate concentration. The dependence of surface diffusivities on pressure, temperature and composition is much more complicated than those of the molecular and Knudsen diffusivities, because of all the complexities of porous medium geometry, surface structure, adsorption equilibrium, mobility of adsorbed molecules, etc.

3.4.3 Models of Mass Transport in Porous Media

3.4.3.1 Dusty Gas Model and Binary Friction Model

Equations 3.15, 3.17 and 3.19 provide the flux relationships in the limiting regimes. There remains the problem of finding the flux relationships in intermediate situations, where the pore size is comparable to the mean free path and the mixture is a multicomponent one. At present, no quantitative kinetic theory exists for flow in the transition region where the dimensions of λ and d_e are comparable. Therefore different simplified models have been developed.

The dusty gas model (DGM) [21] is used most frequently to describe multi component transport in between the two limiting cases of Knudsen and molecular diffusion. This theory treats the porous media as one component in the gas mixture, consisting of giant molecules held fixed in space. The most important aspect of the theory is the statement that gas transport through porous media (or tubes) can be divided into three independent modes or mechanisms:

- free-molecule or Knudsen flow
- viscous flow
- ordinary diffusion.

This model assumes the diffusive flows combine by the additivity of momentum transfer, whereas the diffusive and viscous flows combine by the additivity of the fluxes. To the knowledge of the authors there has never been given a sound argument for the latter assumption. It has been shown that the assumption may result in errors for certain situations [22]. Nonetheless, the model is widely used with reasonably satisfactory results for most situations. Temperature gradients (thermal diffusion) and external forces (forced diffusion) are also considered in the general version of the model. The incorporation of surface diffusion into a model of transport in a porous medium is quite straightforward, since the surface diffusion fluxes can be added to the diffusion fluxes in the gaseous phase.

The flux equation for species i in an n-component mixture can be written as

$$-\frac{1}{RT}\frac{dP_i}{dz} = \sum_{j=1}^{n}\frac{x_j J_i - x_i J_j}{D_{ij}^e} + \frac{J_i}{D_{ik}} + \frac{B_0 P_i}{\mu RT D_{ik}}\frac{dP}{dz} \tag{3.20}$$

Focusing on a binary mixture of A and B the flux equations are

$$\frac{1}{RT}\frac{dP_A}{dz} = \frac{x_B J_A - x_A J_B}{D_{AB}^e} + \frac{J_A}{D_{Ak}} + \frac{B_0 P_A}{\mu RT D_{Ak}}\frac{dP}{dz} \tag{3.21}$$

$$\frac{1}{RT}\frac{dP_B}{dz} = \frac{x_A J_B - x_B J_A}{D_{BA}^e} + \frac{J_B}{D_{Bk}} + \frac{B_0 P_B}{\mu RT D_{Bk}}\frac{dP}{dz} \tag{3.22}$$

Eliminating J_B from Equations 3.21 and 3.22 and assuming $D^e_{AB}=D^e_{BA}$, the flux of A is given by

$$J_A = -CD_A \frac{dx_A}{dz} - x_A \frac{CK_A}{P}\frac{dP}{dz} \tag{3.23}$$

where

$$D_A = \frac{D_{Ak} D^e_{AB}}{D^e_{AB} + x_A D_{Bk} + x_B D_{Ak}}$$

$$K_A = \frac{D_{Ak}(D^e_{AB} + D_{Bk})}{D^e_{AB} + x_A D_{Bk} + x_B D_{Ak}} + \frac{PB_0}{\mu}$$

For the flux of the second component B the suffixes A and B should be interchanged.

Jackson discusses generalizations of the DGM, accounting for the details of the porous structure [19]. The Mean Transport Pore Model is also relevant for the present discussion [23].

As an alternative to the DGM, the Binary Friction Model (BFM) is of interest. This principally new model has been developed recently [22]. The BFM flux equation, contending with Equation 3.20, is:

$$-\frac{1}{RT}\frac{dP_i}{dz} = \sum_{j=1}^{n} \frac{x_j J_i - x_i J_j}{D^e_{ij}} + \left(D_{ik} + \frac{B_0 P}{\mu_{i,f}} \right)^{-1} J_i \tag{3.24}$$

where $\mu_{i,f}$ (which can be interpreted as a viscosity of species i in the mixture) is called the fractional viscosity. This term can be derived from kinetic gas theory and the semi-empirical formula of Wilke [17]:

$$\mu_{i,f} = \frac{\mu_i}{\sum_{j=1}^{n} x_j \Phi_{ij}} \tag{3.25}$$

in which

$$\Phi_{ij} = \frac{1}{\sqrt{8}} \left(1 + \frac{M_i}{M_j} \right)^{-1/2} \left[1 + \left(\frac{\mu_i}{\mu_j} \right)^{1/2} \left(\frac{M_j}{M_i} \right)^{1/4} \right]^2$$

Here μ_i and μ_j are the viscosities of species i and j at system temperature and pressure. Other parameters in Equation 3.24 are the same as those of the DGM. The flux relationship in the form of Equation 3.24 can be explained relatively simply on the basis of momentum transfer considerations. The first term on the right-hand side describes the momentum transfer between species i and other species, the second term describes the

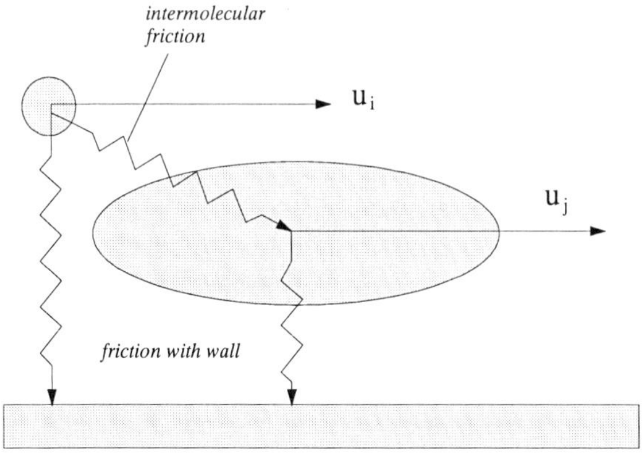

Figure 3.4 Schematic representation of forces acting between molecules, and between molecules and solid wall (adapted from Kerkhof [22]).
Reprinted from Chemical Engineering Journal, **64**, P.J.A.M. Kerkhof, A modified Maxwell-Stefan model for transport through inert membranes: the binary friction model, 319-344, 1996, with kind permission from Elsevier Science S.A., P.O. Box 564, 1001 Lausanne, Switzerland.

momentum transfer to the pore walls, for species *i* as drawn schematically in Figure 3.4. The sum of these terms, multiplied by *RT*, gives the total loss of momentum of the species per unit particle volume, which is equal to the partial pressure gradient, if inertial forces can be neglected.

The difference between the DGM and BFM is in the description of molecule-wall collisions. In contrast to the DGM, the BFM does not separate viscous flow and Knudsen diffusion. According to the BFM they are the limiting forms of the same phenomena, described by the second term in the right-hand side of Equation 3.24. For large Knudsen numbers, that is for small pressures and/or small pore sizes,

$$D_{ik} >> \frac{B_0 P}{\mu_{i,f}}$$

and momentum transfer to the solid phase is characterized through the Knudsen diffusion coefficient. Also in this case,

$$D_{ij}^e >> D_{ik}$$

and Equation 3.24 can be replaced by Equation 3.17. For low Knudsen numbers the fractional viscosity $\mu_{i,f}$ is the determining parameter of the momentum transfer to the pore walls.

For binary mixtures the equations of the BFM can be presented in the form of Equations 3.21 and 3.22, found for the DGM, but the coefficients D_A and K_A are different and are determined by equations

$$D_A = \frac{F_A D_{AB}}{D_{AB} + x_A F_B + x_B F_A}$$

$$K_A = \frac{F_A F_B D_{AB}}{D_{AB}^e + x_A F_B + x_B F_A}\left(\frac{1}{D_{AB}^e} + \frac{1}{F_B}\right)$$

where

$$F_A = D_{Ak} + \frac{B_0 P}{\mu_{A,f}}, \qquad F_B = D_{Bk} + \frac{B_0 P}{\mu_{B,f}}$$

Both the DGM and BFM capture many of the important properties of gas transport in porous media in a qualitatively correct way. They describe the limiting situations of small and large Knudsen number discussed above equally well. However, in some situations the model predictions are qualitatively different. The available experiments for the such questionable situations substantiate the BFM. The examples of the similar and different predictions are given in Chapter 9.

According to the models described above and from available experiments on gas transport in porous media, it follows that a pressure gradient usually exists when a chemical reaction occurs in a porous particle. This pressure gradient can be rather large for catalyst pellets with small pores and the admissibility of the neglect of the total pressure gradient should therefore be checked. It is important that for multicomponent mixtures pressure gradients must exist in catalyst particles even if the reaction proceeds without change in the number of molecules [19].

3.4.3.2 Effective Diffusivity Concept

The DGM and its modifications and generalizations, as well as the BFM, are useful in many practical applications. However, their applicability is restrained due to difficulties in the determination of all the necessary parameters, and by the necessary computational effort. These problems are especially pronounced for multicomponent systems, when viscous flow and surface diffusion are also significant. Therefore, simpler models, containing fewer parameters, are commonly used in practice.

Instead of the detailed description of mass transport inside the porous particles an effective diffusivity D_{eA} is commonly used. For species A in a porous medium this is defined by

$$J_A = -D_{e,A} \frac{dC_A}{dz} \tag{3.26}$$

Where C_A is the concentration of gas A within the pores. Equation 3.25 is known as Fick's law of diffusion. In general, the effective diffusivity is concentration dependent. The total pressure gradient may also have an influence on the values of $D_{e,A}$.

The approach based on the effective diffusivity concept is justified in the region of Knudsen flow, where $D_{e,A} = D_{A,k} + D_{A,surf}$. Equation 3.25 with a constant effective diffusivity also follows from the more general DGM equations in the limiting case of dilute mixtures with one species (B) in considerable excess and a negligible pressure gradient. In this case the other species diffuse independently, as in a Knudsen regime, but with the effective diffusion coefficient governed by the equation

$$\frac{1}{D_{e,A}} = \frac{1}{D_{AB}^{e}} + \frac{1}{D_{Ak}} \tag{3.27}$$

Numerous methods have been proposed with which to estimate the effective diffusivity and to represent pore diffusion with only this parameter [4]. In many cases a good estimate will be found with Equation 3.26. For approximate calculations of D^e_{AB} and D_{Ak} Equations 3.16 and 3.18 can be used. In general, Equation 3.26 should be considered as an approximation in which the effective diffusivity is selected so as to incorporate the various factors that were mentioned earlier. Hence, all relevant transport phenomena are lumped into one single coefficient $D_{e,A}$. Although this approach does not guarantee a high precision in conversion rate calculations, it is adequate for qualitative and quantitative predictions of many properties that are of engineering interest. Its use has been demonstrated by many authors [24]. Whenever possible, it is worth checking the applicability of the concept through calculations using more refined models. In general the effective diffusivity $D_{e,A}$ must be measured experimentally. The most widely used experimental methods are discussed in Chapter 5.

3.4.3.3 Liquid – solid Systems

For liquids, there is no complete theory of multicomponent diffusion yet available. For this reason only rough theoretical approaches, as used for the description of mass transport in the porous particles filled with a liquid are discussed. The effective diffusivity concept just described is the only known approach and

$$D_{e,A} = \frac{\varepsilon_p D_{AB}}{\gamma_p} \tag{3.28}$$

The rate of solute diffusion in liquids in porous materials becomes observably lower than would be expected from Equation 3.27 when the solute molecular size becomes large with respect to the pore size. This phenomena can be important for catalytic processing of heavy liquid petroleum fractions [25]. In the case of liquid transport through finely porous materials, the effective diffusivity can be expressed as [26]

$$D_{e,A} = \frac{\varepsilon_p K_p D_{AB}}{\gamma_p K_r}$$

where K_p is the **equilibrium partition factor** and K_r is the **enhanced drag factor**.

The equilibrium partition factor is defined as the ratio of the concentration of species *A* inside and outside the pores. This concept was first introduced by Ferry in terms of a geometric exclusion effect [25]. Since the center of mass of the molecule, assumed to be a hard sphere, cannot be closer to the pore wall than the distance of the molecular radius, Ferry obtained

$$K_p = (1 - \Lambda)^2 \tag{3.29}$$

where Λ is the ratio of the solute molecule size to the pore diameter. The prediction of Equation 3.29 is based on the assumption that neither the solvent nor the solute is adsorbed on the walls of the pore. Experimental data indicate that this is not always a valid assumption. For example, Satterfield et al. [25] measured a partition factor of

182 for $\Lambda = 0.38$. This high value of K_p must be due to preferential adsorption of the solute.

The enhanced drag coefficient accounts for the increased hydrodynamic drag exerted on the moving solute due to proximity of the walls. Faxen [27] used an approximate solution to obtain

$$K_r^{-1} = 1 - 2.104\Lambda + 2.09\Lambda^3 + 0.95\Lambda^5$$

The theoretical predictions of K_p and K_r are not suitable for cases in which either the solute or solvent are preferentially adsorbed on the walls, or when the wall affects the structure of the liquid in the pore. Unfortunately, at present there is no reliable theory for predicting the influence of these effects.

The above equations are for very small pore diameters, i.e. $\Lambda > 0.2$-0.8, meaning a pore size of the order of 1 nm and less.

3.4.3.4 Material Balance Equations in Steady States

The steady state material balance for component A can be written in the form

$$\frac{dJ_{A,x}}{dx} + \frac{dJ_{A,y}}{dy} + \frac{dJ_{A,z}}{dz} = R_A \tag{3.30}$$

where $J_{A,x}$, $J_{A,y}$, and $J_{A,z}$ are the fluxes of component A in the x, y and z directions, respectively, and R_A is the number of moles of A consumed by chemical reaction per unit volume of catalyst particles and per unit time.

Heterogeneous reactions are basically surface reactions. However, the concentration of A can often be assumed to be nearly constant on the scale of the pore diameter and it is possible to treat these heterogeneous reactions as pseudohomogeneous reactions, as in Equation 3.29. Then, the homogeneous production rate may be defined according to

$$R_A = S\tilde{R}_A$$

where $\tilde{R}_A$ is the production rate per unit surface (which does not depend on the porous structure) and S is the specific internal surface area of the particle per unit porous particle volume.

Combination of the material balance Equation 3.30 and the selected equations, which relate the flux vectors for all species to their concentrations, gives a set of equations to be solved subject to the appropriate boundary conditions at the particle surface. If particles cannot be considered as isothermal, the energy balance and energy flux equations are also required.

When Equation 3.16, describing the flux $J_{A,z}$ in z direction, and similar equations for $J_{A,x}$ and $J_{A,y}$ are inserted into Equation 3.29, we obtain (assuming constant diffusivity)

$$D_{e,A}\left(\frac{d^2C_A}{dx^2} + \frac{d^2C_A}{dy^2} + \frac{d^2C_A}{dz^2}\right) = R_A \tag{3.31}$$

This equation is to be solved with the boundary condition $C_A = C_{A,s}$ at the surface of the particle, where $C_{A,s}$ is the known surface concentration.

The mass conservation equation in rectangular coordinates x, y, and z (Equation 3.31) is suitable for catalyst pellets with a parallelepiped shape, in particular when they can be considered as a plate. For particles with cylindrical surfaces and for spheres, the cylindrical and spherical coordinate systems are more natural. For axisymmetrical concentration distributions in cylindrical coordinates Equation 3.30 becomes

$$D_{e,A}\left[\frac{1}{r}\frac{d}{dr}\left(r\frac{dC_A}{dr}\right)+\frac{d^2C_A}{dz^2}\right]=R_A$$

For a central-symmetrical concentration distribution (i.e. spherical coordinates) we obtain

$$D_{e,A}\frac{1}{r^2}\frac{d}{dr}\left(r^2\frac{dC_A}{dr}\right)=R_A$$

The catalyst may not be completely homogeneous on a pellet scale due to the method of manufacture. Furthermore, the extrusion or pelleting processes may render the material anisotropic, so that properties such as the effective diffusivity are not the same in every direction. Thus, the effective diffusivity may be substantially different in the axial and radial directions for cylindrical particles [4]. This should be accounted for in the balance equations.

The described treatment of mass transport presumes a simple, relatively uniform (monomodal) pore size distribution. As previously mentioned, many catalyst particles are formed by tableting or extruding finely powdered microporous materials and have a bidisperse porous structure. Mass transport in such catalysts is usually described in terms of two coefficients, a effective macropore diffusivity and an effective micropore diffusivity.

For practical purposes the concentration distribution inside particles is not needed. For chemical reactor problems the overall conversion rate (per unit reactor volume or per unit of catalyst mass) and its dependence on conditions in the reactor are of interest. These questions are considered in detail in Chapters 6 and 7.

3.4.3.5 Practical Recommendations on Effective Diffusivity

A simple way to calculate catalyst performance is to use the effective diffusivity. It can be estimated by Equation 3.26 for gases or Equation 3.27 for liquids. The porosity-tortuosity ratio ε_p/γ_p for these calculations can be found with the correlation's between ε_p/γ_p and the porosity ε_p itself. Probst and Wohlfahrt [28] took all available data and checked the proposed correlations. They observed that, for different types of porous systems, different correlations have to be used. They distinguished four systems on the basis of their preparation (Table 3.4).

Table 3.4 Classification of porous systems

	Porous system	
Primary particles	Noncompressed	Compressed
Nonporous	natural packing group A: Equation 3.32	extrudates group B: Equation 3.33
Porous	gel catalyst group C: Equation 3.34	Compressed catalyst group D: Equation 3.35

For group A packing they recommend the correlation

$$\varepsilon_p / \gamma_p = \varepsilon_p^m \tag{3.32}$$

For a packing of glass spheres of different diameters m = 1.43 and for sand m = 1.5. The more the shape of the particles deviate from the sphere, the higher the value of the exponent. For example, for carborundum m = 1.63, for NaCl crystals m = 1.70, for talcum m = 2.80, for diatomaceous earth and kaolin m = 3.00 and for purmice m = 3.10.

Group B particles (produced by compressing or sintering nonporous primary particles, such as crystals) are usually extrudates or sintered pellets. For this group they recommend the correlation originally derived by Hugo [29]:

$$\varepsilon_p / \gamma_p = \varepsilon_p (1.25\varepsilon_p^m - 0.25) \tag{3.33}$$

For m one can assume a value of 1.0, although for fluorite m = 0.85 and for apatite m = 0.91 have been found.

For group C, catalysts produced from gels, a pore structure found that exhibits – even for small macroporosities – low mass transport. Probst and Wohlfahrt propose the correlation

$$\varepsilon_p / \gamma_p = \exp\left(m\left(1 - 1/\varepsilon_p\right)\right) \tag{3.34}$$

For group D the authors recommend a modification of the correlation of Mackie and they have introduced the adjustable parameter m:

$$\varepsilon_p / \gamma_p = \frac{\varepsilon_p^m}{(2 - \varepsilon_p)^{m+1}} \tag{3.35}$$

With this correlation we can describe the most important group of catalysts, namely all metal and metal oxide catalysts (which are produced by compressing powders at different pressures). In most cases, the primary particles are already porous themselves, so bimodal pore-size distributions are obtained. The authors recommend a value of 1.05 when m cannot be determined experimentally. For a large series containing widely varying data, values of m between 0.70 and 1.65 have been observed.

In Table 3.5 are the data and their ranges as found by Probst and Wohlfahrt. It should be realized that the boundaries between the groups are vague. Also, values of m will vary depending on the manufacturing process of the individual catalyst producers.

Table 3.5 The data and their ranges as found by Probst and Wohlfahrt

Group	Recommended m	Range of m	Range of ε_p	Range of ε_p/γ_p
A	1.5	1.43-3.10	0.20-0.85	0.12-0.80
B	1	0.74-1.01	0.20-0.50	0.02-0.18
C	1.8	1.4-2.2	0.35-0.70	0.05-0.35
D	1.05	0.70-1.65	0.05-0.65	0.02-0.35

3.4.4 Heat Transfer in Porous Catalysts

When strongly exothermic or endothermic reactions occur in the catalyst pellet, the temperature cannot be regarded as uniform throughout the catalyst particle. To describe heat transfer through a catalyst particle it is usually considered to be homogeneous. So, the heat flow is described by the conventional heat conduction equation used for isotropic solids or stagnant fluids. When chemical reactions take place, the energy balance equation is

$$\lambda_p\left(\frac{d^2T}{dx^2}+\frac{d^2T}{dy^2}+\frac{d^2T}{dz^2}\right)=\sum_i(-\Delta H_i)R_i \tag{3.36}$$

Where λ_p is the effective heat conductivity of the catalyst particle and $(-\Delta H_i)$ the heat effect of reaction i. The effective conductivity is the quantity which (when multiplied by the gradient of temperature) yields the true rate of energy transport in the porous structure of the solid.

Measurements of λ_p are scarce. The available data are reviewed by Satterfield [2]. The effective thermal conductivity of a porous catalyst can be estimated from the correlation of Russell [30]:

$$\frac{\lambda_p}{\lambda_f}=\frac{\xi\varepsilon_p^{2/3}+1-\varepsilon_p^{2/3}}{\xi(\varepsilon_p^{2/3}-\varepsilon_p)+1-\varepsilon_p^{2/3}+\varepsilon_p}$$

Where ε_p is the porosity, $\xi=\lambda_s/\lambda_f$ and λ_p, λ_s, and λ_f are the thermal conductivities of the pellet, the solid making up the skeleton matrix and the fluid in the pores, respectively.

Thermal conductivities of porous oxide catalysts, which are of major practical interest, (at atmospheric pressure in air) are within the range 0.2 – 0.5 W m^{-1}K^{-1}. The spread of values of the thermal conductivity is remarkably small for the wide range of catalysts studied by several investigators. It does not vary greatly with major differences in void fraction and pore size distribution.

The thermal conductivity may be modeled as two conducting paths with transfer of heat between the two. The fluid contribution is particularly complicated since the pore size distribution of many catalysts is such that, at atmospheric pressure, diffusion takes place in the transition region between the Knudsen and bulk modes. Here, the heat flow through the gas depends significantly on the pressure, although the heat conductivity of a gas is almost independent of pressure. The prediction of the model will vary with the geometry postulated for the two paths, but evidently the effective conductivity is determined primarily by that of the phase with the greater thermal conductivity. Analyzing the fluid heat conductivity's at reaction conditions, Satterfield [2] concluded that the values reported for the effective thermal conductivity of porous catalysts in air usually represent the minimum values to be expected during reaction. Consequently temperature gradients calculated would probably represent maximum values.

3.4.5 Mass Diffusivities

3.4.5.1 Gases

For pressures up to about 1 MPa (or perhaps even higher), the diffusion coefficient for a binary mixture of gases A and B may be estimated from the Fuller, Schettler, and Giddings relationship

$$D_{AB} = \frac{0.00143T^{1.75}}{PM_{AB}^{1/2}\left[(\Sigma_v)_A^{1/3} + (\Sigma_v)_B^{1/3}\right]^2} \tag{3.37}$$

where D_{AB} is the binary diffusion coefficient (cm^2 s^{-1}), T is the temperature (K), M_A and M_B are the molecular weights of A and B (g mol^{-1}), M_{AB} is given by $2[(1/M_A) + (1/M_B)]^{-1}$, and P is the pressure (bar). The term Σ_v is found for each component by addition of the atomic diffusion volumes in Table 3.6 [31]. These atomic parameters were determined by a regression analysis of many experimental data. This correlation cannot distinguish between isomers. Generally, however, Equation 3.36 will predict D_{AB} within 5-10 % accuracy. No composition dependence is shown, because D_{AB} (for gas mixtures) is usually insensitive to this variable. Many other methods are available for prediction of D_{AB}. Reid et al. [31] review these methods and several empirical estimation techniques.

The calculation of the mass diffusivity with Equation 3.36 is illustrated in Example 9.5 in Chapter 9.

3.4.5.2 Liquids

In the absence of a rigorous theory for diffusion in liquids, a number of empirical relationships have been proposed, one of which we mention briefly. For a binary mixture of solute A in solvent B, the diffusion coefficient D^0_{AB} (cm^2 s^{-1}) of A diffusing in an infinitely diluted solution of A in B can be found with the Wilke-Chang correlation:

$$\frac{D_{AB}^0\mu}{T} = 7.4\times10^{-8}\,\frac{(XM)^{0.5}}{V_b^{0.6}} \tag{3.38}$$

where μ is the solution viscosity (cP), T is the temperature (K), X is the association parameter (solvent dependent, Table 3.7), M is the solvent molecular weight (g mol^{-1}), and V_b is the solute molar volume at normal boiling point (Table 3.8) (cm^3 mol^{-1}).

Table 3.6 Atomic diffusion volumes

Atomic and structural diffusion volume increments		Diffusion volumes of simple molecules	
C	15.9	He	2.67
H	2.31	Ne	5.98
O	6.11	Ar	16.2
N	4.54	Kr	24.5
Aromatic ring	-18.3	Xe	32.7
Heterocyclic ring	-18.3	H_2	6.12
F	14.7	D_2	6.84
Cl	21.0	N_2	18.5
Br	21.9	O_2	16.3
I	29.8	Air	19.7
S	22.9	CO	18.0
		CO_2	26.9
		N_2O	35.9
		NH_3	20.7
		H_2O	13.1
		SF_6	71.3
		Cl_2	38.4
		Br_2	69.0
		SO_2	41.8

Table 3.7 Association parameters for various solvents

Solvent	X Parameter
Water	2.6
Methanol	1.9
Ethanol	1.5
Benzene	1.0
Ether	1.0
Heptane	1.0

Table 3.8 Solute molar volumes (V_b) at normal boiling point

Solute	V_b (cc g^{-1})
Air	29.9
Br_2	53.2
Cl_2	48.4
CO	30.7
CO_2	34.0
COS	51.5
H_2	14.3
H_2O	18.9
H_2S	32.9
I_2	71.5
N_2	31.2
NH_3	25.8
NO	23.6
N_2O	36.4
O_2	25.6
SO_2	44.8

This equation is good only for dilute solutions of nondissociating solutes. In engineering work D^0_{AB} is assumed to be a representative diffusion coefficient even for concentrations of A up to 5 – 10 mol %. Note that Equation 3.37 is not dimensionally consistent; the variables must be employed with the specified units (see Example 9.6).

Thus generally, for liquids $D^0_{AB} \neq D^0_{BA}$. Different techniques with which to estimate the infinite dilution diffusion coefficient are described by Reid et al. [31]. Various correlation's (valid for an arbitrary composition of a binary mixture and for electrolytes) are also given. In the Wilke-Chang correlation for D^0_{AB} the effect of temperature has been accounted for by assuming $D^0_{AB} \sim T$. Although this approximation may be valid over small temperature ranges, it is usually preferable to assume that

$$D_{AB} = A\exp\left(\frac{-B}{T}\right)$$

where A and B are constants.

References

1. Lin, K-H., Van Ness, H.C., Abbott, M.M. (1984) in *Perry's Chemical Engineering Handbook* (6th edn). New York: McGraw-Hill, pp. 4-1 – 4-91.

2. Satterfield, C.N. (1991), *Heterogeneous Catalysis in Industrial Practice.* (2nd edn). New York: McGraw-Hill.
3. Wheeler, A. (1951) *Advances in Catalysis*, **3**, 249.
4. Satterfield, C.N. (1970) *Mass Transfer in Heterogeneous Catalysis.* Cambridge: MIT. Press.
5. Richardson, J.T. (1989) *Principles of Catalyst Development.* New York: Plenum Press.
6. Le Page, J.F., Cosyns, J., Courty, P., et al. (1987), *Applied Heterogeneous Catalysis.* Paris: Éditions Technip.
7. Spencer, M.S. (1989) in *Catalyst Handbook* (2nd edn) M.V. Twigg (ed.). London: Wolfe Publishing, pp.17-84.
8. Gregg, S.J., Sing, K.S.W. (1982), *Adsorption, Surface Area, and Porosity.* (2nd edn). New York: Academic Press.
9. Thomas, J.M., Thomas, W.J. (1996) *Principles and Practice of Heterogeneous Catalysis.* Weinheim: VCH.
10. Spencer, D.H.T., Wilson, J. (1976) *Fuel*, **55**, 291.
11. McClellan, A.L., Harnsberger, H.F. (1967) *J. Colloid. Interface Sci.*, **23**, 577.
12. Buyanova, N.E., Zagrafskaya, R.V., Karnaukhov, A.P., Shepelina, A.S. (1983) *Kinetics and Catalysis*, **24**, 1011.
13. Carman, P.C. (1956), *Flow of Gases Through Porous Media.* London: Butterworths.
14. Wit, L.A. de, Scholten, J.J.F. (1975) *J. Catal.*, **36**, 30, 36.
15. Mayer, R.P., Stowe, R.A. (1966) *J. Phys. Chem.*, **70**, 3867.
16. Reimer, L., Pfefferkorn, G. (1973), *Raster-Elektronenmicroskopie.* New York: Springer.
17. Bird, R.B., Stewart W.E., Lightfoot, E.N. (1960) *Transport Phenomena.* New York: Wiley.
18. Kennard, E.H. (1938) *Kinetic Theory of Gases.* New York: McGraw-Hill.
19. Jackson, R. (1977) *Transport in Porous Catalysts.* Amsterdam: Elsevier.
20. Kapoor, A., Yang, R.T., Wong, C. (1989) *Catal. Rev. – Sci. Eng.*, **31**, 129.
21. Mason, E.A., Malinauskas, A.P. (1983) *Gas Transport in Porous Media: The Dusty-Gas Model.* Amsterdam: Elsevier.
22. Kerkhof, P.J.A.M. (1996) *Chem. Eng. J.*, **64**, 319.
23. Schneider, P. (1978) *Chem. Eng. Sci.*, **33**, 1311.
24. Froment, G.F., Bischoff K.B. (1979), *Chemical Reactor Analysis and Design.* New York: Wiley.
25. Satterfield, C.N., Colton, C.K., Pitcher, W.H. (1973) *AIChE J.*, **19**, 628.
26. Luss, D. (1977) in *Chemical Reactor Theory. A Review*: L. Lapidus, N.R. Amundson (eds.). New Jersey: Prentice-Hall, pp. 191-268.
27. Faxen, H. (1959) *Kolloid Z.*, **167**, 146.
28. Probst, K., Wohlfahrt, K. (1979) *Chem.-Ing.-Tech.*, **51**, 737.
29. Hugo, P. (1974) *Chem.-Ing.-Tech.*, **46**, 645.
30. Russell, J. (1935), *Am. Ceram. Soc.*, **18**(1).
31. Reid, R.C., Prausnitz, J.M., Poling, B.E. (1987) *The Properties of Gases and Liquids.* New York: McGraw-Hill.

4 Catalysis and External Transfer Processes

4.1 Mass Transfer in Heterogeneous Systems

Components have to be transported to the catalyst and once they have reached the outer surface of the particle, and possibly into and through the pores of the particle to reach the catalytically active material. This chapter we discusses transport to the outer surface of the particle; transport inside the particle is discussed in Chapter 3.

Several cases can be distinguished. In the case of a hydrogenation, for example, the components to be hydrogenated may be in the liquid phase, in which small catalyst particles are suspended. Here the hydrogen has to pass through the gas to the interface with the liquid, dissolve in the liquid and eventually through the liquid to the catalyst particle. This is a three-phase system with a gas, a liquid and a solid.

In the case that the reactants are already in the phase surrounding the catalyst pellet, the components only have to be transported through this single phase, being either liquid or gas. Reactant A is assumed to be converted according to a first-order reaction. For convenience a chemical reaction rate R_A'' is defined per unit of external surface:

$$R_A'' = \frac{V_p}{A_p}\langle R_A \rangle$$

where R_A'' is the rate based on the external surface area of the pellet (moles of A per square meter of external pellet surface per second), V_p is the pellet volume, A_p is the external surface area of the pellet, and $\langle R_A \rangle$ is the reaction rate based on the pellet volume (moles of A per cubic meter of pellet volume per second). The latter parameter is discussed below in further detail. Corresponding to the reaction rate per unit surface is a surface kinetic constant k_r'' for first-order kinetics:

$$R_A'' = k_r'' C_{A,i} \tag{4.1}$$

where $C_{A,i}$ is the reactant concentration in the direct vicinity of the pellet surface (mol m^{-3}). Since first-order kinetics apply, the dimensions of k_r'' are meters per second. If, for example, a gas-solid reaction is assumed, the stationary conversion rate can be calculated from Equation 4.1 together with the defining equation for the mass-transfer coefficient of A, namely k_g, as

$$N_A = k_g(\overline{C_A} - C_{A,i}) \tag{4.2}$$

and under steady-state conditions

$$R_A'' = N_A \tag{4.3}$$

Elimination of the unknown concentration $C_{A,i}$ gives the overall conversion rate:

$$N_A = \left(\frac{1}{k_r''} + \frac{1}{k_g}\right)^{-1} \overline{C_A}$$

The two asymptotic solutions are (Figure 4.1):

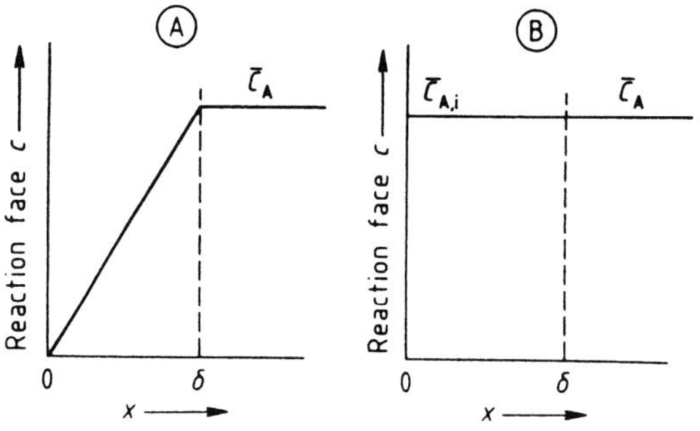

Figure 4.1. Two asymptotic concentration profiles for mass transfer and reaction in series (film theory): A) Fast reaction regime; B) Slow reaction regime.

The **fast reaction regime** for $k_r''/k_g >> 1$. The rate of mass transfer completely limits the conversion rate, which follows from

$$N_A = k_g \overline{C_A}$$

In this regime the conversion rate (hence, reactor capacity) can be enhanced by increasing the mass-transfer rate (e.g., by increasing the fluid velocity in the reactor or decreasing the average particle diameter).

The **slow reaction regime** for $k_r''/k_g << 1$. The chemical reaction rate completely limits the conversion rate, which now follows from

$$N_A = k_r'' \overline{C_A}$$

In this regime, increasing the chemical reaction rate, for example, by increasing the temperature of the reactor can enhance the conversion rate.

The system analyzed above is easily extended to arbitrary kinetics. Langmuir kinetics are discussed here as an example, in which case

$$R_A'' = \frac{k_r'' C_{A,i}}{1 + K C_{A,i}}$$

Combination with Equations 4.2 and 4.3 and elimination of $C_{A,i}$ give

$$N_A = k_g \overline{C_A} \frac{2\beta}{1+\alpha+\beta+\sqrt{(1-\alpha+\beta)^2+4\alpha}} \tag{4.4}$$

with $\alpha = K\overline{C_A}$ and $\beta = \frac{k_r''}{k_g}$

In the slow reaction regime, $k_r''/k_g << 1$ and thus $\beta \to 0$, Equation 4.4 becomes

$$N_A = k_g \overline{C_A} \frac{\beta}{1+\alpha} = \frac{k_r'' \overline{C_A}}{1+K\overline{C_A}}$$

and mass-transfer limitation has no effect on the overall conversion rate.

In the fast reaction regime, where $k_r''/k_g >> 1$ and $\beta \to \infty$, Equation 4.4 yields

$$N_A = k_g \overline{C_A}$$

Hence, due to mass-transfer limitations, apparent first-order kinetics are observed in the fast reaction regime, irrespective of the numerical value of the term $K\overline{C_A}$. The physical explanation is that although the value of the term $K\overline{C_A}$ may be large, in this regime $C_{A,i}$ is much smaller than $\overline{C_A}$, so that $KC_{A,i}$ is much smaller than unity and the adsorption term in Langmuir kinetics becomes negligible. From this, the conclusion can be drawn that mass-transfer limitation may cause an apparent change in the reaction order. Below, more examples illustrating this statement are presented. For other forms of arbitrary kinetics, expressions for the conversion rate analogous to Equation 4.4 are obtained in a similar way. The discussion above is only valid in the case when the chemisorption of A is much more rapid than mass transfer is.

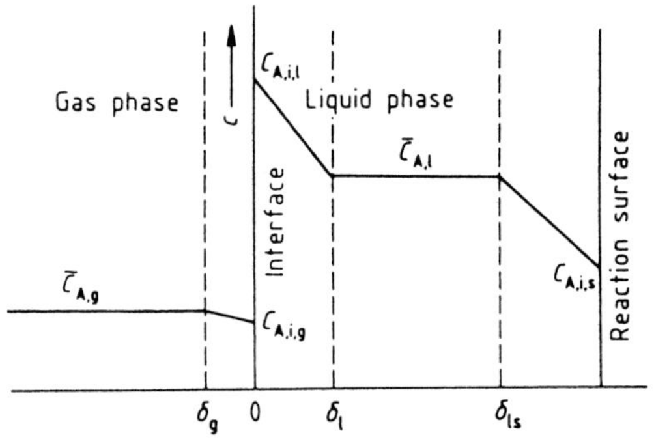

Figure 4.2 Concentration profiles for mass transfer and reaction in series in a gas – liquid – solid system (film theory).

The system discussed is easily extended to a three-phase reactor with mass transfer and reaction in series, for example a gas is absorbed in a liquid in which nonporous particles are suspended. Reaction occurs at the surface of the particles. Examples are the hydrogenation of organic liquids with a solid catalyst and the alkylation of a liquid reactant with gaseous ethylene and solid $AlCl_3$ as catalyst. The concentration profile could be that shown in Figure 4.2. At the gas-liquid interface the assumed equilibrium is given by

$$C_{A,i,l} = mC_{A,i,g} \tag{4.5}$$

where m is called the distribution coefficient. The parameter a_{gl} is defined as the specific gas – liquid surface area (m^{-1}) and a_{ls} is the external specific catalyst particle area (m^{-1}), both per unit volume of dispersion. The term k_g is the gas-phase mass-transfer coefficient for transfer from gas to liquid, and k_l and k_{ls} are the liquid-phase mass-transfer coefficients for transfer from gas to liquid and liquid to solid, respectively (all with dimensions of $m\,s^{-1}$). If first-order kinetics is assumed, then the **overall conversion rate** per unit reactor volume R_{AV} is given by

$$R_{AV} = a_{ls} k_r'' C_{A,i,s}$$

$$= a_{gl} k_g (\overline{C_{A,g}} - C_{A,i,g})$$

$$= a_{gl} k_l (C_{A,i,l} - \overline{C_{A,l}})$$

$$= a_{ls} k_{ls} (\overline{C_{A,l}} - C_{A,i,s})$$

In these equations it is assumed that a steady state has developed and that the consumption of A by reaction equals the fluxes through the boundary layers of the gas phase and of the liquid phase at the gas side and surrounding the catalyst pellet. Combining them with Equation 4.5 and eliminating the unknown concentrations $C_{A,i,g}, C_{A,i,l}, C_{A,i,g}$, and $\overline{C_{A,l}}$ gives

$$R_{AV} = \overline{C_{A,g}} \left(\frac{1}{a_{gl} k_g} + \frac{1}{m a_{gl} k_l} + \frac{1}{m a_{ls} k_{ls}} + \frac{1}{m a_{ls} k_r''} \right)^{-1} \tag{4.6}$$

Reactor capacity per unit volume appears to depend on four resistances in series: the gas-phase transfer resistance, two liquid-phase transfer resistances, and the kinetic resistance. The highest resistance limits the capacity of the reactor. The four resistances have the unit of time and each one individually represents the time constant of the particular process under study. For example, $1/k_g a_{g1}$ is the time constant for the transfer of A from the bulk of the gas through the gas film to the gas-liquid interface. The same holds for the three other resistances. For a first-order reaction in a batch reactor, for example, the concentration after a certain time is given by $C/C_0 = \exp(-t/\tau)$, in which $\tau = 1/k$ is the reaction time constant. For processes in series the individual time constants can be added to find the overall time constant of the total process.

Implicit in the derivation of Equation 4.6 is the assumption that all solid particles are in the bulk of the liquid phase. However, for

$$\frac{1}{a_{ls}k_{ls}}+\frac{1}{a_{ls}k_r''}\ll\frac{1}{a_{gl}k_l}$$

the concentration of A in the bulk of the liquid approaches zero. Then, only the particles in the liquid film at the gas-liquid interface may be active, so that mass transfer and the reaction have become parallel steps instead of being in series. This problem has been studied by Sada and co-workers [1,2].

The discussion above can be extended to bimolecular reactions. Here again a two-phase situation is assumed – a gas-solid reaction on an external pellet surface. If two components take part in a surface reaction, an extra balance for the second component must be considered as well. For

$$A+\nu B\rightarrow \text{products}$$

this gives:

$$N_A=k_{g,A}(\overline{C_A}-C_{A,i})$$

$$N_B=k_{g,B}(\overline{C_B}-C_{B,i})$$

If, for example, the above reaction occurs according to the following kinetics:

$$R_A''=k_r''C_{A,i}C_{B,i} \tag{4.7}$$

A material balance gives

$$N_A=R_A''$$

$$N_B=\nu N_A$$

Combining these equations and eliminating $C_{A,i}, C_{B,i}$, and N_B gives

$$N_A=k_{g,A}\overline{C_A}\left(\frac{2\beta_B}{1+\beta_A+\beta_B+\sqrt{(1-\beta_A+\beta_B)^2+4\beta_A}}\right)^{-1} \tag{4.8}$$

with

$$\beta_A=\nu\frac{k_r''\overline{C_A}}{k_{g,B}}\quad\text{and}\quad\beta_B=\frac{k_r''\overline{C_B}}{k_{g,A}}$$

For the slow reaction regime (i.e. $\beta_A \rightarrow 0$ and $\beta_B \rightarrow 0$) Equation 4.8 simplifies to

$$N_A = k_r'' \overline{C_A C_B}$$

Thus, the mass-transfer resistance is negligible. For the fast reaction regime (i.e. $\beta_A \rightarrow \infty$ and $\beta_B \rightarrow \infty$) Equation 4.8 simplifies to one of the following equations:

For $\nu k_{g,A}\overline{C_A} < k_{g,B}\overline{C_B}$,

$$N_A = k_{g,A}\overline{C_A}$$

hence, mass transfer of the reactant with the lowest mass-transfer rate limits the overall conversion rate.

For $\nu k_{g,A}\overline{C_A} > k_{g,B}\overline{C_B}$

$$N_B = \frac{1}{\nu} k_{g,B}\overline{C_B}$$

Now, B limits the overall conversion rate because it has the lowest mass-transfer rate. Thus, corrected for stoichiometry, the key component with the lowest rate of mass transfer determines the rate of conversion, and the component in excess at the catalyst interface has no influence at all on the rate in this case.

As illustrated for monomolecular reactions, the system analyzed here is easily extended to bimolecular reactions with arbitrary kinetics by modifying the rate equation of R_A'' as given in Equation 4.7, although the algebra involved can become quite tedious.

4.2 Heat and Mass Transfer Coefficients for Flow around Catalyst Particles

4.2.1 Two-phase Reactors

There are many correlations available for heat and mass transfer to particles. These are all have the Nusselt number $Nu = \alpha d_p/\lambda$ (*or the Sherwood number* $Sh = k_g d_p/D$) as a function of the Reynolds number $Re = ud_p/\nu$ and the Prandtl number $Pr = \nu/(\lambda/\rho c_p)$ (or the Schmidt number $Sc = \nu/D$).

Here d_p is a relevant length measure of the catalyst particle, such as its diameter, α and k_g are the heat and the mass transfer coefficients to the particle, u is the average, characteristic velocity of the reaction medium with respect to the particle, and λ, ν, ρ and c_p are the heat conductivity, kinematic viscosity, density and heat capacity of the reaction medium surrounding a particle, either a gas or a liquid. D is the molecular diffusion coefficient of the component transferred to the catalyst particle in the reaction medium.

Gnielinski [3] derived a correlation for single particles, which represents all data measured with sufficient accuracy for technical applications:

$$Nu = Nu_{\min} + \sqrt{Nu_{lam}^2 + Nu_{turb}^2} \tag{4.9}$$

This is valid for $0.1 < Re < 10^7$ and $0.5 < Pr < 2500$ and in which $Nu_{min} = 2$ for spheres and < 2 for other particle shapes and, further,

$$Nu_{lam} = 0.664 Re_{res}^{0.5} Pr^{0.33}$$

$$Nu_{turb} = \frac{0.037 Re_{res}^{0.8} Pr}{1 + 2.44(Pr^{0.67} - 1) Re_{res}^{-0.10}}$$

In these equations Re_{res} is the Reynolds number for a flow field in which both forced as well as free convection occurs:

$$Re_{res} = \sqrt{Re_{forced}^2 + \frac{Gr}{2.5}}$$

With $\mathrm{Re}_{forced} = u_\infty L / \nu$ and for free convection we have

$$Gr = \frac{gL^3}{\nu^2} \times \frac{\rho_\infty - \rho_s}{\rho_s}$$

where u_∞ and ρ_∞ are the velocity and the density of the reaction medium at a large distance from the catalyst pellet, and ρ_s the catalyst surface value. L is the equivalent length of the catalyst particle, given by

$$L = A / l_c$$

Here A is the heat or mass exchanging outer surface face of the particle and l_c the circumference of the projection plane in the direction of the flow. For a sphere with diameter $d_p L = \pi d_p^2 / \pi d_p = d_p$ and for a cylinder with diameter d_p and height H, if the direction of flow is perpendicular to the cylinder axis,

$$L = \frac{2 \times \frac{\pi d_p^2}{4} + \pi d_p H}{2d_p + 2H} = \frac{\pi}{2}\left(\frac{0.5 + H/d_p}{1 + H/d_p}\right) d_p$$

So for very long cylinders ($H >> d_p$) we find $L = \pi d_p / 2$ and for cylinders of equal height and diameter $L = 3\pi d_p / 8$. Also in Nu and Sh the same value for L has to be used.

Further, we should realize that the relations in Equation 4.9 hold for single particles. In a packed bed the heat and mass transfer is enhanced due to the presence of the adjacent particles. For this enhancement Martin [4] derived the following relation:

$$Nu_{packed\ bed} = [1 + 1.5(1 - \varepsilon_b)]Nu_{single\ particle}$$

For a packed bed with $\varepsilon_b = 0.4$ the enhancement is 1.9. In a packed bed u_∞ is the interstitial velocity u_{sup} / ε_b, in which u_{sup} is the superficial velocity. u_{sup} is based on the empty cross-section of the bed and ε_b is the porosity of the packed bed.

We further mention that at low values of the Reynolds number (that is at very low fluid velocities or for very small particles) for flow through packed beds the Sherwood number for the mass transfer can become lower than $Sh_{min} = 2$, found for a single particle stagnant relative to the fluid [5]. We refer to the relevant papers. For the practice of catalytic reactors this is not of interest: at too low velocities the danger of particle runaway (see Section 4.3) becomes too large and this should be avoided, for very small particles suspension or fluid bed reactors have to be applied instead of packed beds. For small particles in large packed beds the pressure drop become prohibitive. Only for fluid bed reactors, like in catalytic cracking, may *Sh* approach a value of 2.

In suspension reactors as well as in fluidized bed reactors for u_∞ the free-falling velocity of a single particle has to be used. The free falling velocity of spheres can be found according to the following derivation. In the steady state the weight of the particle in the fluid is equal to the resistance to flow:

$$\frac{1}{6}\pi d_p^3 g(\rho_p - \rho_f) = \frac{1}{4}\pi d_p^2 C_w \frac{1}{2}\rho_f u_{fall}^2$$

Which leads to

$$u_{fall} = \sqrt{\frac{4(\rho_p - \rho_f) g d_p}{3\rho_f C_w}}$$

where ρ_f is the density of the fluid, d_p is the diameter of the sphere and C_w is the resistance coefficient of a sphere. Kürten et al. [6] derived the following relation for the resistance coefficient:

$$C_w = \frac{21}{Re} + \frac{6}{\sqrt{Re}} + 0.28$$

which is valid for $Re < 10^5$ and where $Re = u_{fall} d_p / \nu$.

4.2.2 Three-phase Reactors

With the above relations for three phase reactors with gas, liquid and solid catalyst phases we can determine the mass and heat transfer coefficients for the transport to and from the catalyst particle, as it is suspended in the liquid phase. The same holds for transfer in the liquid phase surrounding catalyst particles through which gas and liquid flow.

In these cases the gaseous component also has to dissolve in the liquid phase. To determine the resistances for the reactant transfer from gas to liquid we have to know the mass transfer coefficients k_g and k_l at the gas and the liquid sides respectively, the interfacial area a and the solubility m, as can be understood, for example, from Equation 4.6. Most catalyzed gas-liquid-solid reactions are hydrogenations or oxidations. Hydrogen and oxygen are sparingly soluble gases with low solubilities, so that under normal operating conditions the resistance in the gas phase is negligible, compared to the resistance in the liquid phase [7]. Information on a_{gl} and on k_l is therefore of importance. However, data are available only with much less accuracy than the transfer coefficients to catalyst pellets [7].

The mass transfer coefficient at the liquid side of a mobile gas-liquid interface usually ranges from 1 to 4×10^{-4} m s^{-1} in low viscosity liquids. In practice most gas-liquid-solid reactions take place in low viscosity media and with dispersions of gas bubbles with high interfacial areas a. In bubble columns a is of the order of 100-300 m^{2}m^{-3} reactor volume. Higher interfacial areas can be obtained with porous plate gas distributors – not used for industrial reactors – and in agitated tank reactors, where interfacial areas can reach values as high as 1000 m^{2}m^{-3}. If a slurry of fine catalyst particles is present in low concentrations in a bubble column or an agitated reactor, the interfacial area is hardly affected; in the case of higher slurry concentrations, say above 5 vol % of suspended solids, the interfacial area decreases with increasing solids content.

In a trickle flow reactor – a packed catalyst bed with gas and liquid flowing through it – the flow regimes are more complicated. Usually the liquid quietly trickles through the bed. However, at high gas flow rates and cocurrent liquid flow, so-called pulsing flow may also occur. In that case plugs of liquid, in which gas bubbles are dispersed, are alternating rapidly with plugs of gas, in which a fine fog of liquid droplets is present. In the case of trickle flow reactors, which are mostly of the cocurrent down-flow type, the interfacial area between gas and liquid in the trickle flow region ranges from 20 % to 80% of the geometric surface area of the catalyst [8,9]. In the pulsing flow regime the gas-liquid interfacial area can be 5-10 times larger. Unfortunately, at high pressures (above 5 MPa for hydrogenations and above 1 MPa for oxidations) under practical conditions the pulsing flow regime cannot be attained [10].

In packed bubble columns the gas-liquid interfacial area also can be related to the external catalyst surface area, as in trickle flow reactors. However, in packed bubble columns channeling can occur with strongly reduced gas-liquid interfacial areas [11].

4.3 Thermal Behavior of Catalyst Particles and Pellet Runaway

Consider a catalyst particle in a fixed bed through which gas is flowing, and assume that the reaction takes place at the catalyst surface and, accordingly, that the net heat of reaction is released at this surface. An example of such a system is the hydrogenation of acetylene in an ethylene stream, according to

$$C_2H_2 + 2H_2 \rightarrow C_2H_6$$

A pseudosteady state may be assumed to exist in the catalyst bed, and the temperature of a catalyst particle may be thought to be constant. In such a case, all of the heat produced by the reaction must be transported from the catalyst surface to the surroundings by convection to the gas stream. In this way the feedback of heat is established from the reaction site to the gaseous reactant. Under the above conditions the molar flux N_A of hydrogen (which we call reactant A) to the surface of the particle, multiplied by the heat liberated per molar unit of A converted $(-\Delta H_r)_A$ must be equal to the heat flux coming from the surface:

$$HPR = N_A(-\Delta H_r)_A = HWR \tag{4.10}$$

Here HPR stands for heat production rate and HWR for heat withdrawal rate. Usually, the value of N_A depends on the texture of the particle and on the chemical and physical rate parameters. At a low value of T_i (the temperature of the solid), N_A is mainly determined by the rate of the chemical surface reaction. As T_i increases, a situation is reached where the reactant is converted at the external particle surface at such a high rate that its concentration at the surface is zero and N_A is entirely determined by mass transfer. According to the discussion in Section 4.1 we can write

$$N_A = \frac{k_r^{"} k_g C_A}{k_r^{"} + k_g} \tag{4.11}$$

With a constant gas velocity, the mass transfer coefficient k_g does not greatly depend on temperature, so that N_A will be practically constant for a certain value of $\overline{C}_A$ and at high temperatures. Accordingly, the left-hand term of Equation 4.10, i.e. the chemical heat production rate per unit external area depends on T_i in the manner indicated in Figure 4.3, curves 1a and 1b. The equation of these sigmoid curves is given by

$$HPR = \frac{(-\Delta H_r)_A \overline{C}_A}{(1/k_\infty^{"})e^{E/RT_i} + (1/k_g)}$$

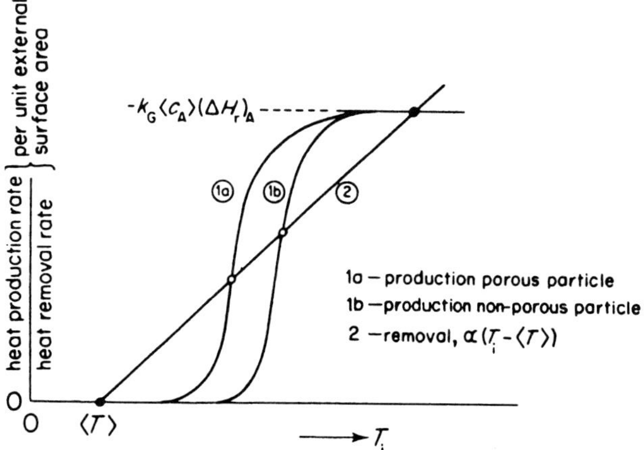

Figure 4.3 Heterogeneous exothermic reaction; possible solutions to Equation 4.10; (•) stable; (o) unstable. *Chemical Reactor Design and Operation*, Westerterp, K.R., Van Swaaij, W.P.M., Beenackers, A.A.C.M., Copyright John Wiley & Sons Limited. Reproduced with permission.

The *HPR* curve is a sigmoid for the following reasons. At low particle temperature the system is in the slow reaction regime; hence the conversion rate equals the reaction rate. In this regime the conversion rate and thus the heat production rate are low. As the pellet temperature increases, the conversion rate and thus *HPR* initially increase with temperature according to the assumed Arrhenius relation. If the pellet temperature becomes still higher, above a certain temperature the reaction rate is so high that mass transfer becomes rate limiting. The pellet then operates in the fast reaction regime, and the overall conversion rate equals the mass-transfer rate. Since mass transfer is virtually independent of temperature, in this regime the *HPR* does not increase further with increasing pellet temperature – it becomes constant.

The rate of heat transfer to the gas flow can be described by means of a heat transfer coefficient α_g:

$$HWR = \alpha_g (T_i - \overline{T}) \tag{4.12}$$

where $\overline{T}$ is the average gas temperature. Since we assumed α_g does not contain a contribution due to thermal radiation, it is fairly independent of temperature. Hence, at a constant gas load and for a given value of $\overline{T}$, *HWR* is practically a linear function of T_i, as shown in Figure 4.3, line 2. The points of intersections of a curve 1 and a line 2 in a diagram such as Figure 4.3 represent solutions to Equation 4.10.

In cases of multiplicity (that is more than one solution to Equation 4.10), of the three intersection points only those at the lowest and highest pellet temperature represent stable conditions. These are therefore called the **stable operating points**. The intermediate intersection point is referred to as an unstable operating point.

The stability of these points is determined by the slope of the *HPR* curve at the point of intersection with the *HWR* line. For stable operating points the slope of the *HPR* curve is less than α_g, the slope of the *HWR* line. Therefore, if the temperature of the pellet increases, more heat is removed than is produced, which implies that the pellet temperature must decrease and thus return to the steady-state value. Similarly, a slightly too low pellet temperature causes more heat to be produced than removed, so that the temperature increases. However, at the intermediate intersection point, the situation is reversed: the slope of the heat production curve is higher than α_g. If the pellet temperature becomes slightly too high, more heat is produced than removed, and the pellet temperature increases further until the upper stable operating point is reached. Similarly, if the pellet temperature is a little too low, more heat is removed than produced and the temperature decreases further until the lower stable operating point is reached. Therefore, the intermediate operating point is an unstable one.

The value of the upper stable operating point is fairly insensitive to variations in gas velocity and the degree of conversion; this is valid not only for one piece of catalyst but equally well for a large collection of particles. This can be understood as follows. The mass and heat transfer coefficients are related by the Chilton-Colburn analogy:

$$Nu/Sh = (Pr/Sc)^{1/3}$$

or:

$$\frac{\alpha_g}{k_g} = \rho_g C_{p,g} Le^{2/3}$$

where ρ_g (kg m^{-3}) is the density of the gas, $C_{p,g}$ (J kg^{-1}K^{-1}) is its specific heat, and *Le is* the Lewis number, defined as

$$Le = \frac{a_g}{D_{A,g}}$$

Here, $a_g = \lambda_g / \rho_g C_{p,g}$ is the thermal diffusion coefficient of the gaseous bulk phase (m^2 s^{-1}) and $D_{A,g}$ (m^2 s^{-1}) the molecular diffusion coefficient of A in the gas.

It is often convenient to make equations dimensionless, because in this way we obtain the dimensionless numbers relevant to describe the system under study. The following dimensionless numbers are introduced for convenience:

$$\theta_p = \frac{T_p}{\Delta T_{ad}} Le^{2/3}$$

$$\theta_g = \frac{\overline{T}}{\Delta T_{ad}} Le^{2/3}$$

$$\xi(\theta_p) = \frac{k_g A_p}{k_p(T_p) V_p} \tag{4.13}$$

Substitution of the above numbers in Equations 4.10 – 4.12 gives

$$\theta_p - \theta_g = \frac{1}{1+\xi(\theta_p)} \tag{4.14}$$

Further, the assumption is made that the mass transfer coefficient k_g is temperature independent and that the dependence of the reaction rate constant k_p (T_i) on temperature is of the Arrhenius type, then

$$k_p(T_i) = k_{p,0} \exp\left(-\frac{E_a}{RT_i}\right)$$

Substitution in Equation 4.13 gives

$$\xi(\theta_p) = \xi_0 \exp\left(\frac{An}{\theta_p}\right) \tag{4.15}$$

with

$$\xi_0 = \frac{k_g A_p}{k_{p,0} V_p}$$

The term An is a modified Arrhenius number defined as

$$An = \frac{E_a}{\Delta T_{ad} R} Le^{2/3}$$

being the dimensionless ratio of the activation temperature E_a/R and the maximum temperature difference between the pellet and the gas. Substituting Equation 4.15 into 4.14 yields

$$\theta_p - \theta_g = \left[1 + \xi_0 \exp\left(\frac{An}{\theta_p}\right)\right]^{-1} \tag{4.16}$$

From this equation, θ_p and thus T_i can be obtained for any given values of θ_g, ξ_0, and An. After calculation of $k_p\ (T_i)$, the relationship

$$k_g a_p (\bar{C}_A - C_{A,i}) = k_p (T_i) C_{A,i}\, V_p$$

yields $C_{A,i}$, and hence the conversion rate $k_p\ (T_i)C_{A,i}$ is obtained.

The right-hand term of Equation 4.16 is a dimensionless pellet heat production rate (*HPR*) scaled between 0 and 1, being the ratio of the actual to the maximum possible heat production rate. The left-hand term can be regarded as a dimensionless pellet heat withdrawal rate (*HWR*). The heat withdrawal rate equation is given by $\theta_p - \theta_g$, which is represented by a straight line with a slope of unity and an intercept with the θ_p axis equal to θ_g. For *exothermic reactions* ΔT_{ad} and thus also θ_p, θ_g and An are positive. For *endothermic reactions* ΔT_{ad} is negative and hence θ_p, θ_g and An are also. As a result, only a single operating point can be obtained (Figure 4.4), which means that for fixed values of θ_g, An, and ξ_0 the value of θ_p is also fixed. Thus, for endothermic reactions, multiplicity cannot occur.

In Figure 4.5, several heat production rate curves and one heat withdrawal rate line are plotted versus the dimensionless pellet temperature θ_p. The parameter values are $An = 3$ and $\theta_g = 0.3$, whereas ξ_0 is varied. Initially $\xi_0 = 0.1$, and the pellet conversion rate is low. Then the gas velocity is decreased. Subsequently the mass transfer coefficient k_g decreases and thus ξ_0. For $\xi_0 = 0.024$ a critical ξ_0 value is passed and the region of multiplicity begins. This multiplicity is illustrated in Figure 4.5 for $\xi_0 = 0.01$. The pellet remains at the lower stable operating point, until ξ_0 is decreased further to 0.0035. Then the pellet is forced to leave the lower stable operating point and jumps to the higher stable operating point; this switch of operating points is sudden. The pellet is said to exhibit *runaway* or, alternatively, *ignition* of the pellet occurs. The pellet temperature in Figure 4.6 switches from $\theta_p = 0.34$ to $\theta_p = 1.26$. If ξ_0 is further decreased the pellet simply remains at the higher stable operating point, as illustrated for $\xi_0 = 0.001$.

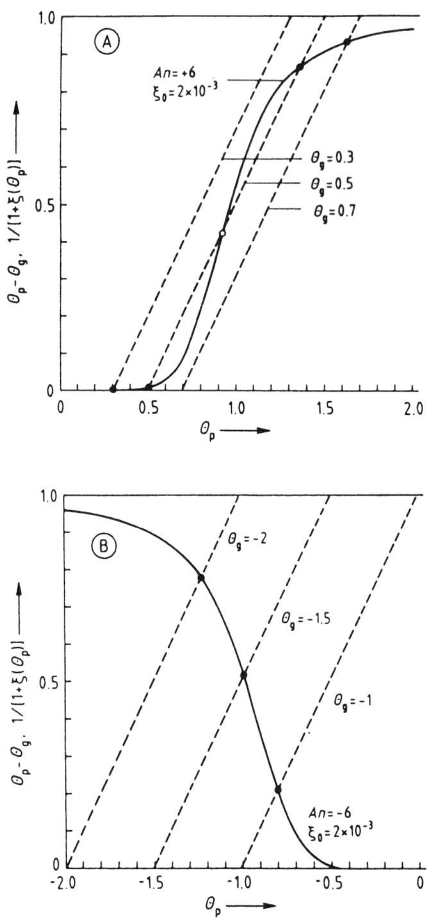

Figure 4.4 Relative heat production and heat withdrawal rates versus dimensionless pellet temperature θ_p: A) exothermic reactions, θ_p, θ_g and $An > 0$; B) endothermic reactions, θ_p, θ_g and $An < 0$; (•) stable operating point; (o) unstable operating point.

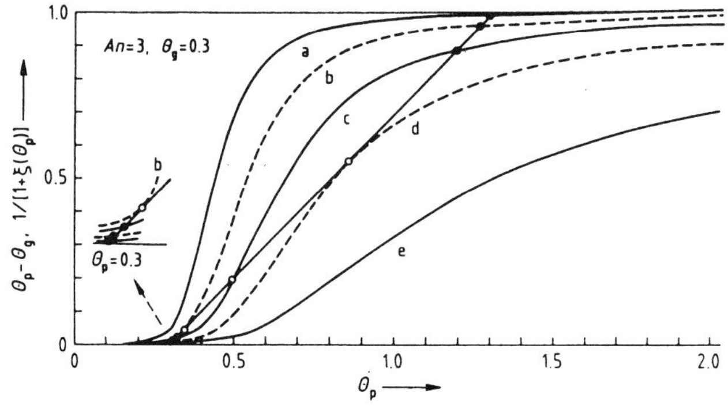

Figure 4.5 Relative heat production rate curves versus dimensionless pellet temperature θ_p for $An = 3$, $\theta_g = 0.3$, and several values of ξ_0: (a) $\xi_0 = 0.001$; (b) $\xi_0 = 0.0035$; (c) $\xi_0 = 0.01$; (d) $\xi_0 = 0.024$; (e) $\xi_0 = 0.1$.

If the velocity and thus ξ_0 are increased again, the pellet reenters the region of multiplicity for $\xi_0 = 0.0035$. In this case the pellet remains at the higher stable operating point

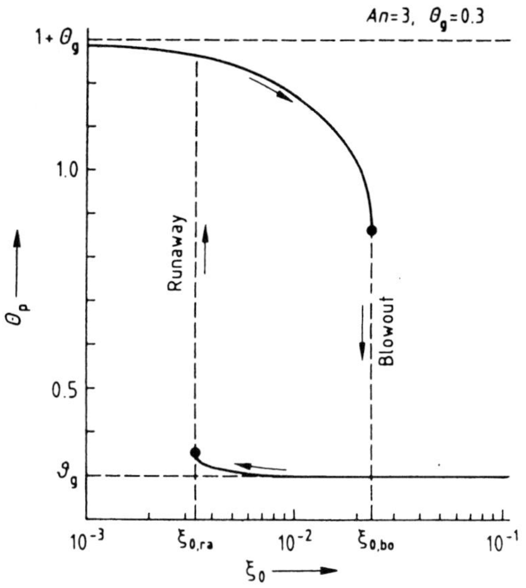

Figure 4.6 Dimensionless pellet temperature θ_p versus ξ_0 for An = 3 and θ_g = 0.3. Note the hysteresis, runaway, and blowout.

until $\xi_0 = 0.024$ is reached. Then the pellet is forced to switch suddenly to the lower operating point. This phenomenon is referred to as *blowout*, or the reaction in the pellet is said to be *extinguished*. For the example given, the pellet temperature drops from θ_p = 0.86 to θ_p = 0.30.

For decreasing ξ_0 values a different line is found than when increasing ξ_0 values as shown in Figure 4.6. This is referred to as *hysteresis*. Hysteresis is commonly encountered when multiplicity of steady states is important.

The critical values of ξ_0 for runaway and blowout $\xi_{0,bo}$, can be found by solving Equation 4.16 together with

$$\frac{\partial}{\partial \theta_p}\left[1+\xi_0 \exp\left(\frac{An}{\theta_p}\right)\right]^{-1} = 1 \tag{4.17}$$

for which case the *HPR* curve touches the *HWR* line. Solving Equations 4.16 and 4.17 and eliminating θ_p gives

$$\xi_{0,bo} = \frac{2+2\theta_g + An - \sqrt{An(An-4\theta_g - 4\theta_g^2)}}{An - 2\theta_g + \sqrt{An(An-4\theta_g - 4\theta_g^2)}}$$

$$\times \exp\left(-\frac{2An(1+An)}{An+2\theta_g An + \sqrt{An(An-4\theta_g - 4_g^2)}}\right) \tag{4.18}$$

and

$$\xi_{0,ra} = \frac{2+2\theta_g + An - \sqrt{An(An-4\theta_g-4\theta_g^2)}}{An-2\theta_g-\sqrt{An(An-4\theta_g-4\theta_g^2)}}$$

$$\times \exp\left(-\frac{2An(1+An)}{An+2\theta_g An-\sqrt{An(An-4\theta_g-4\theta_g^2)}}\right) \tag{4.19}$$

The above equations are useful in the design of packed bed reactors. For highly exothermic reactions, adiabatic temperature increases are of the order of 1000-2000 °C, which implies that at the higher operating point the catalyst temperature is so high that the catalyst would be destroyed. However, increasing the pellet size is economically attractive, so as to decrease the pressure drop over the packed bed and therefore the costs of compression. However, with increasing particle size the value of ξ_0 decreases and therefore the runaway limit given by Equation 4.19 is approached. Since runaway cannot be allowed, Equation 4.19 gives the maximum tolerable particle size for a given reaction, a given gas temperature, and a given gas velocity. In Figure 4.7, $\xi_{0,ra}$ is plotted versus θ_g for several values of An. A multiplicity and a uniqueness region can be distinguished. In this last region the slope at the point of inflection of the *HPR* curve is lower than that of the *HWR* line, which is 1, so multiplicity can never occur. To find the boundary line mentioned, Equation 4.16 must be solved together with Equation 4.17, and

$$\frac{\partial^2}{\partial\theta_p^2}\left[1+\xi_0 \exp\left(\frac{An}{\theta_p}\right)\right]^{-1} = 0$$

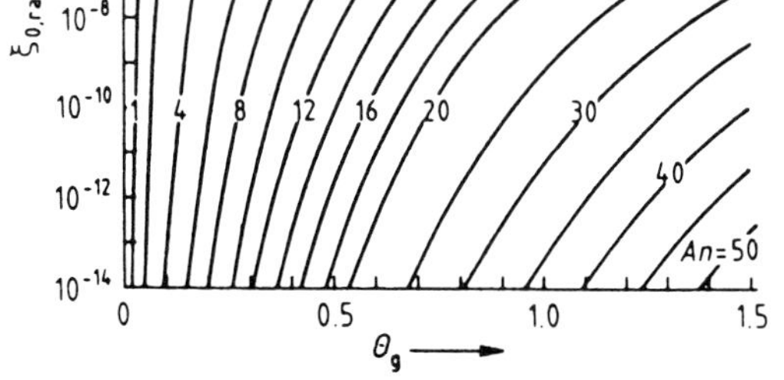

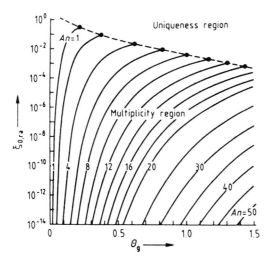

Figure 4.7 Value of $\xi_{0,ra}$ for which runaway occurs versus dimensionless gas temperature θ_g.

which states that the derivative of the *HPR* curve is unity at the inflection point. Solving the equations and eliminating θ_p and *An* gives

$$\xi_0 = \frac{1+\theta_g}{\theta_g} e^{-2(1+\theta_g)}$$

This is the boundary line equation giving the dashed line in Figure 4.7.

All pellets in a reactor cannot always be in the uniqueness region. Then, one must ensure that all pellets work at the lower operating point (see Figure 4.6). Once the particle diameter has been fixed and safe operating conditions have been chosen for the design (e.g. in Figure 4.7, the point given by $\theta_g = 0.3$ and $\xi_0 = 0.07$, see also Figures 4.5 and 4.6) the diagram can be used further to determine operating adjustments. For example, lowering the throughput to 25% of its design value would certainly lead to (near) runaway, because now

$$\xi_{0,new} = \xi_{0,old} \sqrt{u_{new} / u_{old}} = 0.5 \xi_{0,old} = 0.035$$

In the first instance, θ_g should be lowered to prevent pellet runaway. This may lead to too low a temperature and hence too low a conversion. In this case the reactant concentration in the feed may be decreased to increase *An*, so that θ_g can be increased again to achieve sufficiently high reaction rates.

This reasoning also indicates the importance of gas flow rates. If, for example, in a tube of a multitubular reactor, a deposit is formed that decreases the gas velocity because of plugging, at some point pellet runaway will occur irreparably and, consequently, full runaway of that tube [12].

Multiplicity studies for other reaction kinetics proceed along the same lines; simple analytical solutions often cannot be obtained, so numerical approximations must be used.

4.4 Criteria for the Absence of Transport Limitations

The question remains as to when the various diffusion effects really influence the conversion rate in fluid-solid reactions. Many criteria have been developed in the past for the determination of the absence of diffusion resistance. In using the many criteria no more information is required than the diffusion coefficient D_A for fluid phase diffusion and $D_{e,A}$ for internal diffusion in a porous pellet, the heat of reaction and the physical properties of the gas and the solid or catalyst, together with an experimental value of the observed global reaction rate $\langle R_J \rangle$ per unit volume or weight of solid or catalyst. For the time being the following criteria are recommended. Note that intraparticle criteria are discussed in much greater detail in Chapter 6.

The *intraparticle* concentration and temperature gradients in a porous particle can always be neglected, when the pore effectiveness factor η is close to 1. Assuming that η

should be 1 ± 0.05, Mears [13,14] derived for simple irreversible reactions of the *n*th order the following criterion. If

$$\frac{\langle R_A \rangle d_p^2}{4C_{A,s}D_{e,A}} < \frac{1}{n - \varepsilon\alpha} \tag{4.20}$$

then the internal resistances can be neglected. However, if $n - \varepsilon\alpha$ approaches 0, the asymptotic method of Petersen [15] has to be used. In this relationship n is the order of the reaction, $\varepsilon = E_a/RT$, and $\alpha = (-\Delta H)_A D_{e,A} C_{A,s} / \lambda_p T_s$. In Chapter 6 more generalized criteria are discussed.

To check whether the particle can be considered as isothermal – whether there are important concentration gradients or not – the criterion of Anderson [16] can be used. If

$$\frac{|(-\Delta H)_A| \langle R_A \rangle d_p^2}{4\lambda_p T_s} < 0.75 \frac{T_s R}{E_a} \tag{4.21}$$

then the porous particle can be considered as being isothermal.

The intraparticle transport effects, both isothermal and nonisothermal, have been analyzed for a multitude of kinetic rate equations and particle geometries. It has been shown that the concentration gradients within the porous particle are usually much more serious than the temperature gradients. Hudgins [17] points out that intraparticle heat effects may not always be negligible in hydrogen-rich reaction systems. The classical experimental test to check for internal resistances in a porous particle is to measure the dependence of the reaction rate on the particle size. Intraparticle effects are absent if no dependence exists. In most cases a porous particle can be considered isothermal, but the absence of internal concentration gradients has to be proven experimentally or by calculation (Chapter 6).

In fluid-solid systems the interparticle gradients – between the external surface of the particle and the adjacent bulk fluid phase – may be more serious, because the effective thermal conductivity of the fluid may be much lower than that of the particle. For the interparticle situation the heat transfer resistances, in general, are more serious than the interparticle mass transfer effects; they may become important if reaction rates and reaction heats are high and flow rates are low. The usual experimental test for interparticle effects is to check the influence of the flow rate on the conversion while maintaining constant the space velocity or residence time in the reactor. This should be done over a wide range of flow rates and the conversion should be measured very accurately.

For the requirement that the conversion rate at the conditions of the particle surface should not differ by more than 5% of that under the conditions in the bulk of the reaction mixture flowing around the particle, Mears [13,14] derived the following criteria. If

$$\frac{< R_A > d_p}{2k_g C_{A,s}} < \frac{0.15}{n} \tag{4.22}$$

the mass transfer resistance over the gas film can be neglected, and if

$$\frac{|(-\Delta H)_A| < R_A > d_p}{2\alpha_g T_s} < 0.15 \frac{RT_s}{E_a} \tag{4.23}$$

the heat transfer resistance over the gas film can be neglected

From the above it follows that in most practical situations a model, that takes into account only an intraparticle mass and an interparticle heat transfer resistance will give good results. However, in experimental laboratory reactors, which usually operate at low gas flow rates, this may not be true. In the above criteria the heat and mass transfer coefficients for interparticle transport also have to be known. These were amply discussed in Section 4.2.

In Example 9.4.5. the criteria above are applied to the ethylene oxidation reactions:

$$C_2H_4 + \frac{1}{2}O_2 \rightarrow C_2H_4O \quad \text{and} \quad C_2H_4 + 3O_2 \rightarrow 2CO_2 + 2H_2O$$

For the reaction conditions chosen it is shown, that pore diffusion limitations must occur, and that the heat conductivity of the silver containing particles is high enough, so that the particles can be considered to be isothermal. Further it is shown that there must be quite a difference in temperature between the catalyst particles and the flowing gas, so that a particle runaway study must be made.

References

1. Sada, E., Kumazawa, H., Butt, M.A., Sumi, T. (1977b) *Chem. Eng. Sci.*, **32**, 970.
2. Sada, E., Kumazawa, H., Butt, M.A. (1977a) *Chem. Eng. Sci.*, **32**, 972.
3. Gnielinski, V. (1975) *Forsch. Ing. Wes.*, **41**(5), 145.
4. Martin H. (1978) *Chem. Eng. Sci.*, **33**, 913.
5. Nelson, P.A., Galloway, T.R. (1975) *Chem. Eng. Sci.* **30**, 1.
6. Kürten, H. Raasch, J., Rumpf, H. (1966) *Chem.-Ing.-Techn.*, **38**, 941.
7. Westerterp, K.R., Van Swaaij, W.P.M., Beenackers, A.A.C.M. (1987) *Chemical Reactor Design and Operation*. Chichester: Wiley.
8. Wammes, W.J.A., Middelkamp, J., Huisman, W.J., deBaas, C.M., Westerterp, K.R. (1991a) *AIChE J.*, **37**, 1849.
9. Wammes, W.J.A., Middelkamp, J., Huisman, W.J., deBaas, C.M., Westerterp, K.R. (1991b) *AIChE J.*, **37**, 1855.
10. Wammes, W.J.A., Westerterp, K.R. (1991) *Chem. Eng. Technol.*, **14**, 406.
11. Molga, E.J., Westerterp, K.R. (1997) *Ind. Eng. Chem. Res.*, **36**, 622.
12. Wijngaarden, R.J., Westerterp, K.R. (1992) *Chem. Eng. Sci.*, **47**, 1517.
13. Mears, D.E. (1971a) *J. Catal.*, **20**, 127.
14. Mears, D.E. (1971b) *Ind. Eng. Chem. Proc. Des. Dev.*, **10**, 541.
15. Petersen, E.E. (1972) *Chemical Reaction Analysis*, New Jersey: Prentice Hall, Englewood Cliffs.
16. Anderson, J.B. (1963) *Chem. Eng. Sci.*, **18**, 147.
17. Hudgins, R.R. (1968) *Chem. Eng. Sci.*, **23**, 93.

5 Experimental Methods

A quantitative description of heterogeneous catalytic reactors for design, scaling-up, control or optimization purposes requires several parameters. Some of them, including the effective diffusivity and some parameters for the transport models and also the intrinsic chemical rate, should be determined in special experiments.

5.1 Measurement of Diffusion in Porous Solids

The prediction of the parameter values for mass transport through porous materials is too difficult, because we do not know how to take into account the complicated pore geometry as it is in reality. Thus, data for the effective diffusivity or effective molecular and Knudsen diffusivities and the permeability are still more accurately determined experimentally. Experiments of this kind also provide valuable information on the porous structure, such as the average pore size and pore size distribution.

Several experimental techniques have been developed for the investigation of the mass transport in porous catalysts. Most of them have been employed to determine the effective diffusivities in binary gas mixtures and at isothermal conditions. In some investigations, the experimental data are treated with the more refined dusty gas model (DGM) and its modifications. The diffusion cell and gas chromatographic methods are the most widely used when investigating mass transport in porous catalysts and for the measurement of the effective diffusivities. These methods, with examples of their application in simple situations, are briefly outlined in the following discussion. A review on the methods for experimental evaluation of the effective diffusivity by Haynes [1] and a comprehensive description of the diffusion cell method by Park and Do [2] contain many useful details and additional information.

5.1.1 Diffusion Cell Method

A diffusion cell method was introduced by Wicke and Kallenbach [3] and therefore is called the Wicke-Kallenbach method. The experimental set up (Figure 5.1) is called the Wicke-Kallenbach cell. It is the most popular method for different diffusion measurements.

A typical experimental arrangement used for diffusion measurements consists of a cylindrical solid pellet (or multiple pellets in parallel) placed between two gas chambers with the flat surfaces exposed to the gases. Two gas streams of known flow rates and dif-

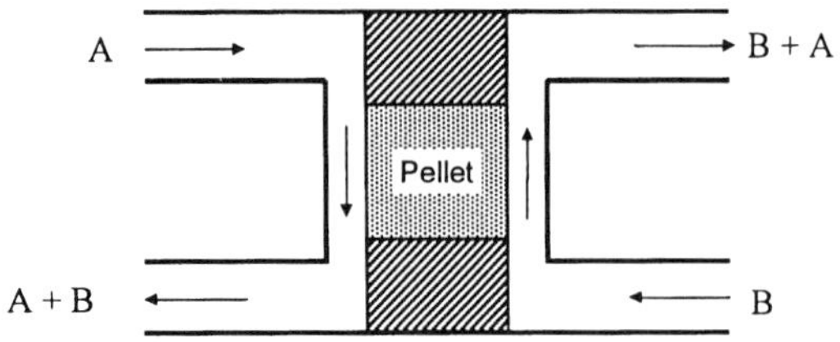

Figure 5.1 Schematic drawing of the Wicke-Kallenbach diffusion cell. Gases A and B may be pure gases or gas mixtures.

ferent inlet compositions flow through the chambers past the solid surfaces (Figure 5.1). The concentrations of the components penetrating through the solid are measured at the outlet of one or both chambers. From these concentrations (and the known inlet compositions, the flow rates and particle dimensions) the transport parameters are calculated. Various cell configurations have been employed [2]. The temperatures and pressures, which can be conveniently used with this technique, depend on the ability to mount the pellet leak-free in the diffusion cell, in view of the very small gas flow rates to be measured. A major distinction can be made between steady-state and unsteady-state methods, which often provide supplementary information.

5.1.2 Steady-state Methods

5.1.2.1 Measurements of Effective Diffusivity

In common practice the rates of the countercurrent flows of the two gases A and B through a pellet are measured. Mixtures of different compositions are passed along the opposite sides of the pellet. For measurements of the effective diffusivity, no pressure gradient is allowed across the pellet unless all pores are sufficiently small and only Knudsen flow occurs. The flux of the component A at a uniform total pressure is determined by the equation

$$J_A = -D_{e,A} C \frac{dx_A}{dz} \tag{5.1}$$

where $D_{e,A}$ is the effective diffusivity of A, C the total molar concentration, x_A is the mole fraction of A, and z the distance across the pellet.

At steady state the flux J_A is independent of time. It also does not depend on the distance z in view of the mass conservation. Provided the effective diffusivity is independent of the concentration and position inside the pellet, integration of Equation 5.1 with respect to z gives

$$J_A = -D_{e,A} C x_A + k_{int} \tag{5.2}$$

The integration constant k_{int} may be eliminated by the use of the boundary conditions

$$z = 0, \quad x_A = x_{A,1} \tag{5.3}$$

$$z = L, \quad x_A = x_{A,2} \tag{5.4}$$

which specify the gas composition $x_{A,1}$ and $x_{A,2}$ at the pellet surface. In that case the following expression for the molar flux of species A is obtained:

$$J_A = CD_{e,A}\frac{x_{A,1} - x_{A,2}}{L} \tag{5.5}$$

To find the effective diffusivity from Equation 5.5, the concentration driving force

$$\Delta C_A = C(x_{A,1} - x_{A,2}) = C_{A,1} - C_{A,2}$$

and the flux J_A through the pellet are measured. The flux can be calculated from the measured gas flow rate Φ_v through one of the chambers and the difference in concentrations of the inlet and outlet streams of this chamber. For example, the material balance over the second chamber gives

$$J_A\frac{\pi d_p^2}{4} = \Phi_{v2}C(x_{2,out} - x_{2,inl}) = \Phi_{v2}\Delta C_{A,2} \tag{5.6}$$

where d_p is the pellet diameter and the subscripts *inl* and *out* designate the inlet and outlet streams.

Combination of Equations 5.5 and 5.6 gives the effective diffusivity:

$$D_{e,A} = \frac{4\Phi_{v2}\Delta C_{A,2}}{\pi d_p^2}\frac{L}{C_{A,1} - C_{A,2}} \tag{5.7}$$

To be valid, Equation 5.7 requires the following conditions. First, the gas velocities along the pellet surfaces should be sufficiently high to:

- avoid resistance to mass transfer from the bulk to the surface;
- provide a sufficiently uniform composition over the chamber (such that the concentration of each gas at the surface of the pellet is close to that in the main flow).

Second, the gas flow rates should not be too high to be able to measure adequately the difference in the compositions of the inlet and outlet streams, $\Delta C_{A,2}$. An alternative method is to provide perfect mixing by a stirrer in each chamber [4]. Examples of the application of the just described technique can be found in many papers [3,5,6].

5.1.2.2 Application of the DGM

The main advantage of the diffusion cell method is the simplicity of the experimental data treatment. This permits analysis of different transport models to describe experiments with pure gases and with binary and ternary gas mixtures [7,8]. As an example consider the application of the DGM for the counter diffusion experiments with binary gas mixtures. In contrast to the measurements described previously, experiments with and without a pressure difference are now of interest. If the pressure gradient can be neglected (the relative pressure drop should be much lower than the relative mole fraction difference [9]) and without surface diffusion, the flux of species A is given by:

$$J_A = -\frac{D_{Ak} D^e_{AB} C}{D^e_{AB} + x_A D_{Ak} + (1 - x_A) D_{Bk}} \frac{dx_A}{dz} \tag{5.8}$$

in which the effective molecular diffusivity $D^e{}_{AB}$ and the effective Knudsen diffusivities of the components D_{Ak} and D_{Bk} are determined by Equations 3.16 and 3.18.

Integration across the length of the pellet and the use of the boundary conditions of Equations 5.3 and 5.4 yields

$$J_A = \frac{C D^e_{AB} D_{Ak}}{L(D_{Bk} - D_{Ak})} \ln \frac{(D_{Bk} - D_{Ak}) x_{A,1} + D_{Ak} + D^e_{AB}}{(D_{Bk} - D_{Ak}) x_{A,2} + D_{Ak} + D^e_{AB}} \tag{5.9}$$

Comparing Equations 5.5 and 5.9 gives the effective diffusivity of A (for the diffusion cell experiment) through the parameters of the DGM:

$$D_{e,A} = \frac{D^e_{AB} D_{Ak}}{(x_{A,1} - x_{A,2})(D_{Bk} - D_{Ak})} \ln \frac{(D_{Bk} - D_{Ak}) x_{A,1} + D_{Ak} + D^e_{AB}}{(D_{Bk} - D_{Ak}) x_{A,2} + D_{Ak} + D^e_{AB}} \tag{5.10}$$

Since the Knudsen diffusivities are interrelated,

$$D_{Bk} = D_{Ak} \sqrt{\frac{M_A}{M_B}}$$

only two coefficients, $D^e{}_{AB}$ and D_{Ak}, should be found. For most practical situations D_{Ak} is independent of the total pressure and $D^e{}_{AB} = D^e{}_0 \times P_0/P$, where P_0 is some reference pressure, (say 1 bar) and $D^e{}_0$ is the value of $D^e{}_{AB}$ at this pressure. Thus, the measurements of the flux J_A and the mole fractions $x_{A,1}$, and $x_{A,2}$ in the chambers at various pressure levels permit the determination of both parameters $D^e{}_0$ and D_{Ak}, independent of the pressure. It is suitable to present experimental data in terms of the effective diffusivity, determined by Equation 5.10, rather than the flux. Measurements at a fixed total pressure but at different temperatures also provide data necessary for model testing and for the determination of its parameters, since the theoretical dependence of $D^e{}_{AB}$ and D_{AK} on the temperature is known, see Equations 3.18 and 3.37. Note that the measurement of the flux of just one component is sufficient for the determination of all model parameters. In practice, however, measurement of both fluxes is highly desirable: inconsistencies in the parameters obtained for the different data sets would indicate an experimental error or model deficiency.

There is an important difference between the use of Equations 5.1 and 5.8 with which to treat experiments. When Equation 5.8 is used (and the DGM is appropriate) its parameters $D^e{}_0$ and D_{Ak} should not depend on the concentrations in the chambers and, for convenience, pure gases can be supplied to the chambers. This is not the case when only the effective diffusivity $D_{e,A}$ is measured. As Equation 5.10 shows, the effective diffusivity may depend on the concentrations. Therefore, Equation 5.1 is applicable only to situations where the concentration difference across the pellet is small.

5.1.2.3 Permeability Measurements

Experiments with different pressures in the chambers provide data for simultaneous determination of the diffusivities in and permeability B_0 of a porous particle (Chapter 3). Allawi and Gunn [8] have described such experiments and the evaluation procedure, accounting for the variation of viscosity with composition. The permeability of the porous particle can also be found in separate experiments on single gas flow through the particle under a pressure difference. In this case the molar flux is given by the equation

$$J_A = -\frac{1}{RT}\left(D_{Ak} + \frac{PB_0}{\mu}\right)\frac{dP}{dz} \tag{5.11}$$

The Knudsen diffusivity and the permeability can be found with the same procedure as has been described for the determination of the effective diffusivity. Integration of Equation 5.11 and accounting for the boundary conditions

$$z = 0, \quad P = P_1 \tag{5.12}$$

$$z = L, \quad P = P_2 \tag{5.13}$$

gives

$$J_A = \left(D_{Ak} + \frac{\overline{P}B_0}{\mu}\right)\frac{P_1 - P_2}{RTL} \tag{5.14}$$

where $\overline{P} = (P_1 + P_2)/2$ is the average pressure. Measured values of the flux and the pressures in the chambers give the values of D_{AK} and B_0/μ.

One should be aware of the condition $Kn = \lambda/d_e \sim 1$, that is the mean free path of the gas molecules is comparable to the average pore size. A graph of $J_A RTL/(P_1 - P_2)$ against the mean pressure $\overline{P}$ should give a straight line as Equation 5.14 predicts. This is confirmed experimentally for the case of Knudsen numbers much lower than 1. However in fine-pore material or at low mean pressures, a pronounced departure from linearity has been found by many authors [8,10]. As an example, the experimental data for forced flow of helium and nitrogen through silica gel are presented in Figure 5.2 [10]. This figure shows the measured gas flow rates per unit pressure difference

$$F = J_A S_{pellet} RT/(P_1 - P_2)$$

as a function of mean pressure $\overline{P}$, where S_{pellet} is the cross-sectional area of the pellet. This old problem has existed since the famous experiments of Knudsen on single gas flow through a capillary at different pressures, when a minimum was found in the de-

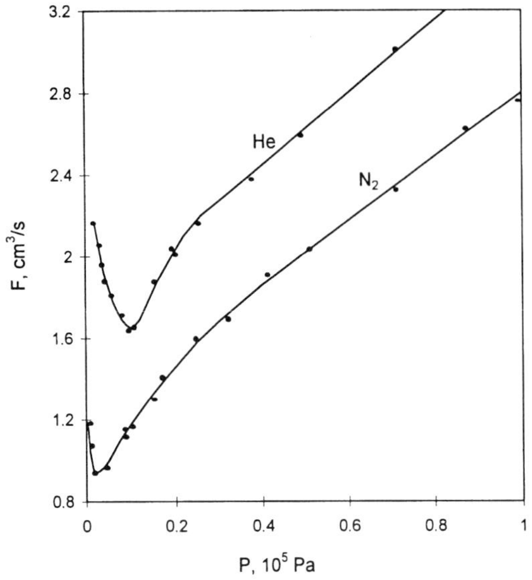

Figure 5.2 Experimental dependence of permeability upon mean pressure for helium and nitrogen through silica gel (adapted from Grachev [10]).

pendence of $J_ART L/(P_1 - P_2)$ on the average pressure, similar to that shown in Figure 5.2 [11]. Different explanations of this effect have been proposed by others [12-14]. Unfortunately, neither theory can give a satisfactory quantitative description of the experimental data found.

5.1.3 Unsteady-state Methods

The available transport models are not reliable enough for porous material with a complex pore structure and broad pore size distribution. As a result the values of the model parameters may depend on the operating conditions. Many authors believe that the value of the effective diffusivity $D_{e,A}$, as determined in a Wicke-Kallenbach steady-state experiment, need not be equal to the value which characterizes the diffusive flux under reaction conditions. It is generally assumed that transient experiments provide more relevant data. One of the arguments is that dead-end pores, which do not influence steady state transport but which contribute under reaction conditions, are accounted for in dynamic experiments. Experimental data confirming or rejecting this opinion are scarce and contradictory [2]. Nevertheless, transient experiments provide important supplementary information and they are definitely required for bidisperse porous material where diffusion in micro- and macropores is described separately with different effective diffusivities.

A variety of diffusion cells have been developed for transient measurements. The experimental arrangements and the data analysis methods for the determination of effective diffusivities are described in reviews [1,2] and in the papers cited therein. Interesting applications of a diffusion cell with one compartment closed have been described recently for the investigation of the dynamics of ternary gas mixtures [15,16]. In the last papers the DGM and the mean transport pore model have been used to describe the experiments [17].

In the usual dynamic method a pure reference or carrier gas B flows across both sides of a pellet. A pulse or step input of tracer A is introduced into the first gas chamber, and the response of the tracer A as a function of time is measured at the exit of the second chamber. Two basic techniques with step and pulse tracer injections for the determination of the effective diffusivity are described.

The mass balance equation for component A is

$$\varepsilon_p \frac{\partial C_A}{\partial t} + \frac{\partial J_A}{\partial z} = 0 \tag{5.15}$$

This equation describes the change of the molar concentration of A with respect to time t at a fixed point inside the pellet pores, this change resulting from the motion of A. Since the pores occupy only a fraction of the pellet volume and the flux J_A relates to the unit area of the pellet, the porosity ε_p of the pellet is accounted for in Equation 5.15. The derivation of the Equation 5.15 makes it essential that neither chemical reaction or adsorption of A occurs. When Equation 5.1 for the flux is inserted into Equation 5.15, we get, for constant diffusivity,

$$\varepsilon_p \frac{\partial C_A}{\partial t} = D_{e,A} \frac{\partial^2 C_A}{\partial z^2} \tag{5.16}$$

5.1.3.1 Step Injection

In this method the concentration of A in both chambers is initially equal to zero, and after that is kept at a constant C_{A1} in the first chamber. If the flow rates in both chambers are large enough, so that the resistance to mass transfer is only located inside the pellet, and the concentration of the tracer in the second chamber is much lower than C_{A1}, the appropriate initial and boundary conditions are:

$$C_A(z,0) = 0 \tag{5.17}$$

$$C_A(0,t) = C_{A1} \tag{5.18}$$

$$C_A(L,t) = 0 \tag{5.19}$$

The analytical solution of Equation 5.16 with these boundary conditions is well known [18]. From this solution the total amount of A, that has entered in the second chamber after a period of time t can be calculated with:

$$Q(t) = -A_p D_{e,A} \int_0^t \frac{\partial C_A(L,t)}{\partial z} dt \tag{5.20}$$

where A_p is the cross-sectional area of the pellet. Note that this solution cannot be matched with the measured response curve because of the approximate boundary condition of Equation 5.19. A more general boundary condition should be used to describe the exit concentration. For a time $t >> L^2/D_e$ the solution, found from Equation 5.20 can be replaced by a more simple expression:

$$Q = \frac{D_{e,A} C_{A1} A_p}{L} \left(t - \frac{L^2 \varepsilon_p}{6 D_{e,A}} \right) \tag{5.21}$$

We may rewrite the last equation as

$$Q = J_A^S A_p (t - t_d) \tag{5.22}$$

where

$$J_A^S = \frac{D_{e,A} C_{A1}}{L} \tag{5.23}$$

is the steady state flux, as determined by Equation 5.5 with $x_{A,2} = 0$, and

$$t_d = \frac{L^2 \varepsilon_p}{6 D_{e,A}} \tag{5.24}$$

Equation 5.22 shows that Q is determined by an equation describing the steady flow of A through the pellet with a time lag t_d. To determine t_d one may plot any quantity proportional to the amount of diffused gas Q, rather than this quantity itself, versus time. The intercept on the t axis will give t_d and $D_{e,A} = L^2\varepsilon_p/(6t_d)$. Note that with this experimental procedure the effective diffusivity can also be found from Equation 5.23, if the steady-state flux is measured.

The described method above is known as the time lag method. It was proposed first by Grachev et al. [19] as a modification of Barrer's method for the determination of the permeability or Knudsen diffusivity [20]. His measurements were conducted with a single gas imposing a pressure drop across the sample. Note that, for effective diffusivity measurements, equal total pressures must be maintained on the two sides of the diffusion cell unless Knudsen diffusion prevails.

5.1.3.2 Pulse Injection

In another version of a dynamic experiment proposed first by Gibilaro et al. [21] an impulse of the tracer is introduced into the carrier gas flowing through the first chamber, and the response $C_A(L,t)$ in the second chamber is recorded. A suitable analysis is based on the normalized first moment of the response curve, which is defined as

$$\mu_1 = \frac{\int_0^\infty t C_A(L,t) dt}{\int_0^\infty C_A(L,t) dt} \tag{5.25}$$

Dogu and Smith [22,23] have given the theoretical basis for the dynamic pulse response technique. If the experimental conditions are identical to those described for the step injection and the injection time is much lower than the diffusion time $L^2/D_{e,A}$, the theoretical first moment is

$$\mu_1 = \frac{L^2 \varepsilon_p}{6D_{e,A}} \tag{5.26}$$

The values of μ_1 determined from Equation 5.25 may than be used in Equation 5.26 to determine the effective diffusivity. Note that the first moment of Equation 5.26 coincides with the time delay in the step injection experiments (Equation 5.24).

5.1.4 Chromatographic Techniques

One of thc popular methods for evaluating effective diffusivities in heterogeneous catalysts is based on gas chromatography. A carrier gas, usually helium, which is not adsorbed, is passed continuously through a column packed with catalyst. A pulse of a diffusing component is injected into the inlet stream and the effluent pulse recorded. The main advantages of this transient method are its applicability to particles of arbitrary shapes, and that experiments can be carried out at elevated temperatures and pressures. Haynes [1] has given a comprehensive review of this method.

Compared to the transient diffusion cell technique, an additional equation to describe the tracer propagation in the particle bed is required for this chromatographic method. This makes the analysis of the experimental data more difficult.

Models with varying degrees of complexity have been employed to analyze the experimental results by a variety of techniques. The most comprehensive models include terms to account for axial dispersion in the packed bed, external mass transfer, intraparticle diffusion in both macropore and micropore regions of the pellet and a finite rate of adsorption. Of the several methods of analysis, the most popular ones are based on the moments of the response curve. The first moment of the chromatogram is defined by Equation 5.25 in which the concentration now is taken at the outlet of the column. The second central moment is calculated from equation

$$\sigma^2 = \frac{\int_0^\infty (t-\mu_1)^2 C_A(t)dt}{\int_0^\infty C_A(t)dt} \tag{5.27}$$

For a short pulse input of a nonadsorbable tracer and a catalyst characterized by a unidisperse pore structure, the following relationships are obtained for the first moment and second central moment of the effluent curve of the bed [24]:

$$\mu_1 = \frac{L}{v}\left[\varepsilon_b + (1-\varepsilon_b)\varepsilon_p\right] \tag{5.28}$$

$$\sigma^2 = \frac{2LD_z}{v^3}\left[\varepsilon_b + (1-\varepsilon_b)\varepsilon_p\right] + \frac{2L(1-\varepsilon_b)\varepsilon_p{}^2 r_p}{3vk_g} + \frac{2L(1-\varepsilon_b)\varepsilon_p^2 r_p^2}{15vD_{e,A}} \tag{5.29}$$

where L is the bed length, D_z is the axial dispersion coefficient, v is the superficial gas velocity, ε_b is the bed porosity, r_p is the particle diameter, and k_g is the interface mass transfer coefficient.

Equation 5.29 shows that the variance of the response curve is separable into contributions from the axial dispersion, and from the external and internal mass transfer. Measurements at different velocities lead to an estimate of all transport coefficients.

Unfortunately the theory, when used for data interpretation, is based on many simplifications that are not always valid. The success of the method depends on the significance of the axial dispersion and the interfacial mass transfer and on the accuracy of the description of these effects. A comparative study of different measuring technique by [25] has shown, for the chromatographic method, that experimental uncertainties may lead to significantly broader confidence limits for the diffusion coefficients than for measurements in single pellets in a diffusion cell.

The method can be applied to investigate the bidisperse pore structures, which consist of small microporous particles formed into macroporous pellets with a clay binder. In such a structure there are three distinct resistances to mass transfer, associated with diffusion through the external fluid film, the pellet macropores, and the micropores. Haynes and Sarma [24] developed a suitable mathematical model for such a system.

As an alternative to the chromatographic pulse technique a method based on a steady-state sinusoidal varying input concentration can be used. This method, first proposed by Deisler and Wilhelm [26], is an improvement over pulse input chromatography in certain systems because simpler, less accurate measurements are required and the modeling process is mathematically less complicated [27].

5.2 Measurement of Reaction Rates

5.2.1 Introduction

The evaluation of catalyst effectiveness requires a knowledge of the intrinsic chemical reaction rates at various reaction conditions and compositions. These data have to be used for catalyst improvement and for the design and operation of many reactors. The determination of the real reaction rates presents many problems because of the speed, complexity and high exo- or endothermicity of the reactions involved. The measured conversion rate may not represent the true reaction kinetics due to interface and intraparticle heat and mass transfer resistances and nonuniformities in the temperature and concentration profiles in the fluid and catalyst phases in the experimental reactor. Therefore, for the interpretation of experimental data the experiments should preferably be done under reaction conditions, where transport effects can be either eliminated or easily taken into account. In particular, the concentration and temperature distributions in the experimental reactor should preferably be described by plug flow or ideal mixing models.

A variety of laboratory reactors have been developed for the determination of the kinetics of heterogeneous reactions, all with specific advantages and disadvantages. Several reviews of laboratory reactors are available [28-33]. The evaluations of the available methods in these reviews are different because of the variation of chemical reactions and catalysts investigated and the different viewpoints of the authors. It is impossible to choose a best kinetic reactor because too many conflicting requirements need to be satisfied simultaneously. Berty [34] discussing an ideal kinetic reactor, collected 20 requirements as set forward by different authors. From these requirements it is easy to conclude, that the ideal reactor, that can handle all reactions under all conditions, does not exist. For individual reactions, or for a group of similar reactions, not all requirements are equally important. In such cases it should be possible to select a reactor that exhibits most of the important attributes.

The most important methods for the determination of kinetics of catalyzed reactions are described here. We emphasize the problems and pitfalls in obtaining reliable reaction rates. The many "diagnostic tests" are briefly discussed and some warnings are given to limitations of commonly used laboratory reactors. Finally, it is worth noting that reaction rates can be expressed per unit mass of catalyst, per unit catalytic surface, per unit external particle area or per unit volume of the reactor, fluid or catalyst. For chemical reactor design it is best to express reaction rates in terms of unit catalyst volume.

5.2.2 Laboratory Reactors for Determination of Kinetics

5.2.2.1 Fixed Bed Reactor

Continuously operated, fixed bed reactors are frequently used for kinetic measurements. Here the reactor is usually a cylindrical tube filled with catalyst particles. Feed of a known composition passes though the catalyst bed at a measured, constant flow rate. The temperature of the reactor wall is usually kept constant to facilitate an isothermal reactor operation. The main advantage of this reactor type is the wealth of experience with their operation and description. If heat and mass transfer resistances cannot be eliminated, they can usually be evaluated more accurately for packed bed reactors than for other reactor types. The reactor may be operated either at very low conversions as a differential reactor or at higher conversions as an integral reactor.

5.2.2.2 Differential Reactor

In the differential flow reactor, the residence time is short so that the conversion remains small, usually a few per cent. This can be achieved in short beds and/or with high feed flow rates. Since the conversion is small all pellets operate approximately under the same conditions and variation of the volumetric flow rate can be neglected. The observed or apparent conversion rate follows directly from the measured inlet and outlet concentrations via a material balance over the bed:

$$\langle R_A \rangle = \frac{\Phi_v (C_{A,0} - C_A)}{(1-\varepsilon_b) V_r} \tag{5.30}$$

where $\langle R_A \rangle$ is the conversion rate in kmol of A converted per unit of time and per unit catalyst volume, Φ_v the volumetric flow rate ($m^3 s^{-1}$), $C_{A,0}$ and C_A are the molar concentrations of species A at the bed inlet and outlet, ε_b is the volume fraction of the fluid phase in the reactor, and V_r is the volume of the tube.

If we introduce the conversion

$$\zeta_A = \frac{C_{A,0} - C_A}{C_{A,0}} \tag{5.31}$$

then Equation 5.30 becomes

$$\langle R_A \rangle = \frac{\Phi_v C_{A,0} \zeta_A}{(1-\varepsilon_b) V_r} \tag{5.32}$$

The quantity $\langle R_A \rangle$ can be further elaborated if interface mass and heat transfer coefficients are known, and with the theory of mass transfer with reaction inside porous catalysts as treated in Chapters 6 and 7:

$$\langle R_A \rangle = f(k_g, D_{e,A}, R_A, \ldots) \tag{5.33}$$

By varying the flow rate Φ_v and the particle size d_p, the interface heat and mass transfer coefficients k_g vary, so that information can be obtained on the significance and location of the rate controlling resistances – in the pellet or in the fluid phase. Once the resistances are known, variation of reactant concentrations and of temperature gives the information necessary to decide whether a particular kinetic equation $R_A(C_A, C_P, T, \ldots)$ fits the conversion data. At sufficiently high flow rates, it should be possible to minimize the temperature and partial pressure gradients so that the fluid phase temperature and compositions are almost the same as that of the catalyst surface. In the ideal situation the variation of the temperature over the bed is negligible, the measured conversion rates do not depend on transport processes and are governed solely by the intrinsic reaction rate and by the amount of catalyst per unit reactor volume:

$$\langle R_A \rangle = \frac{R_A}{1-\varepsilon_b} \tag{5.34}$$

Combining Equations 5.30 and 5.34 gives the equation for the determination of the reaction rate:

$$R_A = \frac{\Phi_v C_{A,0} \zeta_A}{V_r} \tag{5.35}$$

For an irreversible nth-order reaction, $R_A = k_A C_A^n = k_A C_{A,0}^n (1-\zeta_A)^n$ holds and Equation 5.35 becomes:

$$k_A = \frac{\Phi_v \zeta_A}{V_r C_{A,0}^{n-1} (1-\zeta_A)^n} \tag{5.36}$$

Experiments at different temperatures and inlet concentrations provide the data necessary to obtain the reaction order n and the rate constant k_A.

A disadvantage of the differential reactor is the inaccuracy in the determination of conversion and selectivity due to the small concentration changes. The second difficulty in the treatment of experimental data is caused by possible flow nonuniformities. Since the average residence time is short and the fluid elements moving with different axial velocities do not mix, the simplified Equation 5.30 may not be valid. This is because the reactor operates as a segregated flow reactor rather than a plug flow or ideal mixed reactor, on which Equation 5.30 is based.

5.2.2.3 Integral Reactor

The integral mode of operation has the inherent advantage that large conversions are achieved, so a high precision of chemical analysis is not required. If a plug flow model is assumed to be valid for such a reactor, a material balance for component A over a differential bed length dz gives

$$\Phi_v dC_A = (1-\varepsilon_b)\langle R_A\rangle S_R dz \tag{5.37}$$

where S_R is the cross-sectional area of the reactor. This is a differential equation for C_A and it is subject to the initial condition that specifies the feed composition:

$$z = 0, \quad C_A = C_{A,0} \tag{5.38}$$

Introduction of the conversion (Equation 5.31) yields

$$\Phi_v C_{A,0} d\zeta_A = (1-\varepsilon_b)\langle R_A\rangle S_R dz \tag{5.39}$$

subject to

$$z = 0, \quad \zeta_A = \zeta_{A,0} \tag{5.40}$$

The variation of the fluid density usually can be neglected, so the volumetric flow rate can be considered as independent of the axial coordinate. In this case integration of Equation 5.39 over the bed length gives

$$C_{A,0}\int_0^{\zeta_A(L)} \frac{d\zeta_A}{\langle R_A\rangle} = \frac{LS_R(1-\varepsilon_b)}{\Phi_v} \tag{5.41}$$

where L is the bed length. If the conversion over the bed $\zeta_A(L)$ is small, the dependence of the $\langle R_A\rangle$ on ζ_A can be neglected and Equation 5.41 transforms into the balance equation of the differential reactor, Equation 5.32 with the volume of the bed $V_r = LS_R$. Without any internal and external transport limitations Equation 5.34 holds and the intrinsic reaction rate can be found from:

$$C_{A,0}\int_0^{\zeta_A(L)} \frac{d\zeta_A}{R_A} = \frac{V_r}{\Phi_v} = \frac{\tau_L}{\varepsilon_b} \tag{5.42}$$

where τ_L is the average residence time of the reaction mixture in the reactor.

The reaction rate R_A is not obtained directly from the integral reactor data. As suggested by Equation 5.42, the data, which are necessary for the determination of R_A, may be obtained by changing either the feed rate for a given amount of catalyst or by changing the amount of catalyst for a given feed rate, or both. It is usually easier to change the flow rate rather than the amount of catalyst. The general procedure, therefore, involves measurement of the conversions for a number of feed rates with a constant feed composition, pressure and temperature. Similar series of measurements are made for different feed compositions, temperatures, and pressures. Each series of measurements then provides data of the conversion as a function of the average residence time. The relationship of ζ_A and the average residence time can be differentiated to provide the desired value of $\langle R_A \rangle$. If reaction rate equation is known, its parameters can be found directly from Equation 5.42. For example, for power law kinetics with $R_A = k_A C_A^n = k_A C_{A,0}^n (1-\zeta_A)^n$ Equation 5.42 gives

$$-\frac{1}{k_A}\ln\left(1-\zeta_A(L)\right)=\frac{\tau_L}{\varepsilon_b} \quad \text{for } n=1 \tag{5.43}$$

$$\frac{\left(1-\zeta_A(L)\right)^{n-1}-1}{k_A(n-1)C_{A,0}^{n-1}}=\frac{\tau_L}{\varepsilon_b} \quad \text{for } n\neq 1 \tag{5.44}$$

In general the interpretation of the data is somewhat more complicated than for the differential method. Especially for an unknown complicated kinetic functions, the derivation of the correct reaction rate expression R_A from experimental results using Equations 5.41 and 5.33 is more cumbersome than fitting Equations 5.30 and 5.33. This is especially true for complex reaction networks, as in the isomerization and cracking reactions of crude oil fractions, where the integral method is very laborious with which to derive individual rate constants.

Useful information about reaction kinetics can be obtained from measurement of the reaction rate at the bed inlet, since the composition of the fluid is known. Froment [35] gives a survey of this approach, which has been successfully employed to determine the kinetics of the dehydrogenation of ethanol [36]. The initial reaction rates in the catalyst bed are determined by extrapolating the experimentally obtained reaction rate to a zero residence time:

$$R_A = \varepsilon_b C_{A,0} \left.\frac{d\zeta_A}{d\tau}\right|_{\tau\to 0}$$

An alternative approach is to measure the axial concentration profile in the catalyst bed. The initial reaction rate is found from Equations 5.34 and 5.39:

$$R_A = C_{A,0} \left.\frac{\Phi_v}{S_R}\frac{d\zeta_A}{dz}\right|_{z\to 0}$$

Thus the initial reaction rate can be determined in one, single experiment by differentiating the conversion versus axial position curve at $z = 0$. This method has been used for the selective oxidation of ethene [37].

Initial reaction rates obtained with a pure feed in which only reactants are present can be used for the discrimination between rival kinetic models, i.e. to identify whether adsorption, desorption, or surface reactions are the rate-determining steps. When pure A is fed to an integral reactor, for example, initial rates are observed at the inlet, where the product concentration is still zero. Comparing possible rate equations, which are often simpler in case of absence of products, with experimental data obtained at different concentrations of A, helps to reveal the appropriate [33,35].

The disadvantage of the integral method lies in the poor heat transfer in the fixed catalyst bed, which inhibits isothermal conditions for reactions with moderate or high heat effects. Pronounced concentration and temperature gradients in both axial as well as in radial direction are often observed. For such situations the assumption of plug flow and isothermal conditions is not true and partial differential equations, describing multidimensional concentration and temperature fields in the reactor, should be used instead of Equation 5.37. Reaction rates can be found with such equations, as has been demonstrated for various catalytic reactions, such as carbon monoxide oxidation, the dehydrogenation of cyclohexanol and the butene oxidation to maleic anhydride [38,39]. A most refined treatment, accounting for almost all transport effects, has been executed by Schwedock et al. [40] for the oxidation of methanol to formaldehyde. This approach, although possible, requires good skills in packed bed reactor modeling. Fortunately, there is an easier way to investigate fast exo- and endothermic reactions in integral packed bed reactors: the temperature gradients can be reduced to an acceptable level by diluting the catalyst with inert particles.

5.2.2.4 Recycle Reactors

The advantage of both the differential method (small variations in temperature and composition over the bed) and the integral method with its range of conversions are combined in recycle reactors. Various flow schemes with stationary catalysts as well as with moving catalyst particles contained in rotating baskets are applied. In essence, after passing through either a catalyst bed, a single row of pellets or separate pellets, most of the reaction mixture is recycled. A general scheme of such a reactor is presented in Figure 5.3. A material balance demands that

$$\Phi_v C_{A,0} + \Phi_R C_A = (\Phi_v + \Phi_R) C_A^* \qquad (5.45)$$

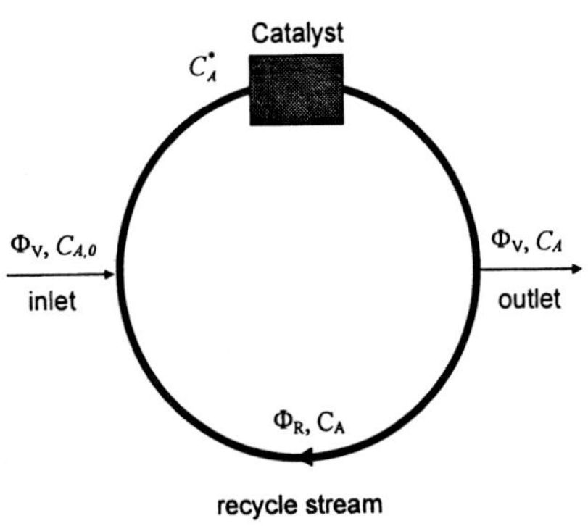

Figure 5.3 Scheme of a recycle reactor.

where $C_{A,0}$, C_A and $C_A{}^*$ are the concentrations in the feed, discharge and recycle streams, respectively. Introducing the conversion per cycle $\zeta_{A,c}$ and the total conversion ζ_A in the reactor, we have

$$\zeta_{A,c} = \frac{C_A^* - C_A}{C_A^*}, \quad \zeta_A = \frac{C_{A,0} - C_A}{C_{A,0}} \tag{5.46}$$

Equation 5.45 can be rewritten as

$$\zeta_{A,c} = \frac{1}{(R_R + 1)\dfrac{1-\zeta_A}{\zeta_A} + 1} \tag{5.47}$$

where R_R is the recycle ratio Φ_R/Φ_v. Equation 5.47 shows that for a finite net conversion ζ_A, the conversion per cycle $\zeta_{A,c}$ approaches zero if the recycle ratio R_R is increased more and more. Thus, at sufficiently high recycle flow rates, the concentration and temperature gradients along the catalyst bed can be kept small in the same way as in the differential reactor. The overall conversion in the reactor can be set at any desirable, easily measurable level by regulating the net flow rate Φ_v. A further benefit of high recycle flow rates lies in the fact that, due to the high fluid velocities past the catalyst particles, any possible interface transport effects can be eliminated. The recycle reactors, therefore, are called gradient-free reactors.

If the recycle rate is sufficiently high, the reaction rate corresponds to the conditions in the reactor exit, which are measured directly. The material balance, based on the feed and effluent concentrations, for the steady state is given by:

$$\frac{\Phi_0 C_{A,0} - \Phi_{out} C_A}{(1-\varepsilon_b)V_r} = R_A(C_A) \tag{5.48}$$

Knowing the inlet and exit compositions, the kinetics of the reaction can be elucidated with Equation 5.48 in the same way as for the differential reactor.

An additional and important advantage of the recycle reactor, compared to the differential packed bed reactor, is that here flow uniformity through the bed is not required, so channeling is not a problem and one layer of catalyst or even separate particles can be used in the reactor. For packed bed reactors, flow nonuniformity would inhibit the application of the plug flow model.

Recycle reactors are often able to cope with fast, complex, exo- or endothermic reactions, and are frequently used in research practice. A great number of different constructions (with external and internal recycling, with stationary and movable catalyst) are described in the literature [32].

External recycle reactor

This reactor is similar to the industrial packed bed reactor in which part of the product stream is recycled to the reactor inlet [41]. As shown in Figure 5.4, a recycle reactor con-

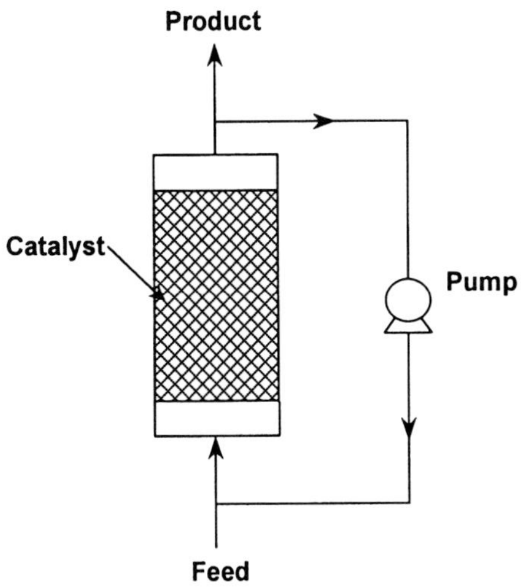

Figure 5.4 External recycle reactor.

sists of a fixed catalyst bed and a circulating pump or ventilator designed to recycle fluid at a rate far in excess to the feed and withdrawal rates. External recycling is usually combined with cooling or heating of the recycle stream to maintain a constant temperature over the reactor and to protect the recycle equipment.

Temkin has proposed the fixed bed reactor with external recycle [42]. Various constructions proposed afterwards differ mainly in the type of circulating system.

The benefit of external recycle reactors results from the ability to control the flow rate through the bed with a relatively high accuracy. Thus, conditions under which the interface heat and mass transfer resistances are negligible, can be established. It is also possible to adjust exactly the physical and chemical conditions around the catalyst in arbitrary regimes.

A disadvantage of the system is the considerable volumes of connecting tubing and the recycle pump. Large free volumes become unacceptable when homogeneous reaction or reactions on the tube walls occur. To use a ventilator that has the required capacity but can only work satisfactorily at relatively low temperatures, the gas from the reactor must be cooled and reheated before being returned to the catalyst bed. This procedure was a problem in a study of the partial oxidation of methanol to formaldehyde, where paraformaldehyde tended to be formed in the cooler parts of the recycle loop [43].

Internal recycle reactor with a stationary packed bed

The difficulties inherent to the external recycle reactor are avoided in an internal recycle reactor. Basically, an internal recycle reactor consists of a basket, in which a variable amount of catalyst can be placed, and an impeller for the internal circulation of the gas in the reactor. One of the most popular recycle reactors is the Berty reactor [29] shown in Figure 5.5. The Berty reactor has a magnetically driven blower and the gas from the turbine flows through a draft tube to the top of the catalyst bed. The flow rate through the bed can be calculated by measuring the pressure drop over the bed, if pressure taps on either side of the catalyst are available. For a catalyst to be studied in this reactor, the

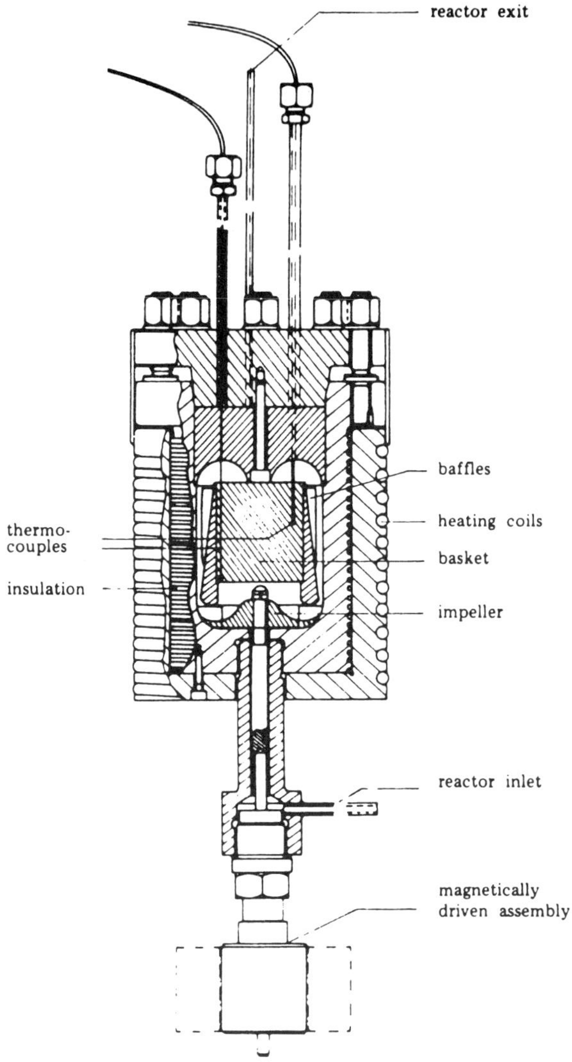

Figure 5.5 Berty internal recycle reactor.
Reprinted from Chemical Engineering Science, **44**, A.N.R. Bos et. al., The kinetics of methanol synthesis on a copper catalysts: an experimental study, 2435-2449, 1989, with kind permission from Elsevier Science Ltd, The Boulevard, Langford Lane, Kidlington OX5 1GB, UK.

pressure drop versus flow relationship has to be measured in separate equipment, or it has to be estimated from available correlations [29]. It should be noted that the flow rate through a packed bed depends on the gas density. Therefore, the calibration measurements should preferably be carried out at the reaction pressure and temperature.

Many authors have proposed reactors with similar basic principles. The best known are those of Garanin et al. [44], Livbjerg and Villadsen [45] and new versions of Berty reactor [34]. Variants of internal recycling reactors have also been proposed by Bennett et al. [43] who tried to decrease the ratio of reactor volume to catalyst volume. In this arrangement the amount of reactant adsorbed increases compared to that in the gas space; as a result the dynamics of the adsorption – desorption processes can be detected through the gas phase measurements.

The advantages of internal recycle reactors for catalyst investigations have been outlined by Berty [46,47]. Like other differential reactors, a single experiment yields the rate directly. Whether mixing is good or not depends largely on the reactor design. In many cases the reactant remains in the reaction zone for a long time, so that homogeneous reactions may become important.

A number of authors have proposed reactors in which mixing occurs by means of a magnetically driven piston reciprocating at high speed, rather than a more usual rotor device [48,49]. It is argued that at high reciprocation rates the vigorous turbulent mixing, produced by rapid changes in the direction of flow of reactants through the stationary catalyst bed, eliminates temperature and concentration gradients more effectively than for flow always in the same direction.

Internal recycle reactors with moving catalyst bed

The concept of a recycle reactor is also fulfilled in a continuous flow reactor with a rotating catalyst phase. The spinning basket reactor is typical for such reactors [28]. The essential features of this reactor are shown in Figure 5.6. Catalyst particles of common commercial sizes (2 mm to 20 mm) are deposited inside wire cages attached to a rotating shaft. Each of the four cages, containing one layer of pellets, behaves as a differential reactor swept through the gas some thousand times per minute. To improve the interface transport baffles are essential in the stirred basket reactor [50].

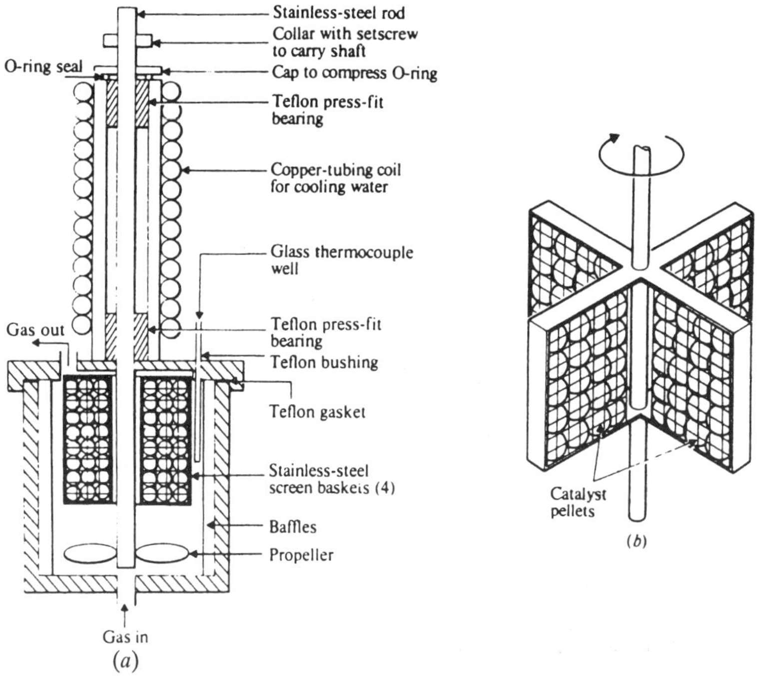

Figure 5.6 The spinning basket reactor: (a) general view ; (b) the catalyst disposition (from Carberry [28]).
J.J. Carberry, Chemical and Catalytic Reaction Engineering, 1976. The material is reproduced with permission of The McGraw-Hill Companies.

Modified types of rotating basket reactors of similar basic principles have been developed [51,52]. The spinning basket reactor apparently can be a good mixer under non-reaction conditions but its effectiveness in reducing interface temperature and concentration differences under reaction conditions is doubted. In this reactor, the true gas velocity relative to the pellets is unknown, because the gas tends to spin along with the basket. Therefore, despite the baffles, an increased speed of rotation does not necessarily reduce the interface temperature and concentration differences to any great extent. This is especially true for smaller catalyst particles and fast reactions.

A novel design for a laboratory reactor to determine the kinetics of fast, heterogeneously catalyzed, gas-phase reactions has been developed recently by Borman et al. [53]. In this reactor perfect mixing is achieved by means of an axial flow impeller in a streamlined enclosure: the reactor has a rounded shape so as to guide the gas flow smoothly through the reactor and to prevent dead zones. To suppress a circular movement of the gas, four baffles are situated on the wall of the reactor. A schematic drawing of the reactor is given in Figure 5.7(a). The diameter of the reactor was chosen such that the cross-sectional area covered by the impeller is equal to that between the impeller and the reactor wall. Thus, compression and expansion

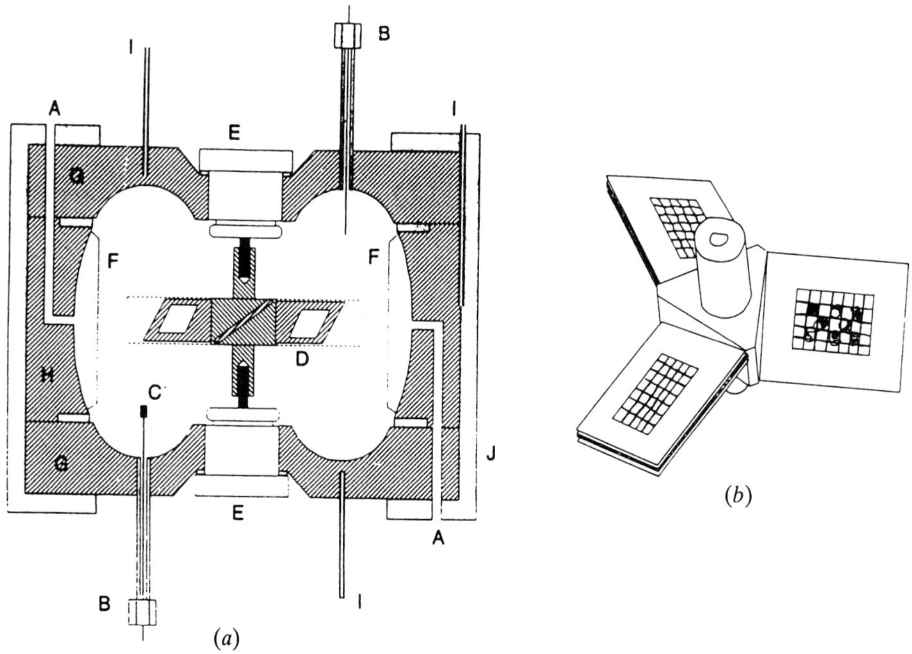

Figure 5.7 The impeller reactor: (a) general view: (A) gas in/outlet; (B) pressure sealed leads for thermocouples; (C) particle fixed on thermocouple; (D) axial flow impeller; (E) impeller bearing; (F) baffles; (G) top/bottom part; (H) center part; (I) thermowell; (J) bronze jacket. (b) axial flow impeller with rectangular openings in the blades in which particles are fixed between gauzes (from Borman et al. [53]).

of the gas is reduced. The catalyst particles are located inside a rectangular opening in the blades of the impeller, as shown in Figure 5.7(b). The main advantage of the new reactor compared to the internal recycle reactor is that mass transfer coefficients at equal impeller tip speeds are 4-6 times higher. Also in the new reactor, the recycle flow rate is independent of the total amount of catalyst contained in the reactor.

5.2.3 Single Pellet Diffusion Reactor

A serious difficulty with most kinetic reactors is to ensure kinetic control has been obtained and transport limitations are negligible. The conditions ensuring kinetic control are usually low temperature and small catalyst particles. These conditions are not easy to satisfy. An elegant method to overcome these problems is the single pellet diffusion reactor. This reactor enables deduction of kinetic information and also allows for the experimental observation of transport effects and their interaction with the chemical kinetic processes on the surface. It is a valuable addition to the more conventional experimental reactor techniques. This reactor was first used by Roiter et al. [54] and later by a number of authors [55,56]. Hegedus and Petersen [57] have given a description of the utility of the single pellet diffusion reactor in reaction engineering studies. In this laboratory reactor (Figure 5.8) the gas flows along the top surface of a pellet of thickness δ. Reactive components diffuse into the pellet and are converted. Reaction products diffuse upwards and then leave the pellet. The space below the pellet is closed and, once stationary conditions have been established, there is no net flow into this space. Then $dC_A/dx = 0$ holds at the bottom surface of the pellet, so that the concentration measured in the space below the pellet represents the concentrations in the center of a pellet of a characteristic length R. Under isothermal conditions the concentration of reactant A can be found by solving

$$D_{e,A}\frac{d^2C_A}{dx^2}=R_A$$

If only internal diffusion resistances are significant, the boundary conditions are

$$x=0 \qquad \frac{dC_A}{dx}=0$$

$$x=L \qquad C_A=C_{A,s}$$

where $C_{A,s}$ is the concentration at the top surface of the pellet. For first-order kinetics, $R_A = k_AC_A$, the solution is

$$\frac{C_A}{C_{A,s}}=\frac{\cosh(\phi x/R)}{\cosh\phi}$$

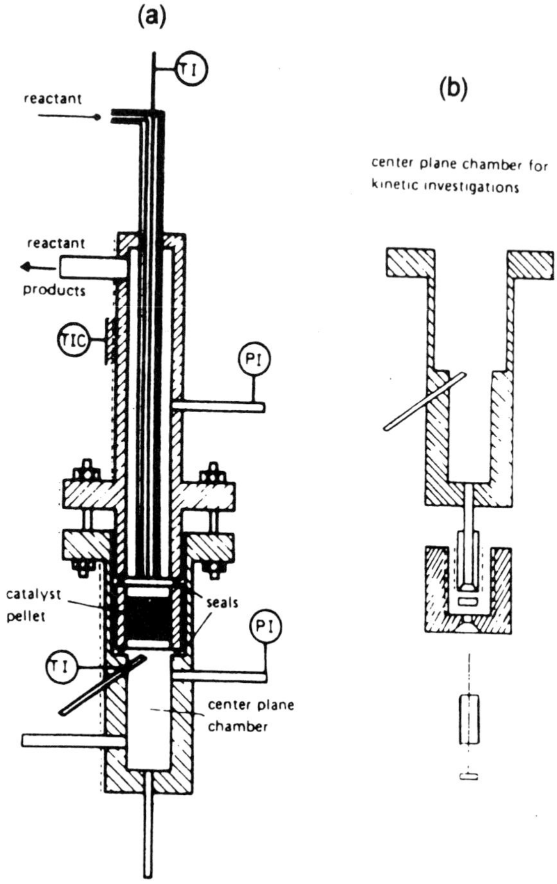

Figure 5.8 Single pellet diffusion reactor: (a) set up for permeability and internal diffusion measurements; (b) part of the cell for measuring kinetics, with sample device (from Christoffel and Schuler [56]).

where $\phi = R\sqrt{k_A / D_{e,A}}$ is the Thiele parameter. The concentration at the bottom surface ($x = 0$) is:

$$\frac{C_{A,b}}{C_{A,s}} = \frac{1}{\cosh \phi}$$

Hence, ϕ follows directly from one, single concentration measurement. Experiments at different pressures and temperatures give the chemical reaction constant and the effective diffusivity. The same procedure can be used for reactions of different orders [57].

The method is useful for kinetics under diffusion controlled conditions, for measuring actual internal diffusivities under reaction conditions, provided the kinetics are known and, above all, to study catalyst deactivation mechanisms and to establish what type of poisoning occurs.

5.2.4 Dynamic Methods

Heterogeneously catalyzed reactions are usually studied under steady-state conditions. There are some disadvantages to this method. Kinetic equations found in steady-state experiments may be inappropriate for a quantitative description of the dynamic reactor behavior with a characteristic time of the order of or lower than the chemical response time ($1/k_A$ for a first-order reaction). For rapid transient processes the relationship between the concentrations in the fluid and solid phases is different from those in the steady-state, due to the finite rate of the adsorption-desorption processes. A second disadvantage is that these experiments do not provide information on adsorption-desorption processes and on the formation of intermediates on the surface, which is needed for the validation of kinetic models. For complex reaction systems, where a large number of rival reaction models and potential model candidates exist, this give rise to difficulties in model discrimination.

To avoid the disadvantages of steady-state investigations, the so-called dynamic technique has been considered [58-60]. Essentially, the dynamic method consists of a sudden change of an operating parameter, such as the reactant concentration or feed temperature. The response of the catalyst to this pulse or step is then measured.

This approach provides a description of the reaction rate in terms of elementary steps which underlie observed kinetics and their parameters and leads to the determination of rates of individual elementary steps of single heterogeneous catalytic reactions.

Various reactors can be used for dynamic experiments. The application of recycle reactors to the investigation of the underlying reaction networks has been demonstrated by Bennett [58], who operated a recycle reactor with the dynamic method by superimposing reactant concentration pulses on the inlet stream. To ensure that the amounts of gaseous components adsorbed on the catalyst are appreciable compared to the amounts of measurable components flowing, the ratio of the catalyst to total reactor volume was made as high as possible. A reactor system, designed for transient experiments, has been described by Bennett et al. [43].

To analyze reaction mechanisms in complex catalytic systems, the application of micropulse techniques in small catalytic packed beds has been used. Christoffel [33] has given an introduction to these techniques in a comprehensive review of laboratory reactors for heterogeneous catalytic processes. Müller and Hofmann [59,61] have tested the dynamic method in the packed bed reactor to investigate complex heterogeneous reactions. Kinetic parameters have been evaluated by a method, which employs concentration step changes and the time derivatives of concentration transients at the reactor outlet as caused by a concentration step change at the reactor inlet.

Acquisition of quantitative information with the dynamic technique is usually more complicated, because of the continuously changing concentrations over the length of the catalytic column and because of the adsorption on the catalyst.

5.2.5 Other Reactor Types

Reactors in which the solid phase is perfectly mixed on a macro scale, such as a stirred tank slurry reactor and the riser reactor with recycle of both phases, are particularly useful for fast catalyst deactivation processes. Notice that the residence time of both phases can be varied independently by introducing an extra recycle flow of one of the phases after the fluid-solid separator. Unfortunately, the fluid phase mass transfer coefficient in these slurry reactors is difficult to vary. Some variation is possible by changing the particle size, but internal transport limitations also change in that case.

The fluidized bed reactor can also handle fast, complex reactions, with mixing and temperature control being especially good when stirring is provided. Unfortunately, the extent of back mixing is difficult to assess so that the residence time distribution of the reactants in the reactor is uncertain. In addition, only small catalyst particles can be used, and attrition, with the consequent breakdown and loss of catalyst, is a problem. Finally, a catalyst bed is adequately fluidized over only a comparatively narrow range of flow rates. More information about kinetic reactors can be found in reviews [33,34,50]. Applications of the basket-type mixed reactor to liquid-solid systems are discussed by Suzuki and Kawazo [62] and by Teshima and Ohashi [63], and the development of a laminar flow, liquid-solid reactor by Schmalzer et al. [64]. In the latter reactor the wall is coated with a catalyst layer.

5.3 General Aspects of Kinetic Investigations

Kinetic research on fluid-solid catalyzed reactions poses many potential pitfalls. The worker in the field therefore must be fully versed with both the theory on mass transfer with reaction and with the characteristics of each available reactor type for kinetic measurements. In kinetic studies of heterogeneous chemical reactions the interpretation of experimental data is greatly facilitated if the composition, pressure and temperature are basically uniform throughout the reactor and mass and heat transfer resistances between the catalyst pellets and the bulk phase and within the catalyst pellets are avoided. Recycle reactors are particularly suitable and popular for kinetic investigations because their construction is suitable to reach almost ideal conditions. These reactors are often called gradient-free, although they are not necessarily such. Whether the ideal conditions are achieved or not depends on the recycle and feed flow rates, the reaction kinetics, and the heat of reaction. It has been pointed out in the literature that recycle reactor data can be seriously influenced by deviations from perfect mixing and/or by heat and mass transport resistances. A too low recycle flow rate can cause an error in the measured reaction rates by imperfect mixing as well as by transport resistances. Bos et al. [65] discuss examples of both types of error in experimental data. Pirjamali et al. [66] discuss a procedure to correct recycle reactor data for inter-particle gradients. Wedel and Villadsen [67] discuss falsification of data due to concentration gradients, thereby introducing the concept of the falsification ratio. Broucek [68] shows that the criterion of Wedel and Villadsen [67] is not always sufficient due to the neglect of temperature gradients. Georgakopoulos and Broucek [69] discuss a method to check whether measured rates are influenced by concentration gradients in the catalyst bed. According to the authors, falsification of the data due to a variation of the concentration throughout the reactor is negligible if no influence is seen of interchanging the inlet and the outlet. Dümpelmann and Baiker [70] have defined a criterion taking into account temperature gradients.

Some of the typical problems in obtaining reliable kinetic data are discussed below.

5.3.1 Criteria for Ideal Mixing in a Recycle Reactor

It is not easy to check directly the uniformity of the temperature and the concentrations inside a recycle reactor. Therefore, in absence of a reaction, transient response techniques are often employed to assess the degree of mixing. Imposition of a pulse of inert tracer on the reactor usually yields a response corresponding to the theoretical prediction under assumption of a perfect mixing condition in the reactor. This is in agreement with the theory: in the recycle reactor a recirculation ratio of the order of 10 is usually enough to achieve ideal mixing conditions in the tracer experiments, as shown in Westerterp et al. p. 181 [41]. On this basis the recycle reactors are regarded as perfect mixed reactors. It is not always realized that a recycle ratio, high enough to give perfect mixing behavior in tracer response experiments, does not assure perfect mixing under reaction conditions.

The experiments with the inert tracer may only show that the time, necessary for the fluid in the reactor to be well mixed, is much smaller than the average residence time. When a chemical reaction takes place, an additional time-scale, the time constant of the chemical reaction, appears. This time characterizes the reaction rate and can be defined as the time in which the reaction proceeds to a certain conversion, say 50%. For many practical heterogeneous catalytic reactions, the reaction time is so short that reactants entering the reactor may be converted without being mixed, for example, during the first cycle. For such fast reactions, of course, the reactor cannot be considered as gradient-free, whatever the recirculation ratio is.

To elucidate the pitfalls in the use of recycle reactors we will discuss a general criterion, the adjustment of which ensures that all reaction takes place in a narrow concentration and temperature range very close to the discharge conditions.

A schematic representation of the recycle reactor is given in Figure 5.3. Assume that a function of the concentration of species A and the temperature, $R(C_A, T)$, can represent the reaction rate per unit volume of the catalyst bed. As described in Section 5.2.2.4 the measured reaction rate is

$$\frac{\Phi_0 C_{A,0} - \Phi_{out} C_A}{(1-\varepsilon_b)V_r} = \langle R_A \rangle \tag{5.49}$$

and represents the average reaction rate. Under the assumption of ideal circumstances the average reaction rate is replaced by the rate at the discharge condition $R(C_A,T)$, and Equation 5.48 is used instead of 5.49. For the assumption to be true the difference between these two rates should not exceed the allowed accuracy:

$$\left| \frac{\langle R_A \rangle - R_A(C_A,T)}{R_A(C_A,T)} \right| < \delta \tag{5.50}$$

Here the vertical bars designate an absolute value. A relative accuracy of 5-10 % for the reaction rate measurements is satisfactory for practical purposes, so $\delta = 0.05 - 0.1$.

In the following derivation we are only interested in small variations of the concentration and the temperature over the bed. Hence, the reaction rate function can be represented as a linear function of the coordinates or state variables C_A and T. With this in mind, the average reaction rate can be written as

$$\langle R_A \rangle = \frac{R_A(C_A^*, T^*) + R_A(C_A, T)}{2} \tag{5.51}$$

where C_A^* and T^* are the concentration and temperature at the bed inlet. Substituting Equation 5.51 into 5.50 gives

$$\left| \frac{R_A(C_A^*, T^*) - R_A(C_A, T)}{2R_A(C_A, T)} \right| < \delta \tag{5.52}$$

Equation 5.52 has a clear meaning: the difference in the inlet and outlet reaction rates must be small. To apply this criterion the inlet concentration C^* and temperature T^* should be known. If they are not measured, as in most recycle reactors, they can be estimated through the measured outlet parameters and the measured net conversion in the reactor, as follows. From the definition of the conversion per cycle (Equation 5.46),

$$C_A^* = \frac{C_A}{1 - \zeta_{A,c}} \tag{5.53}$$

and the conversion per cycle can be calculated from Equation 5.47, which is reproduced here:

$$\zeta_{A,c} = \frac{1}{(R_R + 1)\dfrac{1 - \zeta_A}{\zeta_A} + 1} \tag{5.54}$$

The relationship between the inlet and outlet temperature depends on the heat exchange between the fluid flowing through the bed and the surrounding fluid and/or the reactor parts. The rate of this heat exchange depends on the reactor design and is difficult to predict theoretically. The two limiting situations of isothermal and adiabatic operations can be considered in the evaluation of the reactor performance. Under isothermal conditions,

$$T - T^* = 0 \tag{5.55}$$

When the catalyst bed operates adiabatically the temperature change per cycle is

$$T - T^* = \frac{(-\Delta H)(C_A^* - C_A)}{\rho_g c_p} = \frac{(-\Delta H)C_A}{\rho_g c_p} \frac{\zeta_{A,c}}{1 - \zeta_{A,c}} \tag{5.56}$$

The temperature change over the bed in Equation 5.56 can be related to the adiabatic temperature rise

$$\Delta T_{ad} = (-\Delta H)C_{A,0} / (\rho_g c_p)$$

as

$$T-T^{*}=\Delta T_{ad}\frac{C_{A}}{C_{A,0}}\frac{\zeta_{A,c}}{1-\zeta_{A,c}}=\Delta T_{ad}\frac{(1-\zeta_{A})\zeta_{A,c}}{1-\zeta_{A,c}} \tag{5.57}$$

In a real reactor the variation of the temperature over the bed must be between those predicted by the two above equations. Equations 5.52 – 5.55 and 5.57 can be used to check whether ideal mixing conditions are achieved or not. For practical purposes these equations can be simplified because only small concentration and temperature variations are considered. In this case the inlet reaction rate can be approximated by a linear Taylor expansion:

$$R(C_{A}^{*},T^{*})=R(C_{A},T)+\frac{\partial R(C_{A},T)}{\partial C_{A}}(C_{A}^{*}-C_{A})+\frac{\partial R(C_{A},T)}{\partial T}(T^{*}-T) \tag{5.58}$$

Substitution of Equation 5.58 into 5.52 gives rise to

$$\frac{1}{2R_{A}}\left|\frac{\partial R_{A}}{\partial C_{A}}(C_{A}^{*}-C_{A})+\frac{\partial R_{A}}{\partial T}(T^{*}-T)\right|<\delta \tag{5.59}$$

where the reaction rate and its derivatives are taken at the outlet conditions. Combining Equation 5.59 with Equations 5.53 and 5.54 shows that ideal mixed conditions will prevail in the reactor provided two inequalities are satisfied:

$$\frac{C_{A}}{2R_{A}}\left|\frac{\partial R_{A}}{\partial C_{A}}\right|\frac{\zeta_{A,c}}{1-\zeta_{A,c}}<\delta \tag{5.60}$$

$$\frac{1}{2R_{A}}\left|C_{A}\frac{\partial R_{A}}{\partial C_{A}}-T_{ad}(1-\zeta_{A})\frac{\partial R_{A}}{\partial T}\right|\frac{\zeta_{A,c}}{1-\zeta_{A,c}}<\delta \tag{5.61}$$

These equations correspond to isothermal and adiabatic operation respectively. Using the total conversion from Equation 5.54, Equations 5.60 and 5.61 can be rewritten in terms of measurable quantities:

$$\frac{C_{A}}{2R_{A}}\left|\frac{\partial R_{A}}{\partial C_{A}}\right|\frac{\zeta_{A}}{1-\zeta_{A}}<\delta(1+R_{R}) \tag{5.62}$$

$$\frac{1}{2R_{A}}\left|C_{A}\frac{\partial R_{A}}{\partial C_{A}}-\Delta T_{ad}(1-\zeta_{A})\frac{\partial R_{A}}{\partial T}\right|\frac{\zeta_{A}}{1-\zeta_{A}}<\delta(1+R_{R}) \tag{5.63}$$

The found inequalities show that the minimum recirculation ratio depends strongly on the conversion achieved and the kinetics of the reaction. As an example we will apply the obtained criteria to find the necessary recirculation ratio when a reaction described by the power low rate equation, $R_A = k_A C_A^n$ with $k_A = k_{A,0}\exp(-E_a/RT)$, is carried out in a reactor. Equations 5.62 and 5.63 become, respectively,

$$\frac{|n|\zeta_A}{2(1-\zeta_A)} < \delta(1+R_R) \tag{5.64}$$

$$\frac{1}{2}\left|\frac{n\zeta_A}{(1-\zeta_A)} - \frac{\Delta T_{ad}}{T}\frac{E_a}{RT}\zeta_A\right| < \delta(1+R_R) \tag{5.65}$$

In practice R_R is large compared to 1, so that the criteria can be simplified to

$$\frac{|n|\zeta_A}{2\delta(1-\zeta_A)} < R_R \tag{5.66}$$

$$\frac{1}{2\delta}\left|\frac{n\zeta_A}{(1-\zeta_A)} - \frac{\Delta T_{ad}}{T}\frac{E}{RT}\zeta_A\right| < R_R \tag{5.67}$$

The value of $\varepsilon = E_a/RT$ lies in the range 1 – 30. For some commercial processes, values are given in Westerterp et al. p. 457 [41]. Values of the relative adiabatic temperature rise $\Delta T_{ad}/T$ for typical reaction mixtures are of the order of 1. It can be seen from Equation 5.67 that if ΔT_{ad} and n are of the same sign, the temperature and concentration variation will have an opposite effect on the required value of R_R and thus partly compensate each other. If ΔT_{ad} and n are of opposite sign, the nonuniformities in the reaction rate caused by the two effects will intensify each other.

The criteria found for power law reactions show that the recirculation ratios of an order of 10, which are usually sufficient to observe ideal mixing behavior in tracer experiments, may not be satisfactory under reaction conditions. For example, assume that isothermal conditions in the bed have been achieved and the error due to a concentration nonuniformity should not exceed 5% ($\delta = 0.05$). In this case the required recirculation ratio is determined from

$$R_R > \frac{10|n|\zeta_A}{1-\zeta_A} \tag{5.68}$$

Thus, if the total reactor conversion is 90% the minimum recycle ratio for a first-order reaction should be at least 90 and for a second order reaction 180. This simple example demonstrates that large conversions in recycle reactors are as unfavorable as very small conversions in a packed bed differential reactor. If one is interested in the reaction rate at small concentrations of the reactants, these concentrations in the recycle reactor should be achieved by small or moderate inlet concentrations but not through a large conversion. Large recirculation ratios can be achieved by decreasing of the bed depth.

A possible channeling in the bed is not dangerous if good mixing in the external loop is achieved.

It is worth mentioning that errors in the treatment of the measured reaction rates can be significantly decreased if the reaction rate were evaluated at the average rather than at the outlet concentrations and temperature. Unfortunately, the measurements at the bed inlet are difficult.

5.3.2 Errors in the Determination of Kinetic Parameters

The determination of the chemical reaction rate is based on measurements of the concentrations, temperature and flow rates. Experimental errors in the measurements of these quantities are inevitable.

The importance of correct and accurate temperature measurement follows immediately from a consideration of the reaction rate sensitivity with respect to temperature, as considered in Chapter 2. Although the temperature measurements themselves may be rather accurate, the error of a few degrees due to an incorrect position of the thermocouple can be crucial. Such errors may arise, for example, if the outlet bed temperature is measured at some distance from the outlet or in the reactor exit tube.

To elucidate the significance of the accuracy of concentration and flow rate measurements, two ideal situations in the reactor which are usually assumed to occur in the experimental data collection are considered:

- ideal mixing, typical for recycle reactors;
- plug flow conditions in a fixed bed reactor.

If the resistances to heat and mass transfer can be neglected, Equation 5.35 or 5.36 is used for the determination of the kinetic parameters when an ideal mixing model holds, whereas Equation 5.42 or Equations 5.43 and 5.44 are used when a plug flow model is valid. For the sake of simplicity a first-order reaction is considered. The governing equations used for the determination of the chemical reaction constant are

$$\text{ideal mixing} \quad k_A = \frac{\zeta_A}{\tau(1-\zeta_A)} \tag{5.69}$$

$$\text{plug flow} \quad k_A = -\frac{1}{\tau}\ln(1-\zeta_A) \tag{5.70}$$

where, in both equations, $\tau = V_r/\Phi_v$. An error in the calculation of k_A is determined by the errors in the residence time and the conversion. Its relative value can be found from:

$$\delta_k = \frac{\Delta k_A}{k_A} = \frac{k_A(\tau+\Delta\tau,\ \zeta_A+\Delta\zeta_A) - k_A(\tau,\zeta_A)}{k_A} \approx \frac{1}{k_A}\left(\frac{\partial k_A}{\partial \tau}\Delta\tau + \frac{\partial k_A}{\partial \zeta_A}\Delta\zeta_A\right) \tag{5.71}$$

where $\Delta\tau$ and $\Delta\zeta_A$ are the absolute errors in the determination of the residence time and the conversion respectively. Using Equations 5.69 and 5.70 the errors in the determination of k_A by the two methods are:

$$\text{ideal mixing} \quad \delta_k = \frac{\Delta\tau}{\tau} + \frac{\Delta\zeta_A}{\zeta_A(1-\zeta_A)} \tag{5.72}$$

$$\text{plug flow} \quad \delta_k = \frac{\Delta\tau}{\tau} - \frac{\Delta\zeta_A}{(1-\zeta_A)\ln(1-\zeta_A)} \tag{5.73}$$

Equations 5.72 and 5.73 show that if an error is made in the determination of the residence time the same relative error will be found in the calculation of k_A by both methods. The errors in k_A due to inaccurate conversion measurements depend on the conversion level and are different for the two methods. The significance of accurate conversion measurements is demonstrated in Figure 5.9. This figure presents the relative error in k_A caused by a 1% error in the determination of the conversion, so that $\Delta\zeta_A = 0.01$.

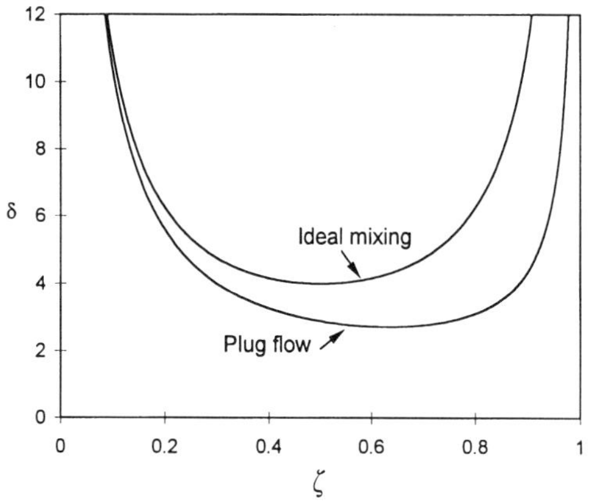

Figure 5.9 Relative errors in the determination of the reaction rate constant due to 1% error in the conversion measurements in ideal mixed and plug flow reactor.

References

1. Haynes, H. W. (1988) *Catal. Rev. Sci. Eng.*, **30**, 563.
2. Park, I.-S., Do, D.D. (1996) *Catal. Rev.-Sci. Eng.*, **38**, 189.
3. Wicke, E., Kallenbach, R. (1941) Kolloid. Z., **97**, 135.
4. Cunningham, R.S., Geankoplis, C.J. (1968) *Ind. Eng. Chem. Fundam.*, **7**, 535.
5. Weisz, P.B. (1957) *Z. Physik. Chem., Neue Folge*, **11**, 1.
6. Henry, J.P., Chennakesavan, B., Smith, J.M. (1961) *AIChE J.*, **7**, 10.
7. Feng, C.F., Kostrov, V.V., Stewart, W.E. (1974) *Ind. Eng. Chem. Fundam.*, **13**, 5
8. Allawi, Z.M., Gunn, D.J. (1987) *AIChE J.*, **33**, 766.
9. Jackson, R. (1977) *Transport in Porous Catalysts.* Amsterdam: Elsevier.

10. Grachev, G.A., Ione, K.G., Nosyreva, G.N., Malinovckaya, O.A. (1972) in *The Porous Structure of Catalysts and Transport Processes in Heterogeneous Catalysis*: G.K. Boreskov (ed.). Budapest: Akadémiai Kiadó, pp. 205-219.
11. Carman, P.C. (1956) *Flow of Gases Through Porous Media*. London: Butterworths.
12. Pollard, W.G., Present, R.D. (1948) *Phys. Rev.*, **73**, 762.
13. Mason, E.A., Malinauskas, A.P. (1983) *Gas Transport in Porous Media: The Dusty-Gas Model*. Amsterdam: Elsevier.
14. Eu, B.C. (1989) *Phys. Rev.*, **A 40**, 6395.
15. Arnošt, D., Schneider, P. (1995) *Chem. Eng. J.*, **57**, 91.
16. Capek, P., Hejtmánek, V., Šolwvá, O., Klusácek, K., Schneider, P. (1997) *Catalysis Today*, **38**, 31.
17. Schneider, P. (1978) *Chem. Eng. Sci.*, **33**, 1311.
18. Carslaw, H.S., Jaeger, J.C. (1959) *Conduction of Heat in Solids* (2nd. edn). Oxford: Clarendon Press.
19. Grachev, G.A., Ione, K.G., Barshev, A.A. (1970) Kinetics and Catalysis, **11**, 445.
20. Barrer, R.M. (1953) *J. Phys. Chem.*, **57**, 35.
21. Gibilaro, L.G., Gioia, F., Greco, G. (1970) *Chem. Eng. J.*, **1**, 85.
22. Dogu, G., Smith, J.M. (1975) *AIChE J.*, **21**, 58.
23. Dogu, G., Smith, J.M. (1976) *Chem. Eng. Sci.*, **31**, 123.
24. Haynes, H.W., Sarma, P.N. (1973) *AIChE J.*, **19**, 1043.
25. Baiker, A., New, M., Richardz, W. (1982) *Chem. Eng Sci.*, **37**, 643.
26. Deisler, P.F., Wilhelm, R.H. (1953) *Ind. Eng. Chem.*, **45**, 1219.
27. Boniface, H.A., Ruthven D.M. (1985) *Chem. Eng. Sci.*, **40**, 2053.
28. Carberry, J.J. (1964) *Ind. Eng. Chem.*, **56**(11), 39.
29. Berty, J.M. (1974) *Chem. Engng Progr.*, **70**(5), 78.
30. Weekman, V. W. (1974) *AIChE J.*, **20**, 833.
31. Sunderland, P. (1976) *Trans. Instn. Chem. Engns*, **54**, 135.
32. Jankowski, H., Nelles, J., Adler, R., Kubias, B., Salzer, C. (1978) *Chem. Techn.*, **30**, 441.
33. Christoffel, E.G. (1982) *Catal. Rev.-Sci. Eng.*, **24**, 159.
34. Berty, J.M. (1983) in *Applied Industrial Catalysis*: B.E. Leach (ed.), New York: Academic Press, pp. 41-67.
35. Froment, G.F. (1975) *AIChE J.*, **21**, 1041.
36. Franckaerts, J., Froment, G.F. (1964) *Chem. Eng. Sci.*, **19**, 807.
37. Schouten, E.P.S., Borman, P.C., Westerterp, K.R. (1994) *Chem. Eng. Sci.*, **24**, 4725.
38. Hofmann, H. (1979) *Ger. Chem. Eng.*, **2**, 258.
39. Hofmann, H. (1986) in *Chemical Reactor Design and Technology*: H. I. de Lasa (ed.). Dordrecht: Martinus Nijhoff Publishers, pp.69-105.
40. Schwedock, M.J., Windes, L.C., Ray W.H. (1989) *Chem. Eng. Comm.*, **78**, 45.
41. Westerterp, K.R., Van Swaaij, W.P.M., Beenackers, A.A.C.M. (1987) *Chemical Reactor Design and Operation*. Chichester: Wiley.
42. Temkin, M.I. (1962) *Kinetics and Catalysis*, **3**, 448.
43. Bennett, C.O., Cutlip, M.B., Yang, C.C. (1972) *Chem. Eng. Sci.*, **27**, 2255.
44. Garanin, V.I., Kurkchi, U.M., Minachev, Kh.M. (1967) *Kinet. Catal.*, **8**, 701.
45. Livbjerg, H., Villadsen, J. (1971) *Chem. Eng. Sci.*, **26**, 1495.
46. Berty, J.M. (1979) *Catal. Rev.- Sci. Eng.*, **20**, 75.
47. Berty, J.M. (1984) *Plant/Operations Progress*, **3**, 163.

48. Korneichuk, G.P., Stasevich, V.P., Semenyuk, Yu.V., Filipov, V.I., Girushtin, Yu.G. (1972) *Kinet. Catal.*, **15**, 970.
49. Sunderland, P., El Kanzi, E.M.A. (1974) *Adv. Chem. Ser.*, **133**, 3.
50. Doraiswamy, L.K., Tajbl, D.G., (1974) *Catal. Rev.-Sci. Eng.*, **10**, 177.
51. Brisk, M.L. Day, R.L., Jones, M., Warren, J.B. (1968) *Trans. Instn. Chem. Engrs.*, **46**, T3.
52. Choudhary, V.R., Doraiswamy, L.K. (1972) *Ind. Eng. Chem., Process Des. Dev.*, **11**, 420.
53. Borman, P.C., Bos, A.N.R., Westerterp, K.R. (1994) *AIChE J.*, **40**, 862.
54. Roiter, V.A., Korneichuk, G.P., Leperson, M.G., Stukanowskaia, N.A., Tolchina, B.I. (1950) *Zh. Fiz. Khim.*, **24** (2), 459.
55. Balder, J.R., Petersen, E.E. (1968) J. Catal., **11**, 195.
56. Christoffel, E.G., Schuler, J.C. (1980) *Chem.-Ing.-Tech.*, **52**, 844.
57. Hegedus, L.L., Petersen, E.E. (1974) *Catal. Rev.-Sci. Eng.*, **9**, 245.
58. Bennett, C.O. (1976) *Catal. Rev.-Sci. Eng.*, **13**, 121.
59. Müller, E., Hofmann, H. (1987) *Chem. Eng. Sci.*, **42**, 1695 and 1705.
60. Renken, A. (1990) *Chem.-Ing.-Tech.*, **62**, 724.
61. Müller-Erlwein, E., Hofmann, H. (1988) *Chem. Eng. Sci.*, **43**, 2245.
62. Suzuki, M., Kawazoe, K. (1975) *J. Chem. Eng. Jpn.*, **8**, 79.
63. Teshima, H., Ohashi, Y. 1977, *J. Chem. Eng. Jpn.*, **10**, 70.
64. Schmalzer, D.K., Hoelscher, H.E., Cowherd, C. (1970) *Can. J. Chem. Eng.*, **48**, 135.
65. Bos, A.N.R., Bootsma, E.S., Foeth, F., Sleyster, H.W.J., Westerterp, K.R. (1993) *Chem. Eng. Proc.*, **32**, 53.
66. Pirjamali, M., Livbjerg, H., Villadsen, J. (1973) *Chem. Eng. Sci.*, **28**, 328.
67. Wedel, S., Villadsen, J. (1983) *Chem. Eng. Sci.*, **38**, 1346.
68. Broucek, R. (1983) *Chem. Eng. Sci.*, **38**, 1349.
69. Georgakopoulos, K., Broucek, R. (1987) *Chem. Eng. Sci.*, **42**, 2782.
70. Dümpelmann, R., Baiker, A. (1992) *Chem. Eng. Sci.*, **47**, 2665.

6 Calculation of Effectiveness Factor

This chapter presents generalized and approximate formulae, which enable calculation of the conversion in solid catalyst materials in cases where both reaction and diffusion influence the overall conversion rate. The conversion mainly depends on the following three aspects:

- Micro properties of the catalyst pellet, the most important being:
  - pore size distribution
  - pore tortuosity
  - diffusion rates of the reacting components in the gas phase
  - diffusion rates of the reacting components under Knudsen flow.
- Macro properties of the catalyst pellet, especially:
  - size and shape of the catalyst pellet
  - possible occurrence of anisotropy of the catalyst pellet.
- Reaction properties, such as:
  - reaction kinetics
  - number of reactants involved
  - complexity of the reaction scheme under consideration.

The micro properties cannot be determined easily. Moreover, due to the complexity of diffusion of reactants in a solid catalyst matrix, models taking the micro properties into account tend to be very complex and inaccurate [1]. Therefore, the micro properties are usually accounted for by a lumped parameter, the so-called effective diffusion coefficient D_e. For solid catalyst particles this approach has proved to be very useful, provided that the particles can be regarded as homogeneous on a micro scale. However, if this is not the case, such as for zeolite in an amorphous matrix, care should be taken in using this approach [2,3].

Here it is assumed that it is possible to use the concept of an effective diffusion coefficient without making too large an error. Hence the effect of micro properties will not be studied here and it is assumed the value of D_e is known. The discussion is restricted to the impact of the macro properties and reaction properties on the effectiveness factor. Furthermore only simple reactions are discussed. Generalized formulae are provided that enable calculation of effectiveness factor for varying properties of the catalyst or the reacting system.

6.1 Literature Survey

Doraiswamy and Sharma [4] have given an extensive literature survey covering effectiveness factors. Some of the topics raised by them are highlighted below.

Thiele was one of the first to use the concept of an effectiveness factor [5]. To calculate this factor he introduced a dimensionless number, nowadays called the Thiele modulus. Thiele defined the effectiveness factor as the amount of a certain component that is actually converted, divided by the amount that could have been converted if retardation by diffusion had not occured. The dimensionless modulus he used was defined as

$$\phi_T = R\sqrt{\frac{R(C_{A,s})}{D_e C_{A,s}}} \tag{6.1}$$

where R is the distance from the center of the catalyst pellet to the surface (Figure 6.1), $R(C_{A,s})$ is the conversion rate of component A for surface conditions, D_e is the effective diffusion coefficient of component A and $C_{A,s}$ is the concentration of component A at the outer surface of the catalyst pellet. A plot of the effectiveness factor versus the modulus ϕ_T is shown in Figure 6.2. As the Thiele modulus increases, the reaction becomes more limited by diffusion and thus the effectiveness factor decreases. For very high values of the Thiele modulus the effectiveness factor is inversely proportional to the Thiele modulus.

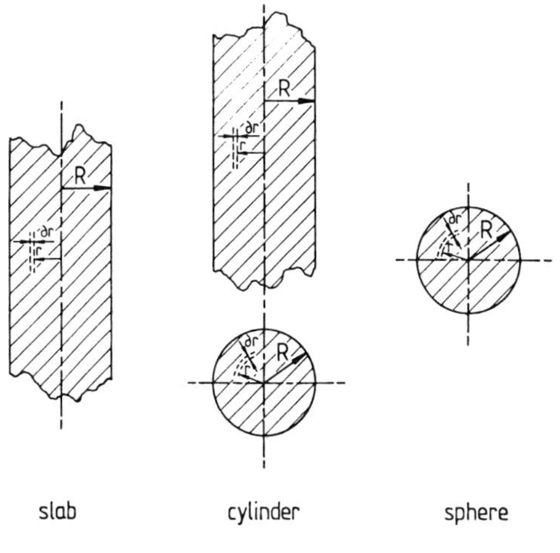

Figure 6.1 Geometry of an infinite slab, cylinder and sphere.

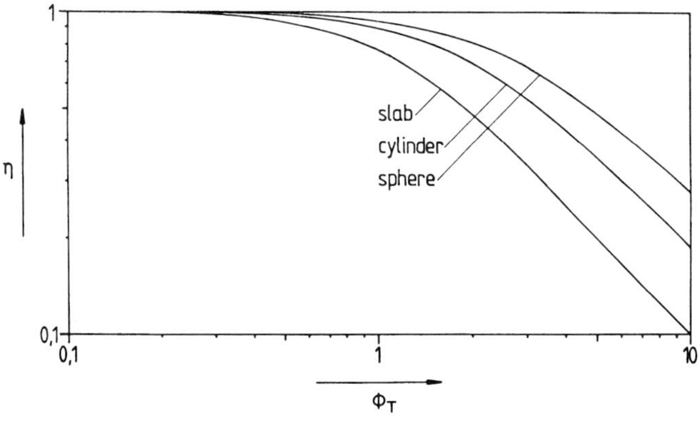

Figure 6.2. Effectiveness factor η versus the Thiele modulus ϕ_T for first-order kinetics in an infinite slab, infinite cylinder and sphere.

It can be seen that the Thiele modulus may be regarded a measure for the ratio of the reaction rate to the rate of diffusion. However, many different definitions are used in the literature, as described below.

In the past, a number of attempts have been made to generalize the definition of the Thiele modulus. Aris [6] noticed that all the Thiele moduli for first-order reactions were of the form:

$$\phi_1 = X_0 \sqrt{\frac{k}{D_e}} \tag{6.2}$$

with k as the reaction rate constant and X_0 a characteristic dimension. Aris showed the curves of η versus ϕ_1 could be brought together in the low η region for all the catalyst shapes, if X_0 is defined as

$$X_0 = \frac{V_P}{A_P} \tag{6.3}$$

where V_p and A_p are the volume and external surface area, respectively, of the catalyst. In Figure 6.3 η is potted versus ϕ_1 for several shapes. It can be seen that the curves coincide both in the high and low η region. In the intermediate region the spread between the curves is largest. Kasaoka and Sakata [7] have observed that this spread is even larger for ring-shaped catalyst pellets.

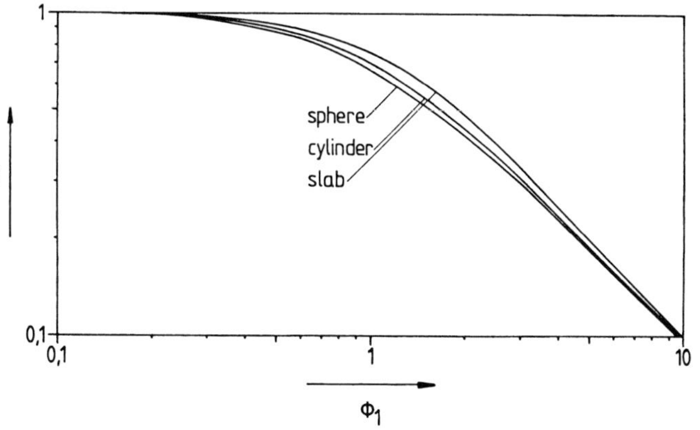

Figure 6.3 Effectiveness factor η versus shape-generalized Thiele modulus ϕ_1 of Aris for first-order kinetics in an infinite slab, an infinite cylinder and a sphere.

Generalizations for the reaction kinetics have also been made. Petersen [8] has shown that for a sphere a generalized modulus can be postulated for nth-order kinetics:

$$\phi_s = \sqrt{\frac{n+1}{2}} R \sqrt{\frac{k}{D_e} C_{A,s}^{\;n-1}} \tag{6.4}$$

Using this generalized modulus the effectiveness factor in the low η region (or high ϕ_s region) can be calculated from

$$\eta = \frac{3}{\phi_s} \tag{6.5}$$

Petersen states that a generalization of the Thiele modulus for the reaction order is also possible for other shapes. For an infinite slab (or plate) he suggested, for the low η region, the effectiveness factor could be calculated from

$$\eta = \frac{1}{\phi_P} \tag{6.6}$$

with ϕ_P being a generalized modulus which follows from the following empirical correlation

$$\phi_P = \frac{n+2.5}{3.5} R \sqrt{\frac{k}{D_e} C_{A,s}^{\;n-1}} \tag{6.7}$$

This correlation should hold within 6 %.

Rajadhyaksha and Vaduseva [9] introduced modified Thiele moduli for a sphere for nth-order kinetics, and for Langmuir-Hinshelwood kinetics assuming the rate equation

$$R\left(C_A\right) = \frac{kC_A}{1+KC_A} \tag{6.8}$$

- For n^{th}-order kinetics

$$\phi_s = \sqrt{n}\, R \sqrt{\frac{k}{D_e} C_{A,s}^{n-1}} \tag{6.9}$$

- For Langmuir-Hinshelwood kinetics

$$\phi_s = \frac{1}{1+KC_{A,s}} \sqrt{\frac{KC_{A,s}}{\ln\left(1+KC_{A,s}\right)}} R \sqrt{\frac{k}{D_e}} \tag{6.10}$$

It should be noticed that the modified moduli given by Equations 6.4 and 6.9 are not in agreement with one another.

A general expression for the modified Thiele modulus for an infinite slab was derived by Bischoff [10]:

$$\phi_P = R\left(C_{A,s}\right) \times \left[2 \int_0^{C_{A,s}} D_e(C_A) R\left(C_A\right) dC_A \right]^{-\frac{1}{2}} \tag{6.11}$$

If the effective diffusion coefficient D_e is independent of the concentration C_A, then for nth-order kinetics Equation 6.11 yields

$$\phi_P = \sqrt{\frac{n+1}{2}}\, R \sqrt{\frac{k}{D_e} C_{A,s}^{\;n-1}} \tag{6.12}$$

It should be noticed that again there is a discrepancy, this time between Equations 6.7 and 6.12.

Other attempts have been made to arrive at modified Thiele moduli for different forms of reaction kinetics. For example, Valdman and Hughes [11] have proposed a simple approximate expression for calculating the effectiveness factor for Langmuir-Hinshelwood kinetics of the type

$$R(C_A)=\frac{kC_A}{(1+KC_A)^2} \tag{6.13}$$

In addition to several empirical correlation's, various numerical approximations have also been presented [12-14]. Even generalized numerical estimation procedures are given, such as the collocation method of Finlayson [15]. Ibanez [16] and Namjoshi et al. [17] have defined transformations for several forms of kinetics, which can be used to make numerical approximation easier.

Another aspect concerns catalyst particles with intraparticle temperature gradients. In general the temperature inside a catalyst pellet will not be uniform, due to the heat effects of the reaction occurring inside the catalyst pellet. The temperature inside the catalyst can be related to the concentration with (see for example [4]):

$$\frac{T}{T_s}=\frac{(-\Delta H)D_e C_{A,s}}{\lambda_P T_s}\left(1-\frac{C_A}{C_{A,s}}\right) \tag{6.14}$$

in which T_s is the surface temperature, $(-\Delta H)$ the reaction enthalpy and λ_p the heat conductivity of the pellet. If the term α is defined as:

$$\alpha=\frac{(-\Delta H)D_e C_{A,s}}{\lambda_P T_s} \tag{6.15}$$

and the dependency of the conversion rate on the temperature is of the Arrhenius type, we can write [4]:

$$\begin{aligned}\frac{k}{k_s}&=\exp\left\{\frac{E_a}{RT_s}\times\left(\frac{1}{T}-\frac{1}{T_s}\right)\right\}\\&=\exp\left\{+\frac{E_a}{RT_s}\times\frac{\alpha\left(1-\frac{C_A}{C_{A,s}}\right)}{1+\alpha\left(1-\frac{C_A}{C_{A,s}}\right)}\right\}\end{aligned} \tag{6.16}$$

where k_s is the reaction rate constant at surface conditions, E_a is the energy of activation and R the ideal gas constant. By defining

$$\varepsilon=\frac{E_a}{RT_s} \tag{6.17}$$

Equation 6.16 can be written as

$$\frac{k}{k_s} = \exp\left(+\alpha\varepsilon \times \frac{1-\dfrac{C_A}{C_{A,s}}}{1+\alpha\left(1-\dfrac{C_A}{C_{A,s}}\right)} \right) \tag{6-18}$$

Since the conversion rate depends on α and ε, the effectiveness factor will be determined by three parameters, namely α, ε and a Thiele modulus. This is illustrated in Figure 6.4 [18]. For values of α larger than zero (exothermic reaction) an increase in the effectiveness factor is found, since the temperature inside the catalyst pellet is higher than the surface temperature. For endothermic reactions ($\alpha < 0$) a decrease of the effectiveness factor is observed.

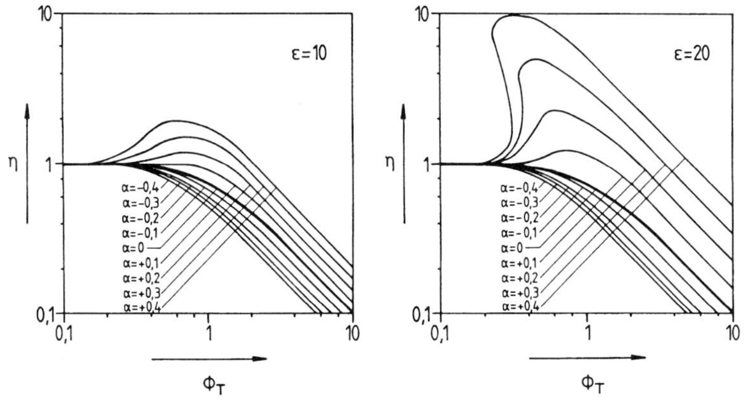

Figure 6.4 Effectiveness factor η versus the Thiele modulus for first order kinetics in an infinite slab and for several values of α and $\varepsilon = E_a/RT_s$.

Criteria, which determine whether or not intraparticle behavior may be regarded as being isothermal, have been reviewed by Mears [19], who gave as a criterion for isothermal operation

$$\alpha\varepsilon < 0.05 \times n \tag{6.19}$$

with n the reaction order. The temperature gradient inside the pellet must be taken into account if this criterion is not fulfilled. For a nonisothermal catalyst many asymptotic solutions and approximations have been derived by various authors [9,20-22].

For bimolecular reactions with nonisothermal pellets, only numerical solutions are presented in literature [23]. Karanth et al. [24] defined the parameter:

$$\xi = \frac{1}{\nu_B}\frac{D_{e,B}C_{B,s}}{D_{e,A}C_{A,s}} - 1 \tag{6.20}$$

for bimolecular reactions of the type:

$$A + \nu_B B \rightarrow \nu_P P + \nu_Q Q \tag{6.21}$$

They proposed calling component A the key reactant. Then the parameter ξ provides a measure for the influence of the so-called nonkey reactant B. Doraiswamy and Sharma [4] suggest pseudo-first order behavior is justifiable when the key reactant A is limiting, hence if ξ is positive. They do not give criteria which guarantee whether or not this statement holds.

If the gas reacting inside the catalyst pellet is nondiluted the following phenomena occur:

- The effective diffusion coefficient may become a function of the reactant concentration and thus it will depend upon the place inside the catalyst pellet.
- Volume changes due to the reaction may become considerable. This may lead to intraparticle pressure gradients, which will influence the effectiveness factor because:
  - the pressure influences both the conversion rate and the effective diffusion coefficient;
  - the pressure gradients give rise to viscous flow, which will influence the concentration profile in the catalyst.

For this problem numerical solutions have also been presented in the literature. The basis of all such solutions is the dusty gas model [25]. Haynes [26] showed viscous flow in the catalyst pores could be neglected. Nir and Pismen [32] made calculations for the case when a pressure gradient exists across the catalyst pellet and presented some results for several Thiele moduli. The results of all these studies cannot be generalized to more complex reactions, each case has to be treated separately.

Many other phenomena have also been modeled, such as the impact on the effectiveness factor of:

- Bimodal pore size distribution, e.g. for zeolite in an amorphous matrix with distinct micro and macro pores [2,3];
- Pore shape [27,28];
- Nonhomogeneity of the catalyst surface [29,30].

As this discussion is restricted to homogeneous catalyst pellets, the above-mentioned phenomena are not discussed in further detail.

This literature survey leaves the impression of an enormous but chaotic quantity of publications, because all the models and correlations focus on a single rather specific problem. A number of attempts have been made to integrate some of the models and correlation's, but without success. Also, the authors tend to use their own tailor-made definitions rather than generalized ones. As a result a large variety of definitions can be found for the same number (Doraiswamy and Sharma [4] use 20 different definitions for the Thiele modulus in their survey). This makes the results of different correlations hard to compare. Finally, some of the formulae which were given contradict each other, which is of course very confusing.

To resist these shortcomings, a generalized theory is presented that covers all the models and correlations mentioned. With this theory it is possible to tackle problems that are more complicated than those discussed here in this survey.

The theory is based upon two newly defined numbers, which we call Aris numbers. This recognizes that, to our knowledge, Aris was the first who substantially contributed to a generalized theory for effectiveness factors, by postulating his shape-generalized Thiele modulus [6]. Aris also wrote a book which gives an excellent survey of all that has been done in this field [31].

The theory is presented in three parts:

- Definition of three generalized dimensionless numbers:
  - effectiveness factor η
  - zeroth Aris number An_0
  - first Aris number An_1
- Generalized approximations for η based upon An_0 and An_1.
- Illustration of how the numbers An_0 and An_1 can be extended for more complex situations. This will be done in a number of case studies:
  - intraparticle temperature gradients
  - bimolecular reactions
  - intraparticle pressure gradients
  - anisotrope catalyst pellets.

The first two topics are covered here in Chapter 6; the third will be covered in Chapter 7.

6.2 Generalized Definitions

The following discussion uses various concepts, which are first defined. The general definitions given here hold for simple reactions only, and are subsequently elaborated further to cover different situations, shapes, conditions etc. The three most important concepts are:

1. Effectiveness factor

The effectiveness factor η of a reaction is defined as the ratio of the amount of a component (participating in that reaction) converted inside a catalyst pellet to the amount that would have been converted if the conditions on the outer surface were to prevail everywhere inside the catalyst pellet. Hence:

$$\eta = \frac{\iiint_{V_P} R \, dV}{V_P R|_{surface}} \tag{6.22}$$

with $R|_{surface}$ the conversion rate at surface conditions.

2. Zeroth Aris number

The zeroth Aris number An_0 is defined as the number which becomes equivalent to $1/\eta^2$ if the effectiveness factor η goes to zero. Hence:

$$\eta \downarrow 0 \quad \rightarrow \quad An_0 \approx \frac{1}{\eta^2} \tag{6.23}$$

From this it follows that An_0 is defined only if η can get infinitely close to zero, which does not always hold, as we will see.

3. First Aris number

The first Aris number An_1 is defined as the number which becomes equal to $1-\eta^2$ if An_0 goes to zero. Hence:

$$An_0 \downarrow 0 \rightarrow An_1 \approx 1-\eta^2 \tag{6.24}$$

Moreover, we have to anticipate that if the zeroth Aris An_0 is not defined, then the first Aris number An_1 is defined as the number which becomes equivalent to $1-\eta^2$ if the effectiveness factor η goes to unity:

$$\eta \downarrow 1 \rightarrow An_1 \approx 1-\eta^2 \tag{6.25}$$

This will be further discussed in the next section.

6.2.1 Effectiveness Factor η

The concept of an effectiveness factor η for the description of the interaction between diffusion and chemical reaction in solid catalyst particles is well understood. Therefore, the derivation of expressions for the effectiveness factor are not elaborated. The following two points should be appreciated:

- The effectiveness factor as we use it is based on the amount that would have been converted if surface conditions were to prevail inside the catalyst pellet, not the amount that would have been converted at local gas phase conditions. This distinction is important if external mass or heat transfer plays a role.
- In the definitions of η the effectiveness factor of a reaction has been mentioned. It should be realized that for a simple reaction all components must have the same effectiveness factor, which therefore can be called the effectiveness factor of the reaction. For example, for the reaction of Equation 6.21:

$$A+\nu_B B \rightarrow \nu_P P+\nu_Q Q$$

The amount of B that is converted is ν_B times larger than the amount of A converted. The amount of B that would have been converted if surface conditions were to prevail is also ν_B times larger than the amount of A which would have been converted at surface conditions. Hence, it follows that A and B have the same effectiveness factor.

Thus for simple reactions the distinction between the effectiveness factor of a component or a reaction is trivial, since both effectiveness factors are the same. For complex reaction schemes, however, the distinction is quite important.

6.2.2 Zeroth Aris Number An_0

An_0 is the Aris number that brings together all the η curves in the low η region. This is illustrated in Figure 6.5 where η is plotted versus An_0 for first-order kinetics in an infinitely long slab, infinitely long cylinder and sphere; see Table 6.1. An_0, as such, brings

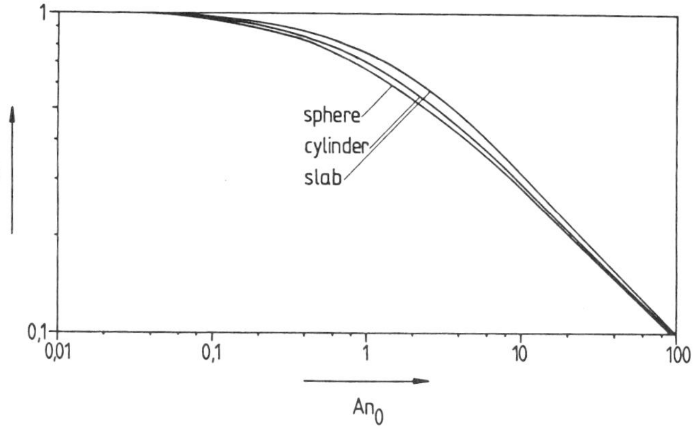

Figure 6.5 Effectiveness factor η versus the zeroth Aris number An_0 for first-order kinetics in an infinite slab, infinite cylinder and sphere.

Table 6.1 Effectiveness factor η as a function of the zeroth Aris number An_0 for first-order kinetics an infinitely long slab, an infinitely long cylinder and a sphere

Geometry	$\eta = \eta(An_0)$
Slab	$\dfrac{\tanh\left(\sqrt{An_0}\right)}{\sqrt{An_0}}$
Cylinder	$\dfrac{I_1\left(2\sqrt{An_0}\right)}{\sqrt{An_0} \times I_0\left(2\sqrt{An_0}\right)}$
Sphere	$\dfrac{1}{\sqrt{An_0}} \times \left\{\dfrac{1}{\tanh\left(3\sqrt{An_0}\right)} - \dfrac{1}{3\sqrt{An_0}}\right\}$

together all the modified Thiele moduli that where introduced in literature, so An_0 is very important from a theoretical point of view. If we define a modified Thiele modulus θ_m as:

$$\theta_m = \sqrt{An_0} \tag{6.26}$$

Then by matter of definition in the low η region, the effectiveness factor η must equal

$$\eta = \frac{1}{\theta_m} \tag{6.27}$$

For the simple reaction

$$A \rightarrow \nu_P P \tag{6.28}$$

it can be shown that the low η Aris number must equal

$$An_0 = \left(\frac{V_P}{A_P}\right)^2 \frac{R(C_{A,s})}{2D_e}\left(\int_0^{C_{A,s}} R(C_A)dC_A\right)^{-1} \tag{6.29}$$

This is not seen easily in considering the defining Equations 6.22 and 6.23, so a mathematical proof of Equation 6.29 is given in Appendix A. Hence the modified Thiele modulus θ_m now becomes

$$\theta_m = \frac{V_P}{A_P}\frac{R(C_{A,s.})}{\sqrt{2D_e}}\left(\int_0^{C_{A,s}} R(C_A)dC_A\right)^{-\frac{1}{2}} \tag{6.30}$$

From this formula it follows that the shape generalization that was introduced by Aris for first-order kinetics holds for arbitrary kinetics. The generalized Equation 6.11 given by Bischoff [10] for an infinite slab holds for arbitrary catalyst geometries, provided that the dimension R is replaced by the characteristic dimension V_p/A_p introduced by Aris. For nth order kinetics Equation 6.30 yields

$$\theta_m = \frac{V_P}{A_P}\sqrt{\frac{n+1}{2}\frac{kC_{A,s}^{\;n-1}}{D_e}} \tag{6.31}$$

Thus, the generalization of Petersen [8] introduced for nth order kinetics and for a sphere,

$$\theta_m^{(1)} = \frac{1}{3}R\sqrt{\frac{n+1}{2}\frac{kC_{A,s}^{\;n-1}}{D_e}} \tag{6.32}$$

holds (see Equations 6.4 and 6.5). However, his approximation for an infinite slab (Equations 6.6 and 6.7)

$$\theta_m^{(2)} = R\frac{n+2.5}{3.5}\sqrt{\frac{kC_{A,s}^{\ n-1}}{D_e}} \tag{6.33}$$

does not. The same can be said for the modified Thiele modulus introduced by Rajadhyaksha and Vaduseva [9] for nth order kinetics in a sphere (Equation 6.9):

$$\theta_m^{(3)} = \frac{1}{3}R\sqrt{n}\sqrt{\frac{kC_{A,s}^{\ n-1}}{D_e}} \tag{6.34}$$

In Figure 6.6 $\theta_m^{(2)}$ and $\theta_m^{(3)}$ are compared with the properly defined θ_m. It can be seen that $\theta_m^{(2)}$ and θ_m agree within 6 % (as claimed by Petersen) only for reaction orders between -0.3 and +2.2. Between $\theta_m^{(3)}$ and θ_m deviations are even larger: the modified Thiele modulus $\theta_m^{(3)}$ can be used within a 10% error range only for reaction orders between +0.7 and +1.6.

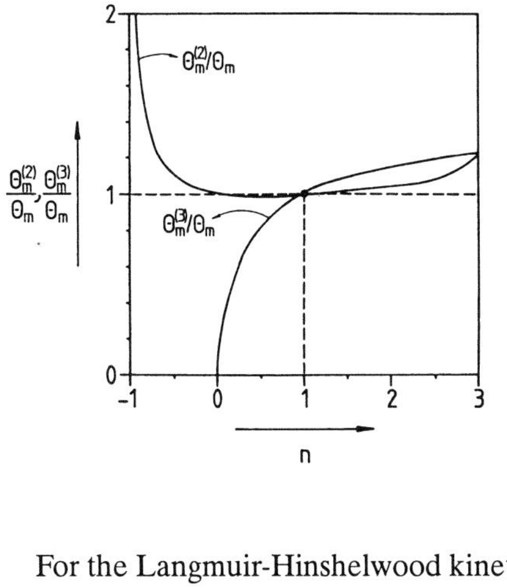

Figure 6.6 Ratio of some Thiele moduli $\theta_m^{(2)}/\theta_m$ and $\theta_m^{(3)}/\theta_m$ as given by Equations 6.31, 6.33 and 6.34 versus the reaction order n.

For the Langmuir-Hinshelwood kinetics of Equation 6.8, namely

$$R(C_A) = \frac{kC_A}{1+kC_A}$$

in a sphere Rajadhyaksha and Vaduseva [9] derived the following modified Thiele modulus (Equation 6.10):

$$\theta_m^{(4)} = \frac{1}{3}R\frac{1}{1+kC_{A,s}}\sqrt{\frac{kC_{A,s}}{\ln(1+kC_{A,s})}}\sqrt{\frac{k}{D_e}} \tag{6.35}$$

From Equation 6.30 we find that the proper modified Thiele modulus for arbitrary catalyst geometries, for this case, is given by

$$\theta_m = \frac{V_P}{A_P}\frac{KC_{A,s}}{1+KC_{A,s}}\sqrt{\frac{1}{2\left[KC_{A,s} - \ln\left(1+KC_{A,s}\right)\right]}}\sqrt{\frac{k}{D_e}} \tag{6.36}$$

Figure 6.7 compares $\theta_m^{(4)}$ and θ_m. It can be seen the use of $\theta_m^{(4)}$ can lead to serious deviations for high values of $KC_{A,s}$. These deviations are higher than 10% for $KC_{A,s} > 2$.

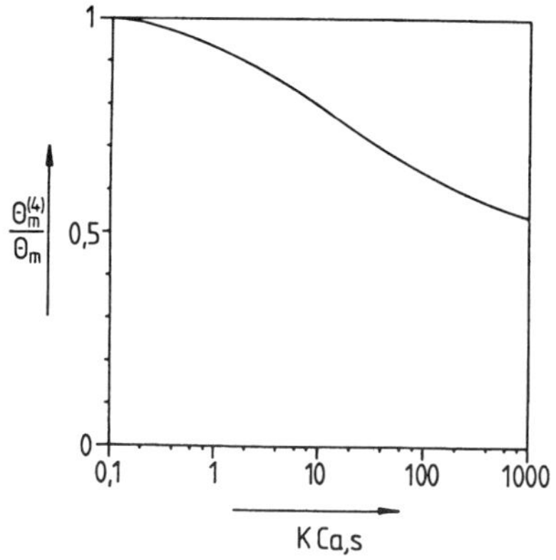

Figure 6.7 Ratio of the modified Thiele moduli $\theta_m^{(4)}/\theta_m$ of Equation 6.35 over Equation 6.36 versus $KC_{A,s}$ for Langmuir-Hinshelwood kinetics.

Valdman and Hughes [11] proposed an approximate relationship for the effectiveness factor for the following form of Langmuir-Hinshelwood kinetics:

$$R(C_A) = \frac{kC_A}{\left(1+KC_A\right)^2} \tag{6.13}$$

From Equation 6.30 the corresponding modified Thiele modulus is obtained as:

$$\theta_m = \frac{V_P}{A_P}\sqrt{\frac{KC_{A,s}}{\left(1+KC_{A,s}\right)^3}}\left\{2\left(\frac{1+KC_{A,s}}{KC_{A,s}}\ln\left(1+KC_{A,s}\right)-1\right)\right\}^{-1/2}\sqrt{\frac{k}{D_e}} \tag{6.37}$$

Therefore, it is not necessary to derive approximate relationships.

In summary, it can be said that the square root of the low η Aris number An_0 is a generalized modified Thiele modulus. If we compare this number with the modified Thiele moduli given in literature we find that serious deviations generally occur for kinetic expressions which are not first order. This is because most of the modified Thiele moduli given in literature are incorrect.

6.2.3 First Aris Number An_1

An_1 is the Aris number that brings together all the η curves in the high η region. This is illustrated in Figure 6.8 where η is plotted versus An_1 for first-order kinetics in an in-

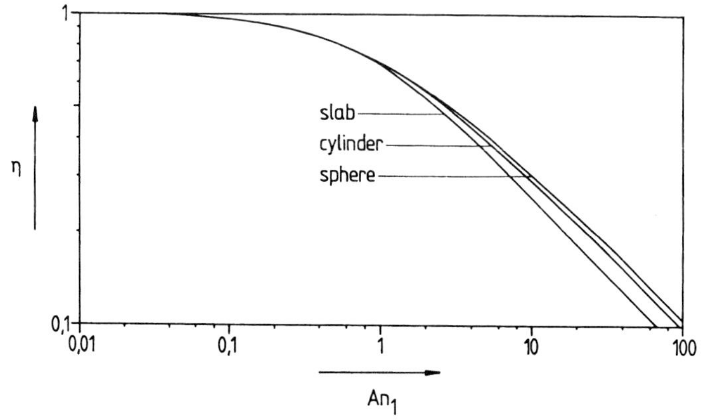

Figure 6.8 Effectiveness factor η versus the first Aris number An_1 for first-order kinetics in an infinitely long slab, an infinitely long cylinder and a sphere.

finite slab, an infinite cylinder and a sphere (Table 6.2 gives the formulae used).

Table 6.2 Effectiveness factor η as a function of the first Aris number An_1 for first-order kinetics in an infinitely long slab, an infinitely long cylinder and a sphere

Geometry	$\eta = \eta(An_1)$
Slab	$\dfrac{\tanh\left(\sqrt{\frac{3}{2}An_1}\right)}{\sqrt{\frac{3}{2}An_1}}$
Cylinder	$\dfrac{I_1\left(2\sqrt{An_1}\right)}{\sqrt{An_1}\,I_0\left(2\sqrt{An_1}\right)}$
Sphere	$\dfrac{1}{\sqrt{\frac{5}{6}An_1}} \times \left\{\dfrac{1}{\tanh\left(\sqrt{\frac{15}{2}An_1}\right)} - \dfrac{1}{\sqrt{\frac{15}{2}An_1}}\right\}$

As such, the high η Aris number is very important from a practical point of view. For chemical reactors in industry, the effectiveness factor will typically range from

70% to 90%. In general, effectiveness factors below 70% are economically unfeasible because the surplus of reactor volume which is needed (larger than (1/0.7)-1 = 43%) bears too heavily upon the investment costs. However, for effectiveness factors lager than 90% it is generally economically worthwhile to increase the size of the catalyst pellets, thereby decreasing the pressure drop across the reactor and thus the compression costs, but also decreasing the effectiveness factor somewhat. In the hot spot area somewhat lower effectiveness factor are often accepted, but if they become too low a different reactor type has to be selected, that can cope with smaller catalyst particles (e.g. a fluid bed).

For the simple reaction:

$$A \rightarrow \nu_P P \tag{6.28}$$

It can be shown (Appendix B) that the first Aris number equals:

$$An_1 = \left(\frac{V_P}{A_P}\right)^2 \frac{\Gamma}{D_e} \times \left.\frac{\partial R(C_A)}{\partial C_A}\right|_{C_A = C_{A,s}} \tag{6.38}$$

This holds for arbitrary reaction kinetics and arbitrary catalyst geometries. For example, for n-th order kinetics this yields

$$An_1 = \left(\frac{V_P}{A_P}\right)^2 n\Gamma \frac{k C_{A,s}^{\,n-1}}{D_e} \tag{6.39}$$

The dimensionless number Γ is a geometry factor. Its value depends on the catalyst geometry only. This is illustrated in Table 6.3 where Γ is given for an infinitely long slab,

Table 6.3 Geometry factor Γ for an infinitely slab, an infinite cylinder and sphere

Geometry	Γ
Slab	$\frac{2}{3}$
Cylinder	1
Sphere	$\frac{6}{5}$

infinitely long cylinder and a sphere. In Appendix B it is shown that Γ can be calculated from

$$\Gamma = -2 \left.\frac{\partial \eta_1(An_0)}{\partial An_0}\right|_{An_0 = 0} \tag{6.40}$$

where η_1 is the function that gives values of the effectiveness factor for a certain value of the zeroth Aris number An_0 for first-order kinetics. Hence for an infinite slab, infinite cylinder and sphere, the functions η_1 are the functions given in Table 6.1. Table 6.3 can be evaluated from Table 6.1 by applying the equations of Table 6.1 to Equation 6.40.

Without any prove it is stated here that the geometry factor Γ falls between the two extremes of $^2/_3$ for the infinitely long slab and of $^6/_5$ for the sphere for almost all practical cases. Thus Γ is almost always close to unity. This holds for any catalyst geometry, hence also for catalyst geometry's commonly found in industry, for example ring-shaped or cylindrical catalyst pellets. For this type of pellet it can be shown (Appendix C) that the geometry factor Γ equals:

$$\Gamma = \frac{64}{\pi^3}\frac{1}{1-\iota^2}\left(1+\frac{\lambda}{1-\iota}\right)^2 \sum_{j=0}^{\infty} \frac{1-\iota^2-2\chi\left(a_j\iota, a_j\right)}{\left(2j+1\right)^4} \tag{6.41}$$

with

$$a_j = \frac{\left(2j+1\right)\eta}{\lambda} \tag{6.42}$$

and

$$\chi(a,b) = \frac{2-b\,Q_0(a,b)-a\,Q_0(b,a)}{b^2\,P_0(a,b)} \tag{6.43}$$

Here, P_0 and Q_0 are the cross-products of modified Bessel functions of the first and second kind:

$$P_0(a,b) = I_0(a)K_0(b) - I_0(b)K_0(a) \tag{6.44}$$

$$Q_0(a,b) = I_0(a)K_1(b) + I_1(b)K_0(a) \tag{6.45}$$

Finally, ι and λ tie down the catalyst geometry of the ring-shaped catalyst pellet (Figure 6.9):

$$\iota = \frac{R_i}{R_u} \tag{6.46}$$

$$\lambda = \frac{H}{R_u} \tag{6.47}$$

Thus for $\iota = 0$ the ring-shaped catalyst pellet becomes a cylindrical catalyst pellet.

Equation 6.41 is illustrated in Figure 6.10. In this diagram ι is plotted versus λ; lines with a constant geometry factor Γ are drawn. The four corners of the diagram represent

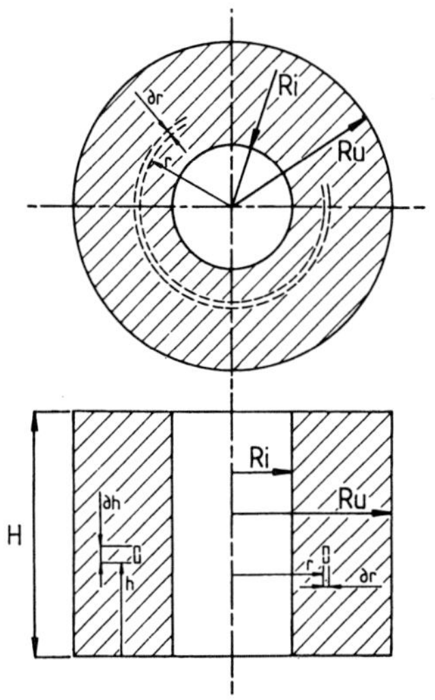

Figure 6.9 Geometry of a ring-shaped catalyst pellet.

the limiting geometries that can be found for a ring-shaped catalyst pellet, and these are shown schematically around the figure.

The upper and right-hand border lines correspond to the geometry of an infinite slab, where from Table 6.3 it follows that the geometry factor Γ equals 2/3. The bottom borderline corresponds to all possible geometries of the cylindrical pellet. In the lower left-hand corner the geometry of an infinitely long cylinder is found, where Γ must equal unity. Finally, the singularity in the upper right-hand corner represents all possible geometries of an infinite beam.

For any ring-shaped catalyst pellet with known values of ι and λ the value of Γ can be obtained from Figure 6.10. The value of the first Aris number An_1 can then be calculated with Equation 6.38.

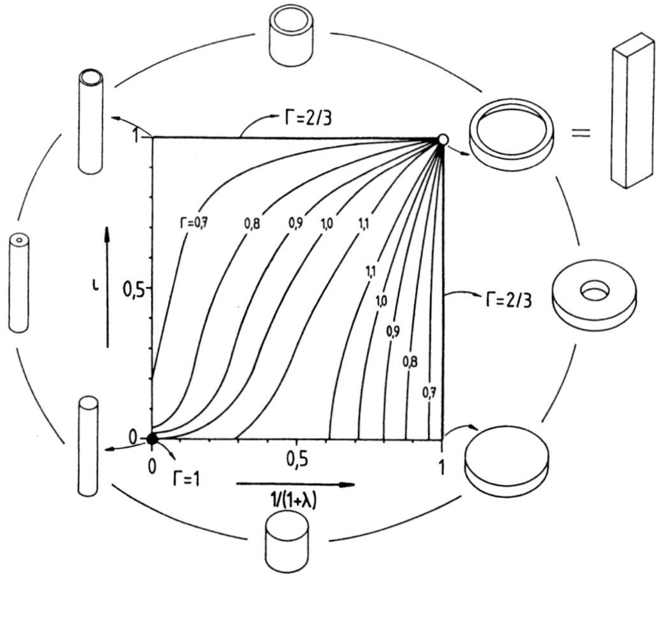

Figure 6.10. Dimensionless inner radius ι versus dimensionless height λ for a ring-shaped catalyst pellet.

6.2.4 Comparison of An_0 and An_1

The Aris numbers An_0 and An_1 are much alike. This is illustrated in Table 6.4 where the formulae for An_0 and An_1 are given for arbitrary kinetics, for n-th order kinetics and for first-order kinetics. In practice, reaction kinetics do not differ too much from first-order kinetics, and hence the values of An_0 and An_1 will remain very close to each other (as also the geometry factor Γ is close to one). In that case both Aris numbers will be roughly equal to the square power of the shape-generalized Thiele modulus of Aris [6].

Table 6.4 Formulae for both the low η Aris number An_0 and high η Aris number An_1 for arbitrary kinetics, nth-order and first-order kinetics

Aris number	Arbitrary kinetics	nth-order kinetics	First-order kinetics
An_0	$\left(\frac{V_p}{A_p}\right)^2 \times \frac{\{R(C_{A,s})\}^2}{2D_{e,A}} \times \left\{\int_0^{C_{A,s}} R(C_A)dC_A\right\}^{-1}$	$\frac{n+1}{2} \times \left(\frac{V_p}{A_p}\right)^2 \times \frac{kC_{A,s}^{n-1}}{D_{e,A}}$	$\left(\frac{V_p}{A_p}\right)^2 \times \frac{k}{D_{e,A}}$
An_1	$\left(\frac{V_p}{A_p}\right)^2 \times \frac{\Gamma}{D_{e,A}} \times \left.\frac{\partial R(C_A)}{\partial C_A}\right\|_{C_A=C_{A,s}}$	$n \times \Gamma \times \left(\frac{V_p}{A_p}\right)^2 \times \frac{kC_{A,s}^{n-1}}{D_{e,A}}$	$\Gamma \times \left(\frac{V_p}{A_p}\right)^2 \times \frac{k}{D_{e,A}}$

Nevertheless, there are some notable differences between An_0 and An_1. As discussed previously, An_0 is important for the low η region, and hence is of interest from a theoretical view-point, whereas An_1 is important for the high η region and thus of interest in practice. There are other differences as well. To illustrate these it is necessary to discuss and understand effectiveness factor plots that are somewhat more exotic than those described so far.

In Figure 6.11 effectiveness factor plots are given for nth-order kinetics in a slab. The lines were calculated with the formulae given in Table 6.5. The effectiveness factor in

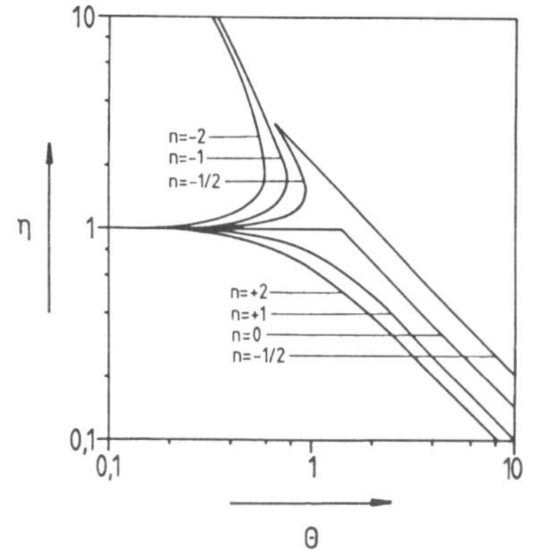

Figure 6.11 Effectiveness factor η versus Thiele modulus θ of Equation 6.48 for nth-order kinetics in an infinitely long slab.

Table 6.5 Effectiveness factor η as a function of the Thiele modulus θ (given by Equation 6.48) for nth-order kinetics in an ininite slab

n	$\eta = \eta(\Theta)$
-2	$\left.\begin{aligned} \Theta &= \frac{1}{\sqrt{2}} \times \frac{\sqrt{y}\sqrt{y-1} + \ln\left(\sqrt{y} + \sqrt{y+1}\right)}{y\sqrt{y}} \\ \eta &= \sqrt{2} \times \frac{\sqrt{y-1}}{\Theta} \end{aligned}\right\} \quad y \in [1, \infty[$
-1	$\left.\begin{aligned} \Theta &= \frac{\sqrt{2}}{y} \times \int_0^{\ln y} e^{x^2} dx \\ \eta &= \sqrt{2} \times \frac{\sqrt{\ln y}}{\Theta} \end{aligned}\right\} \quad y \in [1, \infty[$
-½	$\left.\begin{aligned} \Theta &= \sqrt{6 \times \frac{\eta - 1}{\eta^3}} & \Theta \le \frac{2}{3}\sqrt{2} \\ \Theta &= \frac{2}{\eta} & \Theta > \frac{2}{3}\sqrt{2} \end{aligned}\right\}$
0	$\left.\begin{aligned} \eta &= 1 & \Theta \le \sqrt{2} \\ \eta &= \frac{\sqrt{2}}{\Theta} & \Theta > \sqrt{2} \end{aligned}\right\}$
1	$\eta = \frac{\tanh \Theta}{\Theta}$
2	$\left.\begin{aligned} \Theta &= \sqrt{\frac{y\sqrt{3}}{2} F\left(\phi, \frac{\sqrt{2-\sqrt{3}}}{2}\right)} \\ \phi &= arc\cos\left(\frac{\sqrt{3} + 1 - y}{\sqrt{3} - 1 + y}\right) \\ y &= \frac{1}{\sqrt[3]{1 - \frac{3}{2} \times (\eta\Theta)^2}} \end{aligned}\right\} \quad y \in [1, \infty[\;\wedge\; \phi \in [0, \pi[$

this table is given as a function of θ, which may be regarded as a type of generalized Thiele modulus:

$$\theta = \frac{V_P}{A_P}\sqrt{\frac{R(C_{A,s})}{D_e C_{A,s}}} \tag{6.48}$$

From Figure 6.11 shows that three basically different effectiveness factor plots can be obtained:

- *Case 1* (for nth-order kinetics, $n > 0$):

$$\lim_{C_A \downarrow 0} R(C_A) = 0 \tag{6.49}$$

The concentration profiles inside the slab are as illustrated in Figure 6.12a. The ef-

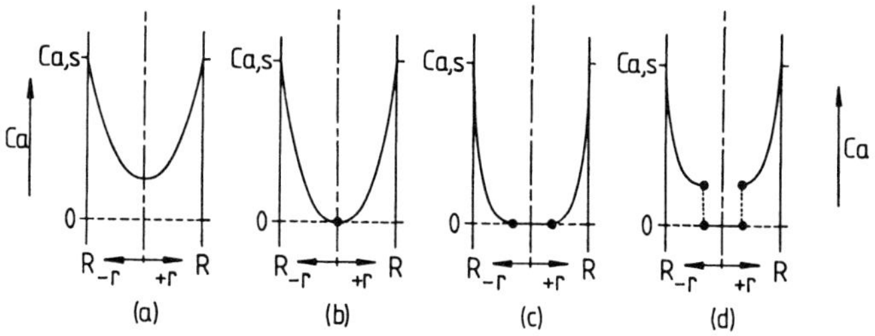

Figure 6.12 Concentration profiles as they can occur inside a catalyst pellet.

fectiveness factor plot is continuous and differentiable for every value of θ and η is always smaller than one.

- *Case 2* (for n^{th}-order kinetics: $-1 < n \leq 0$):

$$\lim_{C_A \downarrow 0}\{C_A R(C_A)\} = 0 \quad \textit{and} \quad \lim_{C_A \downarrow 0} R(C_A) \neq 0 \tag{6.50}$$

For low values of θ the concentration profiles inside the slab are as given in Figure 6.12a. As the value of θ increases, η increases. At a certain point the profile of Figure 6.12b is obtained. For this specific value of θ the effectiveness factor plot is continuous but nondifferentiable. For values of θ higher than this specific value, the concentration profiles are as in Figure 6.12c and the effectiveness factor decreases with increasing θ. For this case in general, the effectiveness factor will become larger than one. It is even possible that multiplicity occurs, as is illustrated in Figure 6.11 for $n = -{}^1/_2$.

- *Case 3* (for n^{th}-order kinetics: $n \leq -1$):

$$\lim_{C_A \downarrow 0}\{C_A R(C_A)\} \neq 0 \tag{6.51}$$

For low values of θ multiplicity occurs. There are two operating points, each one with a concentration profile that follows Figure 6.12a. For a certain critical value of

θ both operating points coincide to a single point and the two concentration profiles coincide to a single profile with the form of Figure 6.12a. Above this critical value of θ the concentration profile collapses and takes the form of Figure 6.12d. Once such a concentration profile is obtained, the effectiveness factor cannot be calculated from a steady state mass balance. In this region the effectiveness factor η can have any real value for a certain value of θ. The value of η when operating in the steady state, can only be calculated from a transient micro mass balance by solving the balance and studying the solution for approaching infinite times.

Figure 6.11 holds for a slab. Similar figures can be obtained for other catalyst geometries. This is illustrated in Figure 6.13 where the effectiveness factor is plotted versus θ for zeroth-order kinetics in an infinite slab, infinite cylinder and a sphere. Figure 6.13 has been constructed on the basis of the formulae given in Table 6.6. Hence, the discussion that follows is not restricted to a slab, but holds for any arbitrary catalyst geometry.

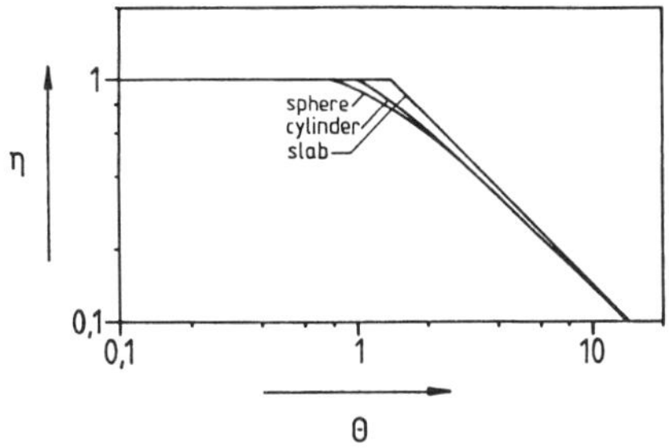

Figure 6.13 Effectiveness factor η versus Thiele modulus θ of Equation 6.48 for zeroth-order kinetics in an infinitely long slab, an infinitely long cylinder and a sphere.

Table 6.6 Effectiveness factor η as a function of the Thiele modulus θ (given by Equation 6.48) for zeroth-order kinetics an infinite slab, an infinite cylinder and sphere

Geometry	$\eta = \eta(An_1)$	
Slab	$\eta = 1$	$\theta \leq \sqrt{2}$
	$\eta = \dfrac{\sqrt{2}}{\theta}$	$\theta > \sqrt{2}$
Cylinder	$\eta = 1$	$\theta \leq 1$
	$\eta + (1-\eta)\ln(1-\eta) = \dfrac{1}{\theta^2}$	$\theta > 1$
Sphere	$\eta = 1$	$\theta \leq \sqrt{\dfrac{2}{3}}$
	$\left(1-\sqrt[3]{1-\eta}\right)^2\left(1+2\sqrt[3]{1-\eta}\right) = \dfrac{2}{3\theta^2}$	$\theta > \sqrt{\dfrac{2}{3}}$

Figure 6.11 shows the following differences between An_0 and An_1:

- An_0 can only have positive values, whereas An_1 can have any real value, thus it may also become negative. This follows directly from the definitions of An_0 and An_1. Since the effectiveness factor cannot become imaginary, An_0 cannot become negative (Equation 6.23). Since the effectiveness factor can have values both larger than one and smaller than one, it follows that An_1 can have both negative and positive values respectively (Equation 6.24). From Equation 6.38 we see that An_1 will become negative if

$$\left.\frac{\partial R(C_A)}{\partial C_A}\right|_{C_A=C_{A,s}} > 0 \tag{6.52}$$

Then, from the definition of An_1 it follows that effectiveness factors larger than one can be expected if criterion 6.52 is adhered to.
At this point the introduction of Aris numbers rather than Thiele moduli is stressed. One could say that a Thiele modulus corresponds to the square root of an Aris number. Hence a negative Aris number will correspond to an imaginary Thiele modulus. Because negative values of An_1 often occur (An_0 can only be positive!) and because we want to avoid working with imaginary numbers, we abstain on the use of Thiele moduli.

- An_1 is always defined for any differentiable function $R(C_A)$. The low η Aris number An_0, however, is defined only when the function $R(C_A)$ fulfils the condition

$$\lim_{C_A \downarrow 0}\{C_A\, R(C_A)\} = 0 \tag{6.53}$$

If this is not the case, no low η region exists, i.e. for reaction orders lower than and equal to minus one (Figure 6.11). This also implies that, in this case, An_0 is undefined. Note that this is in agreement with Equation 6.29 repeated below:

$$An_0 = \left(\frac{V_P}{A_P}\right)^2 \frac{\{R(C_{A,s})\}^2}{2D_e}\left(\int_0^{C_{A,s}} R(C_A)\,dC_A\right)^{-1}$$

If Equation 6.53 is not fulfilled, the integral in the above equation is not defined and thus An_0 is also undefined. However, the number An_1 is not restricted.

6.3 Generalized Approximations

Several formulae have been given for the calculation of the effectiveness factor as a function of one of the Aris numbers An_0 or An_1, or as a function of a Thiele modulus. These formulae can become very complex and, for most kinetic expressions and catalyst geometries, it is impossible to derive analytical solutions for the effectiveness factor, so that we have to resort to methods of numerical approximation. Even then calculation of the effectiveness factor can be very tedious and time consuming as, for example, for ring-shaped catalyst pellets where partial differential equations must be solved. Therefore, there is a need for generalized approximations.

It is possible to give generalized approximations on the basis of the numbers An_0 and An_1. The simplest are:

- For values of the effectiveness factor η close to one

$$\tilde{\eta} = \sqrt{1 - An_1} \tag{6.54}$$

- For values of η close to zero

$$\tilde{\eta} = \frac{1}{\sqrt{An_0}} \tag{6.55}$$

These approximations follow directly from the definitions (Equations 6.23-6.25).

In order to study relationship 6.54 and 6.55, as well as other approximations, the error Δ is defined as

$$\Delta = \left|\frac{\eta - \tilde{\eta}}{1 - \eta}\right| \; for \; \eta > \frac{1}{2}$$

$$\Delta = \frac{|\eta - \tilde{\eta}|}{\eta} \; for \; 0 < \eta < \frac{1}{2} \tag{6.56}$$

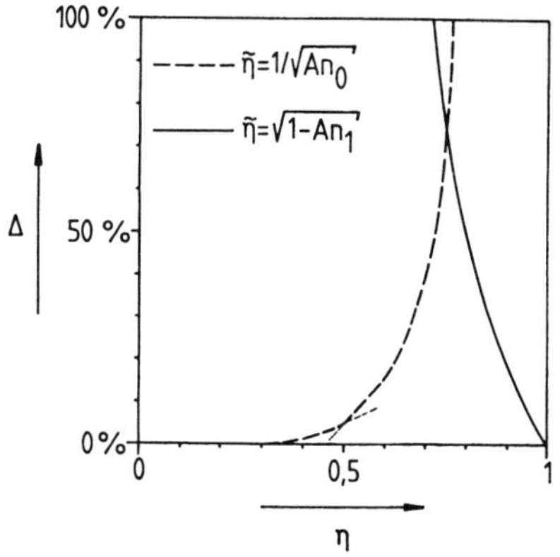

Figure 6.14 Relative error Δ versus effectiveness factor η for first – order kinetics in an infinite slab.

If Δ is plotted versus η two branches occur. For $\eta > ½$, Δ is a measure for the relative error of 1-η. For $0 < \eta \leq ½$, Δ is a measure for the relative error of η. Since for $\eta = ½$ the values of 1-η and η are equal, the two branches meet for $\eta = ½$. Thus if Δ is plotted versus η the resultant line will also be continuous for $\eta = ½$. However, in general the line will be nondifferentiable for $\eta = ½$. This is illustrated in Figure 6.14, where Δ is plotted versus η for first-order kinetics in a slab. Here, η was calculated with the formulae given in Tables 6.1 and 6.2, and $\tilde{\eta}$ was calculated with either approximation 6.54 (the extended line), or approximation 6.55 (the dotted line). For values of the effectiveness factor close to one (as generally in industrial reactors), we are interested in how much the effectiveness factor drops below one. It is desirable to know, for example, whether the effectiveness factor is 89 % or 85 %, since this means an 11 % or 15 % increase in required reactor volume, respectively. Hence it is more appropriate to take the relative error in 1-η as a measure for the relative error. However, for low values of η the relative error in 1-η will always be small (since 1-η or 1-$\tilde{\eta}$ are close to one, whereas η and $\tilde{\eta}$ are much smaller than one). Therefore, for low values of η, the relative error in η is a much better indicator for the relative error. It is logical to switch from the first relative error to the second at η = ½, since for that case both errors are equal.

Figure 6.14 compares the two approximations 6.54 and 6.55. For high values of η approximation 6.54 yields good results and Δ values are low. This is logical, since this approximation is based upon An_1, which is defined for the high η region. Contrary to this, approximation 6.55 yields good results in the low η region, since it is based upon An_0. Results are very poor in the opposite regions, and in the intermediate region the results are unsatisfactory.

An improvement is obtained if we replace the approximations of Equations 6.54 and 6.55 by 6.57 and 6.58, respectively:

$$\tilde{\eta} = \frac{1}{\sqrt{1 + An_1}} \tag{6.57}$$

$$\tilde{\eta} = \frac{1}{\sqrt{1 + An_0}} \tag{6.58}$$

For values of $\tilde{\eta}$ close to one, Equations 6.54 and 6.57 coincide; for $\tilde{\eta}$ close to zero Equations 6.55 and 6.58 coincide. The similarity between Equations 6.57 and 6.58 is notable in that they have exactly the same from.

Equations 6.57 and 6.58 are illustrated in Figure 6.15 where Δ according to Equation 6.56 is plotted versus η for first-order kinetics in a slab. The effectiveness factor η and the approximation $\tilde{\eta}$ were calculated from Tables 6.1 and 6.2 and Equations 6.57 and

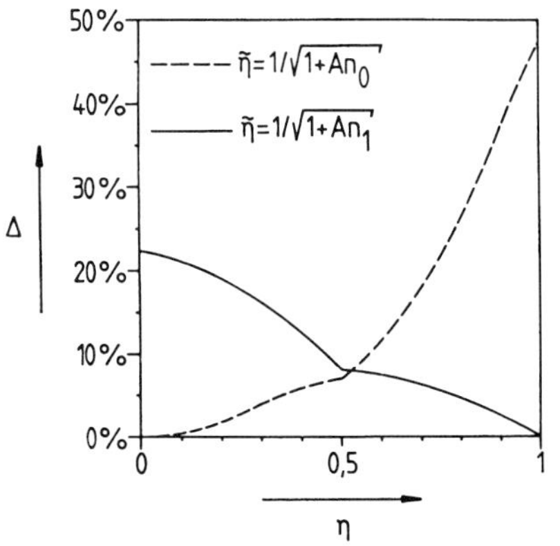

Figure 6.15 Relative error Δ versus effectiveness factor η for first-order kinetics in an infinite slab.

6.58. Figure 6.15 is basically the same as Figure 6.14. With approximations 6.57 and 6.58 much better results are obtained in the intermediate η region, as compared to the results of Equations 6.54 and 6.55 in the same region.

The best approximation is obtained if both An_1 and An_0 are used in one approximation. Good results are obtained from:

$$\tilde{\eta} = \frac{1}{\sqrt{1 + \tilde{\eta} An_1 + (1 - \tilde{\eta}) An_0}} \qquad (6.59)$$

This approximation can be used only if An_1 is positive, which will be true in most cases. Equation 6.59 coincides with 6.57 for η close to one, and with 6.58 for η close to zero. Equation 6.59 can be solved as an implicit relationship. The value of $\tilde{\eta}$ can be easily obtained by using the contraction theorem. This is illustrated in Table 6.7 for $An_1 = An_0 = 1$.

Table 6.7 Illustration of the contraction theorem for the calculation of an approximate value of the effectiveness factor $\tilde{\eta}$ defined by Equation 6.59

new value of $\tilde{\eta}$	calculated value of $\frac{1}{\sqrt{1 + \tilde{\eta} An_1 + (1 - \tilde{\eta}) An_0}}$
1.0000	0.8165
0.8165	0.7926
0.7926	0.7897
0.7897	0.7893
0.7893	0.7893

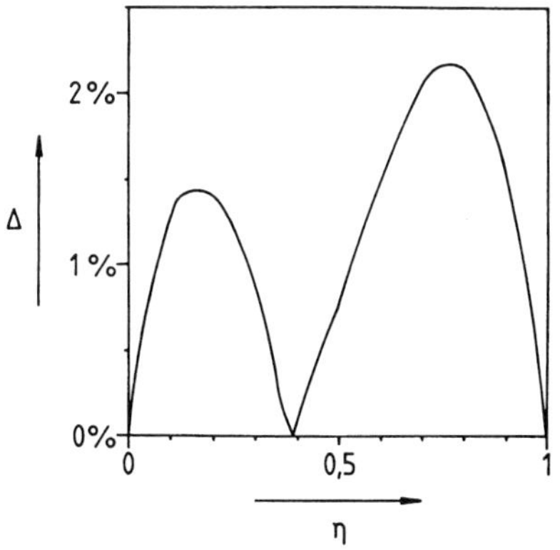

Figure 6.16 Relative error Δ versus effectiveness factor η for first-order kinetics in an infinite slab using the approximation

$$\tilde{\eta} = \frac{1}{\sqrt{1 + \tilde{\eta} An_1 + (1 - \tilde{\eta}) An_0}}$$

The value of Equation 6.59 is demonstrated in Figure 6.16 for first-order kinetics in a slab. It can be seen that Equation 6.59 predicts the effectiveness factor with an error of only (2 % at the most. Further, this equation gives the exact values for η close to one or zero, which must be so by matter of definition. Hence:

$$\lim_{\eta \uparrow 1} \Delta = 0 \tag{6.60}$$

$$\lim_{\eta \downarrow 0} \Delta = 0 \tag{6.61}$$

For a certain value of η a global maximum denoted Δ_{max} will occur for Δ. Thus:

$$\Delta_{max} = \max_{\eta \in]0,1[} \Delta \tag{6.62}$$

For Figure 6.16 Δ_{max} equals 2.2 %.

Up till now the discussion has been restricted to first-order kinetics in a slab. However, if Δ is plotted versus η for other forms of kinetics or other catalyst geometries, curves similar to Figure 6.16 are obtained. For each individual case a different value of Δ_{max} will be obtained. This is illustrated in Table 6.8 where Δ_{max} is given for several forms of kinetics and catalyst geometries. For all cases the value of Δ_{max} amounts to a few per cent only, which is well within the usual accuracy of kinetic data.

Table 6.8 Value of the maximum relative error Δ_{max} for n[th]-order kinetics in an infinite slab, infinite cylinder and sphere if η is approximated using Equation 6.59

Geometry n	Slab	Cylinder	Sphere
½	$\Delta_{max} = 2.6\,\%$	$\Delta_{max} = 5.2\,\%$	$\Delta_{max} = 6.2\,\%$
1	$\Delta_{max} = 2.2\,\%$	$\Delta_{max} = 3.4\,\%$	$\Delta_{max} = 3.9\,\%$
2	$\Delta_{max} = 1.7\,\%$	$\Delta_{max} = 1.4\,\%$	$\Delta_{max} = 1.4\,\%$

Equation 6.59 is especially convenient for complex catalyst geometries. For a ring-shaped catalyst pellet (Figure 6.9), with the formulae given in Appendix C it is possible to calculate Δ_{max} for first-order reactions occurring inside the pellet. This is illustrated in Figure 6.17, which is comparable to Figure 6.10. From this figure it can be concluded that the approximation 6.59 can be used for every geometry of the hollow cylinder without introducing an error larger than 4.4 %. Hence it is possible to estimate the effectiveness factor with one simple formula, whereas otherwise a partial differential equation would have to be solved. Apparently, Equation 6.59 is of great value! The discussion is clarified with Example 9.8.

The approximation of Equation 6.59 can only be used if the number An_1 is positive. If this is not the case, other approximations can be constructed by comparison with a similar but simpler situation. For example, calculation of the effectiveness factor for a given situation for a ring-shaped catalyst pellet can be achieved by calculating the effectiveness factor for the same situation in a slab, and afterwards correcting for the geometry of the ring-shaped catalyst pellet using the Aris numbers (Fig. 6.18). Since the Aris numbers are generalized in every respect, only small errors are introduced with this method (the more similar the complex and the simple situation are, the smaller the error will be).

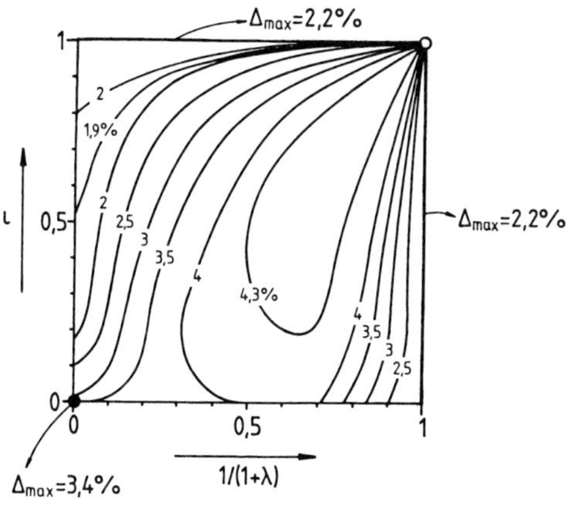

Figure 6.17 Dimensionless inner radius ι versus dimensionless height λ for a ring-shaped catalyst pellet. Lines of constant maximum relative error Δ_{max} are drawn for first-order kinetics using the approximation

$$\tilde{\eta} = \frac{1}{\sqrt{1 + \tilde{\eta} An_1 + (1 - \tilde{\eta}) An_0}}$$

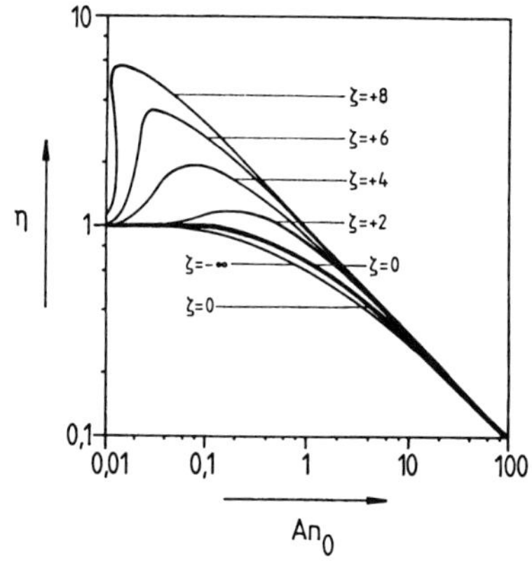

Figure 6.18 Effectiveness factor η versus zeroth Aris number An_0 for first-order kinetics in an infinite slab and several values of ζ.

References

1. Schmalzer, D.K. (1969) *Chem. Eng. Sci.*, **13**, 226.
2. Ors, N., Dogu, T. (1979) *AIChEJ.*, **25**, 723.
3. Kulkarni, B.D., Jayaraman, V.K., Doraiswamy, L.K. (1981) *Chem. Eng. Sci.*, **36**, 943.
4. Doraiswamy, L.K., Sharma, M.M. (1984) *Heterogeneous Reactions: Analysis, Examples, and Reactor Design*, Vol. 1, New York: John Wiley & Sons.
5. Thiele, E.W. (1939) *Ind. Eng. Chem.*, **31**, 916.
6. Aris, R. (1957) *Chem. Eng. Sci.*, **6**, 262.
7. Kasaoka, S., Sakata, Y. (1966) *Kagaku Kogaku*, **30**, 650.
8. Petersen, E.E. (1965) *Chemical Reaction Analysis.* Prentice-Hall, New Jersey: Englewood Cliffs.
9. Rajadhyaksha, R.A., Vaduseva, K. (1974) *J. Catal.*, **34**, 321.
10. Bischoff, K.B. (1974) *Chem. Eng. Sci.,* **29**, 1348.
11. Valdman, B., Hughes, R. (1976) *AIChEJ.*, **22**, 192.

12. Schneider, P., Mitschka, P. (1966) *Chem., Eng. Sci.*, **21**, 455.
13. Roberts, G.W., Satterfield, C.N. (1966) *Ind. Eng. Chem. Fundam.*, **5**, 317.
14. Kao, H.S.P., Satterfield, C.N. (1968) *Ind. Eng. Chem. Fundam.*, **7**, 664.
15. Finlayson, B.A. (1972) *The Method of Weighted Residuals and Variational Principles*, New York: Academic Press.
16. Ibanez, J.L. (1979) *J. Chem. Phys.*, **71**, 5253.
17. Namjoshi, A., Kulkarni, B.D., Doraiswamy, L.K. (1983) *AIChEJ.*, **29**, 521.
18. Weisz, P.B., Hicks, J.S. (1962) *Chem. Eng. Sci.*, **17**, 265.
19. Mears, D.A. (1971) *Ind. Eng. Chem., Process Des. Dev.*, **10**, 541.
20. Patterson, W.R., Cresswell, D.L. (1971) *Chem. Eng. Sci.*, **26**, 605.
21. Carberry, J.J. (1961) *AIChEJ.*, **7**, 350.
22. Liu, S.L. (1970) *AIChEJ.*, **16**, 742.
23. Maymo, J.A., Cunningham, R.E. (1966) *J. Catal.*, **6**, 186.
24. Karanth, N.G., Koh, H.P., Hughes, R (1974) *Chem. Eng. Sci.*, **29**, 451.
25. Kehoe, J.P.G., Aris, R. (1973) *Chem. Eng. Sci.*, **28**, 2094.
26. Haynes, H.W., Jr. (1978) *Can. J. Chem. Eng.*, **56**, 582.
27. Broekhoff, J.C.P., de Boer, J.H. (1968) *J. Catal.*, **10**, 153.
28. Chu, C., Chon, K. (1970) *J. Catal.*, **17**, 71.
29. Goldstein, W., Carberry, J.J. (1973) *J. Catal.*, **28**, 33.
30. Varghese, P., Varma, A., Carberry, J.J. (1978) *Ind. Eng. Chem. Fundam.*, **17**, 195.
31. Aris, R. (1975) *The Mathematical Theory of Diffusion and Reaction in Permeable Catalyst*, Vol. 1, Oxford: Clarendon Press.
32. Nir, A., Pismen, L.M. (1977) *Chem. Eng. Sci.*, **32**, 35.

7 Complex Situations

7.1 Intraparticle Temperature Gradients

The Aris numbers An_1 and An_0 can be applied to nonisothermal catalyst pellets. Three items will be discussed:

- comparison of the three- and two-parameter model;
- formulae for the calculation of Aris numbers;
- negligibility criteria, for determination as to whether or not intraparticle temperature gradients may be neglected.

7.1.1 Two- and Three-parameter Model

As mentioned in Chapter 6, for a reaction rate constant of the Arrhenius type we can write (Equation 6.18):

$$\frac{k}{k_s} = \exp\left(+\alpha\varepsilon \frac{1-\dfrac{C_A}{C_{A,s}}}{1+\alpha\left(1-\dfrac{C_A}{C_{A,s}}\right)} \right)$$

where k_s is the reaction rate constant at the surface temperature. The number ε was given by $\varepsilon = E_a/RT_s$. From Equations 6.14 and 6.15 it follows that α is given by

$$\alpha = \frac{T_m - T_s}{T_s} \tag{7.1}$$

with T_m the maximum temperature that can be obtained inside the catalyst pellet for $C_A = 0$. From Equation 6.18 we see that the conversion rate can be written as

$$R(C_A, T) = R(C_A, T_s)\exp\left(+\alpha\varepsilon \frac{1-\dfrac{C_A}{C_{A,s}}}{1+\alpha\left(1-\dfrac{C_A}{C_{A,s}}\right)} \right) \tag{7.2}$$

Hence, the effectiveness factor is a function of three parameters: α, ε and a Thiele modulus. Therefore we speak of a three-parameter model.

If the assumption is made that

$$\alpha\left(1-\frac{C_A}{C_{A,s}}\right) \ll 1 \tag{7.3}$$

then Equation 7.2 can be rewritten as

$$R(C_A,T)=R(C_A,T_s)\,e^{\varsigma\left(1-\frac{C_A}{C_{A,s}}\right)} \tag{7.4}$$

with

$$\zeta=\alpha\varepsilon=\frac{E_a}{RT_s}\frac{(-\Delta H)\,D_e C_{A,s}}{\lambda_P T_s} \tag{7.5}$$

The number ζ is the dimensionless ratio of the maximum increase in conversion rate due to intraparticle temperature gradients and the conversion rate at surface temperature. This can be seen as follows:

$$\begin{aligned}\zeta=\varepsilon\alpha&=\frac{E_a}{RT_s}\times\frac{T_m-T_s}{T_s}=\left(\frac{\partial R}{\partial T}\right)\bigg|_s\times\frac{T_s}{R_s}\times\frac{T_m-T_s}{T_s}\\&=\frac{1}{R_s}\times(T_m-T_s)\times\left(\frac{\partial R}{\partial T}\right)\bigg|_s\approx\frac{R_m-R_s}{R_s}\end{aligned} \tag{7.6}$$

Here, R_s is the conversion rate for the surface temperature and R_m is the conversion rate that would have been obtained for the maximum temperature that can occur inside the catalyst pellet.

Thus, the effectiveness factor is given by two parameters: ζ and a Thiele modulus. So, a two-parameter model is obtained if assumption 7.3 holds. The error we introduce by using a two- instead of three-parameter model is

$$\begin{aligned}\delta^{2-3}&=\frac{\exp\left\{\alpha\varepsilon\dfrac{1-\dfrac{C_A}{C_{A,s}}}{1+\alpha\left(1-\dfrac{C_A}{C_{A,s}}\right)}\right\}-\exp\left\{\alpha\varepsilon\left(1-\dfrac{C_A}{C_{A,s}}\right)\right\}}{\exp\left\{\alpha\varepsilon\dfrac{1-\dfrac{C_A}{C_{A,s}}}{1+\alpha\left(1-\dfrac{C_A}{C_{A,s}}\right)}\right\}}\\&=1-\exp\left\{\alpha^2\varepsilon\frac{\left(1-\dfrac{C_A}{C_{A,s}}\right)^2}{1+\alpha\left(1-\dfrac{C_A}{C_{A,s}}\right)}\right\}\end{aligned} \tag{7.7}$$

The maximum value for $\delta^{2\text{-}3}$ is obtained for C_A equal to zero, thus $\delta^{2\text{-}3}$ is largest for very low effectiveness factors and in the center of the catalyst pellet. Substituting $C_A = 0$ for a worst case analysis gives

$$\delta^{2-3}_{\max} = \exp\left(\frac{\alpha^2\varepsilon}{1+\alpha}\right) - 1 \tag{7.8}$$

In Table 7.1 values are given for α, ε and $\delta^{2\text{-}3}{}_{\max}$ for some typical industrial processes [1]. $\delta^{2\text{-}3}{}_{\max}$ is generally very small and lower than 10% for all cases. Hence we can use the two-parameter model without introducing a significant error.

Table 7.1 Value of the maximum relative errror $\delta^{2\text{-}3}_{\max}$ that will be made if the two-instead of the three-parameter model is used, for some typical processes and, hence, for some typical values ε and α (data from Hlavacek and Kubicek [1])

Process	ε	α	$\delta^{2\text{-}3}_{\max}$
Ethylene hydrogenation	13.2	2.9×10^{-3}	1×10^{-4}
Benzene hydrogenation	15.9	3.0×10^{-2}	1×10^{-2}
Methane oxidation	21.5	5.6×10^{-3}	7×10^{-4}
Ethylene oxidation	13.5	6.8×10^{-2}	$6.\times10^{-2}$
Naphtalene oxidation	22.2	1.5×10^{-2}	$5.\times10^{-3}$
Acrylonitrile synthesis	9.9	1.9×10^{-3}	3×10^{-5}

From now on the two-parameter model is used because it is almost as accurate as the three-parameter model and it gives a better insight. For example, the curves which were drawn by Weisz and Hicks [2] for different values of α and ε (Figure 6.4) reduce to one. This is illustrated in Figure 7.1 where the effectiveness factor is plotted versus An_0 for several values of ζ and for a first-order reaction occurring in a slab. Notice that all the curves in Figure 7.1 coincide in the low η region, since η is plotted versus An_0. The formulae used for An_0 now follow.

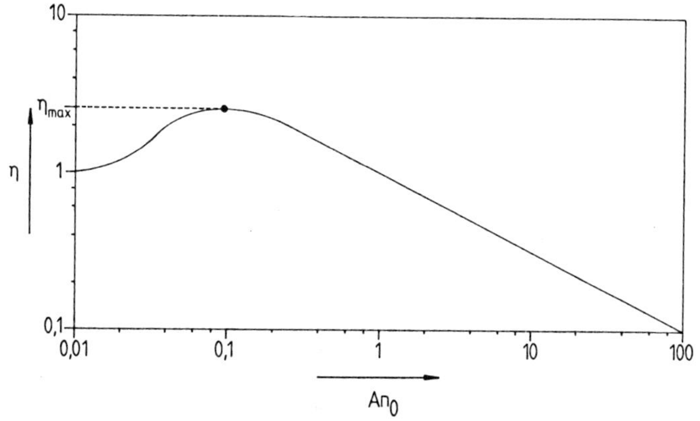

Figure 7.1 Schematic plot of the effectiveness factor η versus the zeroth Aris number An_0 for an exothermic nth-order reaction ($n > 0$ and $\zeta > n$). Notice the occurrence of a maximum for the effectiveness factor.

7.1.2 Aris Numbers

The temperature inside the catalyst can be related to the concentration of component A via:

$$\frac{T}{T_s} = 1 + \alpha\left(1 - \frac{C_A}{C_{A,s}}\right) \tag{7.9}$$

(Equations 6.14 and 6.15). Hence, the conversion rate, which will generally be a function of both the concentration C_A and the temperature T, can be written as a function of the concentration only:

$$R(C_A, T) = R\left(C_A, T_s\left\{1 + \alpha\left[1 - \frac{C_A}{C_{A,s}}\right]\right\}\right) \tag{7.10}$$

As elucidated in Appendices A and B the Aris numbers An_0 and An_1 can be calculated from Equation 6.29 and 6.38, provided the temperature dependency of the conversion rate is included in the formulae. This yields:

$$An_0 = \left(\frac{V_P}{A_P}\right)^2 \frac{\{R(C_{A,s}, T_s)\}^2}{2D_e} \left\langle \int_0^{C_{A,s}} R\left[C_A, T_s\left\{1 + \alpha\left(1 - \frac{C_A}{C_{A,s}}\right)\right\}\right] dC_A \right\rangle^{-1} \tag{7.11}$$

$$An_1 = \left(\frac{V_P}{A_P}\right)^2 \frac{\Gamma}{D_e} \left. \frac{\partial R\left[C_A, T_s\left\{1 + \alpha\left(1 - \frac{C_A}{C_{A,s}}\right)\right\}\right]}{\partial C_A} \right|_{C_A = C_{A,s}} \tag{7.12}$$

These formulae give the Aris numbers for arbitrary reaction kinetics and intraparticle temperature gradients. Thus the dependency of the conversion rate on the temperature can also be of any arbitrary form.

If it is assumed that the temperature dependency is of the Arrhenius type, the two-parameter model is taken, the Equations 7.11 and 7.12 simplify to

$$An_0 = \left(\frac{V_P}{A_P}\right)^2 \frac{\{R(C_{A,s}, T_s)\}^2}{2D_e} \left\{ \int_0^{C_{A,s}} R(C_A, T_S)\, e^{\zeta\left(1 - \frac{C_A}{C_{A,s}}\right)} dC_A \right\}^{-1} \tag{7.13}$$

$$An_1 = \left(\frac{V_P}{A_P}\right)^2 \frac{\Gamma}{D_e} \frac{\partial \left\{ R(C_A, T_s) e^{\zeta\left(1-\frac{C_A}{C_{A,s}}\right)} \right\}}{\partial C_A} \Bigg|_{C_A=C_{A,s}} =$$

$$= \left(\frac{V_P}{A_P}\right)^2 \frac{\Gamma}{D_e} \left\{ \frac{\partial R(C_A, T_s)}{\partial C_A}\Bigg|_{C_A=C_{A,s}} - \zeta \frac{R(C_{A,s}, T_s)}{C_{A,s}} \right\} \tag{7.14}$$

This is elucidated in Chapter 9 (Example 9.9).

The above suggests that the discussion of the Aris numbers for simple reactions also holds for nonisothermal pellets. For example, effectiveness factors larger than one are found if the number An_1 becomes negative. According to Equation 7.14 this is the case if

$$|\zeta| > \frac{C_{A,s}}{R(C_{A,s}, T_s)} \frac{\partial R(C_A, T_s)}{\partial C_A}\Bigg|_{C_A=C_{A,s}} \tag{7.15}$$

For nth-order kinetics Equation 7.15 yields

$$|\zeta| > n \tag{7.16}$$

Thus, if criterion 7.16 is fulfilled there will be a maximum value for the effectiveness factor, provided $n > -1$. Hence, when we increase the particle size, first the conversion will rise until a maximum conversion is reached. This is illustrated schematically in Figure 7.2. For particle sizes larger than this optimum particle size, the conversion decreases again with increasing particle size. Thus, if criterion 7.16 or more general criterion 7.15 is fulfilled, the highest conversion is not obtained for the smallest particle size, but for some optimum particle diameter instead.

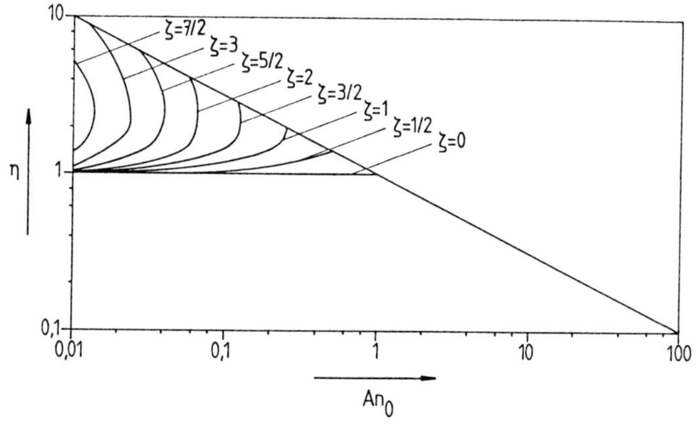

Figure 7.2 Effectiveness factor η versus the zeroth Aris number An_0 for an exothermic zeroth-order reaction in an infinite slab. Lines for several values of ζ ($\zeta \geq 0$) are drawn.

The above is illustrated in Example 9.10. In this specific example the heat effects are negligible. It would be handy to have criteria to determine beforehand whether or not heat effects are negligible. This is discussed further in Section 7.1.3.

It has been shown that approximation 6.59 cannot be used if the number An_1 is zero or negative. It was also stated that approximations can be constructed by comparing a given situation with a similar but simpler case. This is illustrated in Example 9.11.

7.1.3 Negligibility Criteria

Negligibility criteria are derived for both the high and low η region. For the high η region heat effects can be neglected if the number

$$\delta_{\zeta,1} = \left| \frac{\left(1-\eta|_{\zeta}\right)-\left(1-\eta|_{\zeta=0}\right)}{\left(1-\eta|_{\zeta=0}\right)} \right| \tag{7.17}$$

is very low. In this formula $\eta|_{\zeta}$ is the effectiveness factor with heat effects, $\eta|_{\zeta=0}$ is the effectiveness factor that would have been obtained under isothermal conditions, and $\delta_{\zeta,1}$ can be regarded as a relative error if heat effects are neglected. For effectiveness factors close to one,

$$1+\eta|_{\zeta} \approx 1+\eta|_{\zeta=0} \approx 2 \tag{7.18}$$

and we may write

$$\begin{aligned} \delta_{\zeta,1} &\approx \left| \frac{\left(1+\eta|_{\zeta}\right)\left(1-\eta|_{\zeta}\right)-\left(1+\eta|_{\zeta=0}\right)\left(1-\eta|_{\zeta=0}\right)}{\left(1+\eta|_{\zeta=0}\right)\left(1-\eta|_{\zeta=0}\right)} \right| = \\ &= \left| \frac{\left(1-\eta^2\right)_{\zeta}-\left(1-\eta^2\right)_{\zeta=0}}{\left(1-\eta^2\right)_{\zeta=0}} \right| = \left| \frac{An_1|_{\zeta}-An_1|_{\zeta=0}}{An_1|_{\zeta=0}} \right| \end{aligned} \tag{7.19}$$

Using n^{th}-order kinetics we may write (Equation 7-14):

$$An_1|_{\zeta} = \left(\frac{V_P}{A_P}\right)^2 \frac{\Gamma}{D_e} kC_{A,s}^{n-1}(n-\zeta) \tag{7.20}$$

Thus for $\zeta = 0$,

$$An_1|_{\zeta} = \left(\frac{V_P}{A_P}\right)^2 \frac{\Gamma}{D_e} kC_{A,s}^{n-1} \times n \tag{7.21}$$

Substitution of Equations 7.20 and 7.21 into 7.19 gives

$$\delta_{\zeta,1} = \left|\frac{\zeta}{n}\right| \tag{7.22}$$

For relative errors smaller than 10 % (thus $\delta_{\zeta,1} < {}^{1}/_{10}$), the following criterion is obtained:

$$|\zeta| < \frac{1}{10} \times |n| \tag{7.23}$$

In comparing with the criterion of Mears [3] (Equation 6.19):

$$|\zeta| < \frac{1}{20} \times n \tag{7.24}$$

These results are quite similar, although they are arrived at by entirely different reasoning.

With the aid of the number An_0 it is possible to derive criteria for isothermal operation in the low η region. For this region, the relative error introduced by assuming isothermal operation $\delta_{\zeta,0}$ is

$$\delta_{\zeta,0} = \frac{\left|\eta|_{\zeta} - \eta|_{\zeta=0}\right|}{\eta|_{\zeta=0}} \tag{7.25}$$

Since for small values of η, η equals $\dfrac{1}{\sqrt{An_0}}$, Equation 7.25 can be written as

$$\delta_{\zeta,0} = \frac{\left|\dfrac{1}{\sqrt{An_0|_{\zeta}}} - \dfrac{1}{\sqrt{An_0|_{\zeta=0}}}\right|}{\dfrac{1}{\sqrt{An_0|_{\zeta=0}}}} \tag{7.26}$$

From Equation 7.13 it then follows that the error $\delta_{\zeta,0}$ does not depend on the catalyst geometry or a Thiele modulus: like $\delta_{\zeta,1}$ it only depends on the value of ζ and the reaction kinetics. This is illustrated in Figure 7.3 where $\delta_{\zeta,0}$ is plotted versus ζ for zeroth-, first- and second-order kinetics and for exothermic reactions ($\zeta > 0$). The curves were obtained from the formulae given in Table 7.2, which were calculated with the aid of Equations 7.13 and 7.26.

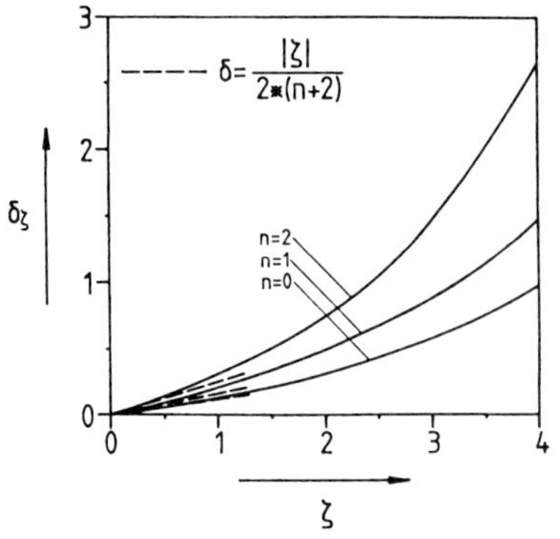

Figure 7.3 Relative error δ_ζ introduced by assuming isothermal operation versus ζ for nth-order kinetics in an infinite slab. The dotted lines are the approximation

$$\delta_\zeta = \frac{|\zeta|}{2(n+2)}$$

Table 7.2 Relative error δ_ζ by assuming isothermal operation as a function of ζ for nth-order kinetics in an infinite slab

n	δ_ζ
0	$\left|\sqrt{\frac{e^\zeta - 1}{\zeta}} - 1\right|$
1/2	$\left|\sqrt{\frac{1}{2/3 \times \zeta}\left(1/2 \times \sqrt{\frac{\pi}{\zeta}} e^\zeta \operatorname{erf}\sqrt{\zeta} - 1\right)} - 1\right|$
1	$\left|\sqrt{\frac{e^\zeta - 1 - \zeta}{1/2 \times \zeta^2}} - 1\right|$
3/2	$\left|\sqrt{\frac{1}{4/15 \times \zeta^2}\left(1/2 \times \sqrt{\frac{\pi}{\zeta}} e^\zeta \operatorname{erf}\sqrt{\zeta} - 1 - 2/3 \times \zeta\right)} - 1\right|$
2	$\left|\sqrt{\frac{e^\zeta - 1 - \zeta - 1/2 \times \zeta^2}{1/6 \times \zeta^3}} - 1\right|$

Typical values of ζ range between zero and one. In Table 7.3 values of ζ are given for some typical industrial processes. For small values of ζ, $\delta_{\zeta,0}$ can be approximated with

$$\delta_{\zeta,0} = \frac{|\zeta|}{n+2} \tag{7.27}$$

This approximation is given by the dotted lines in Figure 7.4.

A criterion for isothermal operation can be obtained by demanding that $\delta_{\zeta,0}$ is smaller than 10 %. This yields

$$|\zeta| < \frac{n+2}{5} \tag{7.28}$$

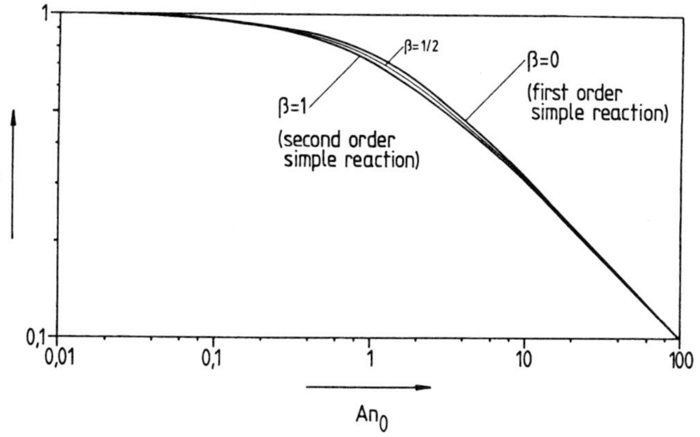

Figure 7.4 Effectiveness factor η versus zeroth Aris number An_0 for a bimolecular reaction with (1,1) kinetics occurring in an infinite slab, and for several values of β.

Table 7.3 Value of ζ for some typical processes and the relative errors $\delta_{\zeta,1}$ and $\delta_{\zeta,0}$ for those processes, where first-order kinetics is assumed

Process	ζ	$\delta_{\zeta,1}$	$\delta_{\zeta,0}$
Ethylene hydrogenation	0.038	0.04	6×10^{-4}
Benzene hydrogenation	0.48	0.48	8×10^{-2}
Methane oxidation	0.12	0.12	2×10^{-2}
Ethylene oxidation	0.92	0.92	1×10^{-2}
Naphtalene oxidation	0.33	0.33	6×10^{-2}
Acrylonitrile synthesis	0.019	0.02	3×10^{-3}

In Table 7-3 values of $\delta_{\zeta,1}$, Equation 7.22, and $\delta_{\zeta,0}$, Equation 7.27 are given for some typical processes, where first order kinetics is assumed. We see that heat effects can play an important role in the high η region, where we have to consider the value of $\delta_{\zeta,1}$. They are as good as negligible for all cases in the low h region, where $\delta_{\zeta,0}$ has to be considered. The high values of $\delta_{\zeta,1}$ implicate that reactor tube lengths can be reduced considerably or alternatively, much larger catalyst pellets can be used decreasing in this way the pressure drop and compression costs.

In Table 7.4 the criteria 7.23 and 7.28 and of Mears (Equation 7.24) are compared for several values of n. In order of increasing stringency we have Equation 7.23 < 7.28 < 7.24 of Mears. This is also illustrated in Table 7.5 where $\delta_{\zeta,1}$ and $\delta_{\zeta,0}$ are given for several values

Table 7.4 Comparison of the negligibility criteria for isothermal operation and nth-order kinetics

n	High η criterion	Low η criterion	Mears' criterion
1/2	$\lvert\zeta\rvert<0.05$	$\lvert\zeta\rvert<0.5$	$\lvert\zeta\rvert<0.025$
1	$\lvert\zeta\rvert<0.1$	$\lvert\zeta\rvert<0.6$	$\lvert\zeta\rvert<0.05$
2	$\lvert\zeta\rvert<0.2$	$\lvert\zeta\rvert<0.8$	$\lvert\zeta\rvert<0.1$

of n, if Mears' criterion is just fulfilled. It can be seen that applying Mears' criterion leads to only very small errors indeed. For most practical cases it is too restrictive (Example 9.12).

Table 7.5 Relative errors $\delta_{\xi,1}$ and $\delta_{\xi,0}$ for several reaction orders n, if Mears' criterion is just fulfilled

n	$\delta_{\xi,1}$	$\delta_{\xi,0}$
1/2	0.05	0.005
1	0.05	0.008
2	0.05	0.013

In summary it can be stated that intraparticle heat effects are more important in the high η region than in the low η region. In the low η region heat effects can almost always be neglected; in the high η region criterion 7.23 can be used to determine whether heat effects are negligible or not.

7.2 Bimolecular Reactions

Consider the bimolecular reaction

$$A + \nu_B B \rightarrow \nu_P P + \nu_Q Q$$

introduced as Equation 6.21. As already stated, all components participating in the reaction have the same effectiveness factor. Of interest are formulae which enable us to calculate the effectiveness factor and then derive neglibility criteria, whether or not the concentration of the nonkey reactant B inside the catalyst pores can be assumed to be uniform. In the latter case, mathematical manipulations can be simplified considerably.

7.2.1 Aris Numbers

The concentrations C_A and C_B inside the catalyst pellet for bimolecular reactions are interrelated in a similar way as the temperature T and C_A for monomolecular, nonisothermal reactions. A micro mass balance yields

$$D_{e,B}\left(C_{B,s} - C_B\right) = \nu_B D_{e,A}\left(C_{A,s} - C_A\right) \tag{7.29}$$

If we introduce the dimensionless number

$$\beta = \nu_B \times \frac{D_{E,A}}{D_{e,B}} \times \frac{C_{A,s}}{C_{B,s}} \tag{7.30}$$

then Equation 7.29 can be written as

$$\frac{C_B}{C_{B,s}} = 1 - \beta(1 - \frac{C_A}{C_{A,s}}) \tag{7.31}$$

The reactant not present in excess is chosen as component A, thus

$$\beta < 1 \tag{7.32}$$

Using Equation 7.31 the conversion rate, which generally is a function of both C_A and C_B, can be written as a function of C_A only:

$$R[C_A, C_B] = R\left[C_A, C_{B,s}\left\{1 - \beta\left(1 - \frac{C_A}{C_{A,s}}\right)\right\}\right] \tag{7.33}$$

Using the same line of reasoning as in Appendices A and B, it can be seen that the numbers An_0 and An_1 follow from:

$$\begin{aligned} An_0 &= \left(\frac{V_P}{A_P}\right)^2 \frac{\{R(C_{A,s}, C_{B,s})\}^2}{2D_{e,A}} \\ &\times \left(\int_0^{C_{A,s}} R\left[C_A, C_{B,s}\left\{1 - \beta\left(1 - \frac{C_A}{C_{A,s}}\right)\right\}\right] dC_A\right)^{-1} \end{aligned} \tag{7.34}$$

and

$$\begin{aligned} An_1 &= \left(\frac{V_P}{A_P}\right)^2 \frac{\Gamma}{D_{e,A}} \times \\ &\left.\frac{\partial R\left[C_A, C_{B,s}\left\{1 - \beta\left(1 - \frac{C_A}{C_{A,s}}\right)\right\}\right]}{\partial C_A}\right|_{C_A = C_{A,s}} \end{aligned} \tag{7.35}$$

This is illustrated in Figure 7.4 where the effectiveness factor is plotted versus the low η Aris number An_0 for a bimolecular reaction with (1,1) kinetics, and for several values of β. β lies between 0 and 1, calculations were made with a numerical method. Again all curves coincide in the low η region, because η is plotted versus An_0. For $\beta = 0$, the excess of component B is very large and the reaction becomes first order in component A. For $\beta = 1$, A and B match stoichiometrically and the reaction becomes pseudosecond order in component A (and B for that matter). Hence the η-An_0 graphs for simple first- and second-order reactions are the boundaries when varying β.

Comparison of Figures 7.5 and 7.1 shows that, in general, the effect of a reaction being bimolecular is far less dramatic than the effect of heat production inside a cat-

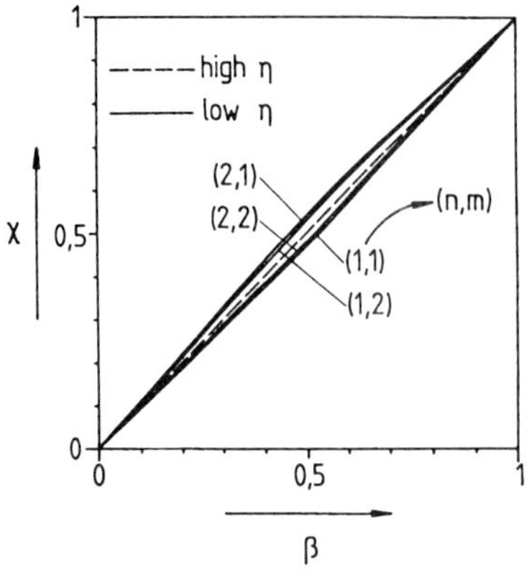

Figure 7.5. The number χ versus the number β for several forms of (n,m) kinetics: (-----) χ versus β for the high η region and all forms of (n,m) kinetics; (_____) χ versus β for the low η region and for several forms of (n,m) kinetics.

alyst pellet. All the conclusions drawn for simple reactions hold for bimolecular reactions as well. This is illustrated with Examples 9.13, 9.14 and 9.15. From these examples it becomes clear that for bimolecular reactions the effectiveness factor can be calculated in the same manner as for simple reactions. Since we can calculate the Aris numbers for both simple and bimolecular reactions, there is no basic difference. However, the formulae we found for bimolecular reactions are rather tedious and cumbersome (e.g. Equations 9.17 and 9.18). This can be overcome by using the approximation

$$\tilde{\eta}\Big|_{\beta} = \tilde{\eta}\Big|_{\beta=0} + \beta\left(\tilde{\eta}\Big|_{\beta=1} - \tilde{\eta}\Big|_{\beta=0}\right) \tag{7.36}$$

Since for $\beta = 0$ and $\beta = 1$ bimolecular reactions can be written as simple reactions, $\tilde{\eta}|_{\beta=0}$ and $\tilde{\eta}|_{\beta=1}$ can be calculated easily. Hence if β is known, $\tilde{\eta}|_{\beta}$ can also be calculated easily.

To investigate whether approximation 7.36 is valid, the number χ is defined as

$$\chi = \frac{\tilde{\eta}\Big|_{\beta} - \tilde{\eta}\Big|_{\beta=0}}{\tilde{\eta}\Big|_{\beta=1} - \tilde{\eta}\Big|_{\beta=0}} \tag{7.37}$$

In the high η region for η close to one, χ can be written as:

$$\begin{aligned}\chi &= \frac{\left(1-\tilde{\eta}\Big|_{\beta}\right)-\left(1-\tilde{\eta}\Big|_{\beta=0}\right)}{\left(1-\tilde{\eta}\Big|_{\beta=1}\right)-\left(1-\tilde{\eta}\Big|_{\beta=0}\right)} \\ &\approx \frac{\left(1+\tilde{\eta}\Big|_{\beta}\right)\left(1-\tilde{\eta}\Big|_{\beta}\right)-\left(1+\tilde{\eta}\Big|_{\beta=0}\right)\left(1-\tilde{\eta}\Big|_{\beta=0}\right)}{\left(1+\tilde{\eta}\Big|_{\beta=1}\right)\left(1-\tilde{\eta}\Big|_{\beta=1}\right)-\left(1+\tilde{\eta}\Big|_{\beta=0}\right)\left(1-\tilde{\eta}\Big|_{\beta=0}\right)} \\ &= \frac{An_1\Big|_{\beta}-An_1\Big|_{\beta=0}}{An_1\Big|_{\beta=1}-An_1\Big|_{\beta=0}}\end{aligned} \tag{7.38}$$

In the low η region χ is given by

$$\chi = \frac{\dfrac{1}{\sqrt{An_0|_\beta}} - \dfrac{1}{\sqrt{An_0|_{\beta=0}}}}{\dfrac{1}{\sqrt{An_0|_{\beta=1}}} - \dfrac{1}{\sqrt{An_0|_{\beta=0}}}} \tag{7.39}$$

For (n,m) kinetics

$$R(C_A, C_B) = kC_A{}^n C_B{}^m \tag{7.40}$$

This can be written as:

$$R\left(C_A, C_{B,s}\left\{1-\beta\left(1-\frac{C_A}{C_{A,s}}\right)\right\}\right) = kC_{B,s}{}^m \times C_A{}^n \left\{1-\beta\left(1-\frac{C_A}{C_{A,s}}\right)\right\}^m \tag{7.41}$$

Substitution of Equation 7.41 into 7.35 yields the following formula for the high η Aris number:

$$An_1 = \left(\frac{V_P}{A_P}\right)^2 \frac{\Gamma}{D_{e,A}} kC_{A,s}{}^{n-1} C_{B,s}{}^m \times (n + \beta m) \tag{7.42}$$

Substitution into Equation 7.38 yields

$$\chi = \beta \tag{7.43}$$

Hence in the high η region the approximation 7.36 is exact.

For the low η region the expressions for χ are not as simple as Equation 7.43. This is illustrated in Table 7.6 where χ is given for several forms of (n,m) kinetics. The for-

Table 7.6 Number χ as a function of the number β for several forms of (n,m) kinetics

m / n	1	2
0	$\chi = \dfrac{1-\sqrt{1-\frac{\beta}{2}}}{1-\sqrt{\frac{1}{2}}}$	$\chi = \dfrac{1-\sqrt{1-\beta+\frac{\beta^2}{3}}}{1-\sqrt{\frac{1}{3}}}$
1	$\chi = \dfrac{1-\sqrt{1-\frac{\beta}{3}}}{1-\sqrt{\frac{2}{3}}}$	$\chi = \dfrac{1-\sqrt{1-\frac{2\beta}{3}+\frac{\beta^2}{6}}}{1-\sqrt{\frac{1}{2}}}$

Table 7.6 Number χ as a function of the number β for several forms of (n,m) kinetics (Continued)

m / n	1	2
2	$\chi = \dfrac{1-\sqrt{1-\dfrac{\beta}{4}}}{1-\sqrt{\dfrac{3}{4}}}$	$\chi = \dfrac{1-\sqrt{1-\dfrac{2\beta}{3}+\dfrac{\beta^2}{6}}}{1-\sqrt{\dfrac{1}{2}}}$
n	$\chi = \dfrac{1-\sqrt{1-\dfrac{\beta}{n+2}}}{1-\sqrt{\dfrac{n+1}{n+2}}}$	$\chi = \dfrac{1-\sqrt{1-\dfrac{2\beta}{(n+2)}+\dfrac{2\beta^2}{(n+2)(n+3)}}}{1-\sqrt{\dfrac{n+1}{n+3}}}$

mulae were obtained from Equation 7.39. On their basis Figure 7.5 was constructed, where χ is plotted versus β for several values of n and m. The dotted line is approximation 7.36. If we consider here the high η region, where $\chi = \beta$ (see Eq. 7-43), the dotted line also represents the value of χ. It can be seen that generally χ does not differ very much from β. Even in the low h region, where the deviations are the largest, they do not become higher than 10 %. The relative error in η», with respect to η˜, will always be smaller than the relative deviation between χ and β, as can be seen as follows:

$$
\begin{aligned}
\frac{\left|\tilde{\tilde{\eta}}\big|_{\beta}-\tilde{\eta}\big|_{\beta}\right|}{\tilde{\eta}\big|_{\beta}} &= \frac{\left|\tilde{\eta}\big|_{\beta}-\tilde{\eta}\big|_{\beta=0}\right|}{\tilde{\eta}\big|_{\beta}} \\
&\times\left|\frac{\tilde{\tilde{\eta}}\big|_{\beta}-\tilde{\eta}\big|_{\beta=0}}{\tilde{\eta}\big|_{\beta=1}-\tilde{\eta}\big|_{\beta=0}}-\frac{\tilde{\eta}\big|_{\beta}-\tilde{\eta}\big|_{\beta=0}}{\tilde{\eta}\big|_{\beta=1}-\tilde{\eta}\big|_{\beta=0}}\right| \\
&\times\left|\frac{\tilde{\eta}\big|_{\beta}-\tilde{\eta}\big|_{\beta=0}}{\tilde{\eta}\big|_{\beta=1}-\tilde{\eta}\big|_{\beta=0}}\right|^{-1} = \frac{\left|\tilde{\eta}\big|_{\beta}-\tilde{\eta}\big|_{\beta=0}\right|}{\tilde{\eta}\big|_{\beta}} \\
&\times\left|\frac{\beta-\chi}{\chi}\right| \approx \frac{\left|\tilde{\tilde{\eta}}\big|_{\beta}-\tilde{\eta}\big|_{\beta=0}\right|}{\tilde{\tilde{\eta}}\big|_{\beta}}\times\left|\frac{\beta-\chi}{\chi}\right|
\end{aligned}
\tag{7.44}
$$

This implies the relative error in η^{χ} will always be much smaller than 10 % and hence the use of approximation 7.36 is reasonable. This is illustrated in Example 9.16.

Now that it is clear how to calculate effectiveness factors for bimolecular reactions, criteria are derived to tell whether the reaction is monomolecular or not.

7.2.2 Negligibility Criteria

Substitution of $C_{B,s}$ for C_B in a kinetic expression is acceptable if, for the high η region

$$\delta_{\beta,1} = \left| \frac{\eta|_\beta - \eta|_{\beta=0}}{1 - \eta|_{\beta=0}} \right| \tag{7.45}$$

is small enough. Similarly, for the low η region:

$$\delta_{\beta,0} = \frac{\left| \eta|_\beta - \eta|_{\beta=0} \right|}{\eta|_{\beta=0}} \tag{7.46}$$

should be small enough. Since approximation 7.36 can be used to estimate $\eta|_\beta$, Equations 7.45 and 7.46 can be written as

$$\delta_{\beta,1} \approx \left| \frac{\tilde{\eta}|_\beta - \tilde{\eta}|_{\beta=0}}{\tilde{\eta}|_{\beta=1} - \tilde{\eta}|_{\beta=0}} \right| \times \left| \frac{\tilde{\eta}|_{\beta=1} - \tilde{\eta}|_{\beta=0}}{1 - \tilde{\eta}|_{\beta=0}} \right| \approx \beta \times \left| \frac{\tilde{\eta}|_{\beta=1} - \tilde{\eta}|_{\beta=0}}{1 - \tilde{\eta}|_{\beta=0}} \right| \tag{7.47}$$

$$\delta_{\beta,0} \approx \left| \frac{\tilde{\eta}|_\beta - \tilde{\eta}|_{\beta=0}}{\tilde{\eta}|_{\beta=1} - \tilde{\eta}|_{\beta=0}} \right| \times \frac{\left| \tilde{\eta}|_{\beta=1} - \tilde{\eta}|_{\beta=0} \right|}{\tilde{\eta}|_{\beta=0}} \approx \beta \times \frac{\left| \tilde{\eta}|_{\beta=1} - \tilde{\eta}|_{\beta=0} \right|}{\tilde{\eta}|_{\beta=0}} \tag{7.48}$$

Substitution of the Aris numbers An_1 and An_0 into Equations 7.47 and 7.48 then yields

$$\delta_{\beta,1} = \beta \times \left| \frac{An_1|_{\beta=1} - An_1|_{\beta=0}}{An_1|_{\beta=0}} \right| \tag{7.49}$$

$$\delta_{\beta,0} = \beta \times \frac{\left| \dfrac{1}{\sqrt{An_0|_{\beta=1}}} - \dfrac{1}{\sqrt{An_0|_{\beta=0}}} \right|}{\dfrac{1}{\sqrt{An_0|_{\beta=0}}}} \tag{7.50}$$

Assuming that substitution of $C_{B,s}$ for C_B is justifiable, provided that a relative error smaller than 10 % is introduced, the following criteria are obtained:

- for the high η region

$$\beta < \frac{1}{10}\left|\frac{An_1\big|_{\beta=0}}{An_1\big|_{\beta=1} - An_1\big|_{\beta=0}}\right| \tag{7.51}$$

- for the low η region

$$\beta < \frac{1}{10}\frac{\dfrac{1}{\sqrt{An_0\big|_{\beta=0}}}}{\left|\dfrac{1}{\sqrt{An_0\big|_{\beta=1}}} - \dfrac{1}{\sqrt{An_0\big|_{\beta=0}}}\right|} \tag{7.52}$$

For the situation of (n,m) kinetics (Equation 7.40) the two criteria become:

$$\beta < \frac{1}{10}\left|\frac{n}{m}\right| \tag{7.53}$$

and

$$\beta < \frac{1}{10}\frac{\sqrt{n+m+1}}{\left|\sqrt{n+m+1} - \sqrt{n+1}\right|} \tag{7.54}$$

respectively. If we assume $n = m = 1$, the criteria yield $\beta < 0.1$ for the high η region and $\beta < 0.55$ for the low η region. Again, the criterion for the high η region is more stringent than that for the low η region.

Criteria for other forms of kinetics are easily derived using the general formulae 7.51 and 7.52. For example, for Langmuir-Hinshelwood kinetics as in Equation 9.16, for the high η region the criterion becomes

$$\beta < \frac{1}{10}\left|\frac{1 + K_A C_{A,s} - K_B C_{B,s}}{1 - K_A C_{A,s} + K_B C_{B,s}}\right| \tag{7.55}$$

and, for the low η region,

$$\beta < \frac{1}{10} \times \left|\frac{\mu_1}{\mu_0} \times \frac{\mu_0 - 1}{\mu_1 - 1} \times \sqrt{\frac{\dfrac{1}{\mu_1} + 1 - \dfrac{2\ln\mu_1}{\mu_1 - 1}}{\dfrac{1}{\mu_0} - 1 + \ln\mu_0}} - 1\right|^{-1} \tag{7.56}$$

where

$$\mu_0 = \frac{1 + K_A C_{A,s} + K_B C_{B,s}}{1 + K_B C_{B,s}} \tag{7.57}$$

$$\mu_1 = 1 + K_A C_{A,s} + K_B C_{B,s} \tag{7.58}$$

The criteria 7.55 and 7.56 are illustrated in Example 9.17.

7.2.3 Multimolecular Reactions

In summary, it can be said that calculating the effectiveness factor for bimolecular reactions does not pose any new problems as compared to simple reactions. Criteria for isothermal operation and criteria which determine whether or not $C_{B,s}$ may be substituted for C_B can easily be determined. From the last criteria it can be seen that the conclusion of Doraiswamy and Sharma [4], that $C_{B,s}$ can be substituted for C_B if the key reactant is limiting, is not always true.

Also worthy of note is that if the conversion rate depends on the concentrations of more than two components, the Aris numbers can be calculated using the same approach as outlined above. For example, if, for the reaction

$$A + \nu_B B \rightarrow \nu_P P + \nu_Q Q$$

C_P occurs in the conversion rate expression, (e.g. because it acts as an inhibitor on the catalyst surface) the conversion rate can be written as

$$R(C_A, C_B, C_P) = R\left[C_A, C_{B,s}\left\{1 - \beta_B\left(1 - \frac{C_A}{C_{A,s}}\right)\right\}, C_{P,s}\left\{1 + \beta_P\left(1 - \frac{C_A}{C_{A,s}}\right)\right\}\right] \tag{7.59}$$

where

$$\beta_B = \nu_B \frac{D_{e,A}}{D_{e,B}} \frac{C_{A,s}}{C_{B,s}} \tag{7.60}$$

$$\beta_P = \nu_P \frac{D_{e,A}}{D_{e,P}} \frac{C_{A,s}}{C_{P,s}} \tag{7.61}$$

Hence An_1 and An_0 can be calculated as

$$An_1 = \left(\frac{V_P}{A_P}\right)^2 \frac{\Gamma}{D_{e,A}} \times \left. \frac{\partial R\left[C_A, C_{B,s}\left\{1-\beta_B\left(1-\frac{C_A}{C_{A,s}}\right)\right\}, C_{P,s}\left\{1+\beta_P\left(1-\frac{C_A}{C_{A,s}}\right)\right\}\right]}{\partial C_A}\right|_{C_A=C_{A,s}} \tag{7.62}$$

$$An_0 = \left(\frac{V_P}{A_P}\right)^2 \frac{\{R(C_{A,s}, C_{B,s}, C_{P,s})\}^2}{2D_{e,A}} \times \left(\int_0^{C_{A,s}} R\left[C_A, C_{B,s}\left\{1-\beta_B\left(1-\frac{C_A}{C_{A,s}}\right)\right\}, C_{P,s}\left\{1+\beta_P\left(1-\frac{C_A}{C_{A,s}}\right)\right\}\right] dC_A\right)^{-1} \tag{7.63}$$

Where it is assumed that A again is the component not in excess, thus:

$$\beta_B < 1 \tag{7.64}$$

If component Q is also important for the conversion rate, the concentration C_Q can be introduced in the formulae for An_1 and An_0 in a similar way as C_P. Hence, it is concluded that calculating the effectiveness factor for multimolecular reactions is basically not very different than for simple reactions (if the proper Aris numbers are used).

7.3 Intraparticle Pressure Gradients

The gas composition may vary considerably across a catalyst pellet. If the concentration of the reactants is high, this may result in two effects:

- The effective diffusion coefficients, which up till now have been assumed constant, become a function of the concentration and, therefore, due to the internal concentration gradients, the effective diffusion coefficients will depend on the distance inside the catalyst pellet.
- Because of volume changes due to the reaction, pressure gradients may occur inside the catalyst pellet. This can give rise to two effects. First, it influences the effective diffusion coefficients, since the gas-phase diffusion coefficients depend on pressure. Second, the pressure gradients affect the concentrations (or more accurately, chemical activities), which determine the reaction rate. Hence pressure gradients must directly influence the effectiveness factor.

Using the dusty gas model [5] analytical solutions are derived to describe the internal pressure gradients and the dependence of the effective diffusion coefficient on the gas composition. Use of the binary flow model (BFM, Chapter 3) would also have yielded almost similar results to those discussed below. After discussion of the dusty gas model, results are then implemented in the Aris numbers. Finally, negligibility criteria are derived, this time for intraparticle pressure gradients. Calculations are given in appendices; here we focus on the results.

7.3.1 Pressure Gradients

If the catalyst pores are very broad, i.e. the Knudsen number is very low and we operate in the so-called 'continuum region', no pressure gradient builds up inside the catalyst pellet and viscous flow does not play a role (see Chapter 3). However, if the catalyst pores are very narrow, the Knudsen number is high and we operate in the 'free molecule region'. In this regime pressure gradients inside the catalyst pellet may become considerable. Furthermore, viscous flow is negligible, because it is overshadowed completely by Knudsen flow. The resistance against viscous flow is proportional to $1/r_p^2$ and Knudsen flow to $1/r_p$, hence for small values of the pore radius r_p the rate of Knudsen flow will be much higher than that of viscous flow.

Thus viscous flow can only be of any importance in the transition region between the continuum region and free molecule region, a feature also concluded by Mason and Malinauskas [5]. Results of Kehoe and Aris [6] and of Haynes [7] suggest that viscous flow can be neglected completely. By neglecting viscous flow, Haynes arrived at the same effectiveness factors as Kehoe and Aris who included viscous flow. This is confirmed in Appendix D, where criteria are derived to determine whether viscous flow can be neglected. There it is proved that these criteria are always fulfilled for all practical cases, so that viscous flow inside the catalyst pores is indeed always negligible.

For the reaction (Equation 6.28)

$$A \rightarrow \nu_P P$$

The gas phase consists of A, P and m components D_i ($i = 1, 2, \ldots, m$). The components D_i are inert (i.e., do not play a role in the reaction). With the aid of the dusty gas model the pressure inside the catalyst pellet can be expressed as a function of the concentration inside the catalyst pellet. Since viscous flow can be neglected, this function becomes (Appendix D)

$$\begin{aligned}&\left(\sqrt{\nu_P}-1\right)\kappa_{A,s}\left(1-\frac{C_A}{C_{A,s}}\right)\\&=\frac{\sigma^2}{2Kn^*}+\sigma\left(1+\frac{1}{Kn^*}\right)+\left\{\left(\sqrt{\nu_P}-1\right)\kappa_{A,s}+1+\frac{1}{2Kn^*}\right\}\times\left\{1-\exp\left(-\frac{\sqrt{\nu_P}+1}{Kn^*}\sigma\right)\right\}\end{aligned} \tag{7.65}$$

where

$$\sigma = \frac{P - P_s}{P_s} \tag{7.66}$$

P is the pressure inside the catalyst pellet, P_s is the pressure on the outer surface, $\kappa_{A,s}$ the mole fraction of A on the outer surface and Kn^* is a modified Knudsen number, defined as

$$Kn^* = \frac{D_{AP}(P_s)}{D_{AK}} \cdot \left[1 + \sum_{i=1}^{m} \left\{ \kappa_{D_i}\Big|_{\kappa_A=0} \left(\frac{D_{AP}(P_s)}{D_{AD_i}(P_s)} - 1 \right) \right\} \right]^{-1} \tag{7.67}$$

In Equation 7.67 D_{AP} (P_s) and D_{AD_i} (P_s) are the Maxwell diffusion coefficients of A in a binary mixture respectively of P and D_i at surface pressure. Neither D_{AP} (P_s) nor D_{AD_i} (P_s) depend on the gas composition or pressure. D_{Ak} is the Knudsen diffusion coefficient of A in the catalyst pores, which is also independent of the gas composition and the pressure. The term $\kappa_{D_i}\Big|_{\kappa_A=0}$ is the mole fraction of D_i that would have been obtained if all of component A had been converted into product P. This is illustrated in Example 9.18.

From Equation 7.67 it follows that the modified Knudsen number Kn^* is not a function of the pressure. Once a certain feed composition is chosen, Kn^* does not depend on the gas composition either.

Consider again Equation 7.65 for the pressure inside the catalyst pellet as a function of the gas composition. Notice that for $\nu_P = 1$ or $\kappa_{A,s} = 0$ or $C_A = C_{A,s}$, which corresponds to no volume changes due to reaction, the number σ equals zero and thus the pressure inside the catalyst pellet equals the surface pressure. Also for very low Knudsen numbers, which corresponds to very broad catalyst pores, the number σ becomes equal to

$$Kn^* \downarrow 0 \quad \Rightarrow \quad \sigma \rightarrow \frac{2(Kn^*)^2}{\sqrt{\nu_P}+1} \left(\sqrt{\nu_P} - 1\right) \kappa_{A,s} \left(1 - \frac{C_A}{C_{A,s}}\right) \tag{7.68}$$

Therefore, in this case as well, the number σ equals zero and again the intraparticle pressure gradient vanishes. So, no pressure will build up inside the pellet either if there is no volume change due to reaction or if the catalyst pores are very broad.

The maximum pressure gradient will occur for very high Knudsen numbers, i.e., for very small catalyst pores. For this case it follows (from Equation 7.65)

$$Kn^* \rightarrow \infty \quad \Rightarrow \quad \sigma \rightarrow \left(\sqrt{\nu_P} - 1\right) \kappa_{A,s} \left(1 - \frac{C_A}{C_{A,s}}\right) \tag{7.69}$$

Hence the maximum pressure difference occurring inside a catalyst pellet will be obtained for $Kn^* \rightarrow \infty$ (small pores) and $C_A = 0$ (low effectiveness factors):

$$\sigma_{max} = \left(\sqrt{\nu_P} - 1\right) \kappa_{A,s} \tag{7.70}$$

This equation is illustrated in Example 9.19.

Equation 7.65 is intractable because of its complexity. Therefore, the following approximation for the relative pressure is used

$$\tilde{\sigma} = \frac{(Kn^*)^2}{(Kn^*)^2 + \{(\nu_P - 1)\kappa_{A,s} + \sqrt{\nu_P} + 2\}Kn^* + \frac{1}{2}(\sqrt{\nu_P}+1)} \qquad (7.71)$$

$$(\sqrt{\nu_P} - 1)\kappa_{A,s}\left(1 - \frac{C_A}{C_{A,s}}\right)$$

This is also derived in Appendix D. In this appendix it is further shown that approximation 7.71 is exact for $Kn^* \downarrow 0$, $Kn^* \rightarrow \infty$, $\nu_P = 1$, $\kappa_{A,s} = 0$ or $C_A = C_{A,s}$. For a given value of ν_P, the relative error

$$\Delta_\sigma = \left| \frac{\tilde{\sigma} - \sigma}{\sigma} \right| \qquad (7.72)$$

will have a maximum value for $\kappa_{A,s} = 1$, $C_A = 0$ and a certain value of $Kn^* \in [0,\infty[$. We denote this value $\Delta_{\sigma,max}$:

$$\Delta_{\sigma,\max} = \max_{Kn^* \in [0,\infty[} \Delta_\sigma \Big|_{\kappa_{A,s}=1 \ \ and \ \ C_A=0} \qquad (7.73)$$

In Figure 7.6 the maximum relative error $\Delta_{\sigma,max}$ is plotted versus ν_P. It can be seen that even in the worst case, Equation 7.71 can be used without introducing a relative error larger than 10 %, provided $\nu_P \in [0,4]$ (which will be adhered to in almost all cases). This means that Equation 7.71 can be used in practice without any problem.

Now that we have derived the intraparticle pressure gradients, we can also determine the effective diffusion coefficient as a function of the gas composition.

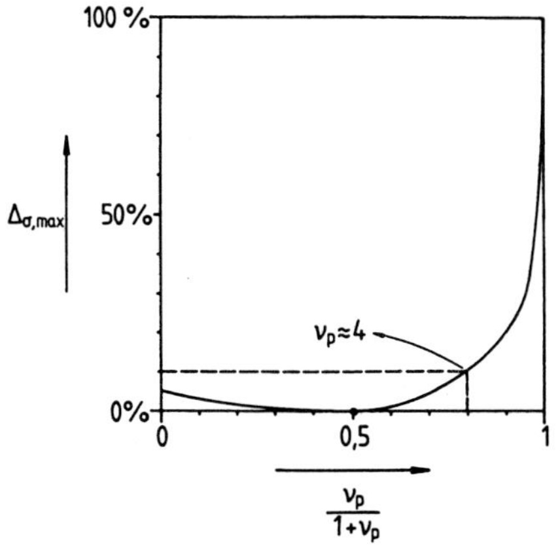

Figure 7.6 Maximum relative error $\Delta_{\sigma,max}$ versus the stoichiometric coefficient ν_P for simple reaction.

7.3.2 Effective Diffusion Coefficient

For the reaction

$$A \rightarrow \nu_P P$$

with the gas phase consisting of A, P and m inert components D_i ($i = 1, 2, \ldots, m$) the dusty-gas model yields (Appendix E)

$$D_{e,A} = \frac{\varepsilon_P}{\gamma_P} D_{Ak} \left(1 + \frac{1 + (\nu_P - 1)\kappa_A}{Kn^*} \times \frac{P}{P_s} \right)^{-1} \tag{7.74}$$

In this equation ε_p is the porosity of the catalyst pellet and γ_p the tortuosity of the catalyst pores as discussed in Chapter 3 (the rest of the symbols are as defined before). From this formula it follows that the effective diffusion coefficient depends on both the gas composition and the pressure. Since we know the pressure as a function of the concentration, Equation 7.74 provides the effective diffusion coefficient as a function of the concentration. If we define

$$\psi = \frac{(Kn^*)^2}{(Kn^*)^2 + \left\{(\nu_p - 1)\kappa_{A,s} + \sqrt{\nu_P} + 2\right\}Kn^* + \frac{1}{2}\left(\sqrt{\nu_P} + 1\right)} \tag{7.75}$$

then substitution of Equation 7.71 into 7.74 yields

$$D_{e,A} = \frac{\varepsilon_P}{\gamma_P} D_{Ak} \left(1 + \frac{1 + \psi\left(\sqrt{\nu_P} - 1\right)\kappa_{A,s}\left(1 - \frac{C_A}{C_{A,s}}\right) + (\nu_P - 1)\kappa_{A,s}\frac{C_A}{C_{A,s}}}{Kn^*} \right)^{-1} \tag{7.76}$$

It is convenient to write the last equation as

$$\frac{D_{e,A}\big|_{C_A}}{D_{e,A}\big|_{C_A=0}} = \left(1 + \gamma \frac{C_A}{C_{A,s}} \right)^{-1} \tag{7.77}$$

where

$$\gamma = \frac{\sqrt{\nu_P} + 1 - \psi}{1 + Kn^* + \psi\left(\sqrt{\nu_P} - 1\right)\kappa_{A,s}} \times \left(\sqrt{\nu_P} - 1\right)\kappa_{A,s} \tag{7.78}$$

The number γ is a measure for the dependence of the effective diffusion coefficient on the concentration. It can vary between minus one (for $\nu_P = 0$, $Kn^* = 0$ and $\kappa_{A,s} = 1$ at the same time) and infinite (for $\nu_P \to \infty$). For ν_P U 1 or small values of $\kappa_{A,s}$, γ will be close to zero and the concentration dependence vanishes.

For any given values of ν_P and $\kappa_{A,s}$, two regimes can be distinguished:

- For small values of the Knudsen number (roughly $Kn^* < 0.3$) we operate in the continuum region and γ reaches its maximum value:

$$\gamma\Big|_{Kn^* \downarrow 0} = (\nu_P - 1)\,\kappa_{A,s} \tag{7.79}$$

Then γ equals the so-called volume-change modulus, as introduced by Weekman and Gorring [8],

- As the value of Kn^* increases, the value of γ decreases and eventually the concentration dependence vanishes. For high enough values of the Knudsen number (roughly $Kn^* > 10$) we operate in the free molecule region and γ equals:

$$\gamma\Big|_{Kn^* \to \infty} = \frac{\sqrt{\nu_P}\left(\sqrt{\nu_P} - 1\right)}{Kn^*}\,\kappa_{A,s} \tag{7.80}$$

The physicochemical reason why two regimes occur is as follows. In the free molecule region, the effective diffusion coefficient is determined by collisions of the molecules with the catalyst pore walls. Hence, the gas composition will not influence the effective diffusion coefficient. However, in the continuum region mutual collisions between the molecules are determining and the dependence of the effective diffusion coefficient on the concentration becomes most pronounced.

7.3.3 Aris Numbers

If the concentration C_A influences the effective diffusion coefficient $D_{e,A}$, the Aris numbers can be calculated from

$$An_0 = \left(\frac{V_P}{A_P}\right)^2 \frac{\{R(C_{A,s})\}^2}{2} \left(\int_0^{C_{A,s}} D_{e,A}(C_A) R(C_A)\, dC_A\right)^{-1} \tag{7.81}$$

$$An_1 = \left(\frac{V_P}{A_P}\right)^2 \frac{\Gamma}{D_{e,A}(C_{A,s})} \left.\frac{\partial R(C_A)}{\partial C_A}\right|_{C_A = C_{A,s}} \tag{7.82}$$

Equation 7.81 can be obtained directly by using the same line of reasoning as in Appendix A. Equation 7.82 cannot be obtained that easily, but is proved for arbitrary reaction kinetics in an infinite slab in Appendix F. Substitution of Equation 7.77 into 7.81 and 7.82 yields

$$An_0 = \left(\frac{V_P}{A_P}\right)^2 \frac{\{R(C_{A,s})\}^2}{2D_{e,A}\big|_{C_A=0}} \left(\int_0^{C_{A,s}} \frac{R(C_A)}{1+\gamma\frac{C_A}{C_{A,s}}} dC_A \right)^{-1} \tag{7.83}$$

$$An_1 = (1+\gamma)\left(\frac{V_P}{A_P}\right)^2 \frac{\Gamma}{D_{e,A}\big|_{C_A=0}} \left.\frac{\partial R(C_A)}{\partial C_A}\right|_{C_A=C_{A,s}} \tag{7.84}$$

For first-order kinetics in a slab, from Equation 7.83 it follows that An_0 can be calculated from:

$$An_0 = R^2 \frac{k}{D_{e,A}\big|_{C_A=0}} \times \frac{\gamma^2}{2\{\gamma - \ln(1+\gamma)\}} \tag{7.85}$$

Similarly, An_1 follows from

$$An_1 = R^2 \frac{k}{D_{e,A}\big|_{C_A=0}} \times \frac{2}{3} \times (1+\gamma) \tag{7.86}$$

In Figure 7.7 the effectiveness factor η is plotted versus the low γ Aris number An_0 for several values of γ. We see that the effect of γ is rather small. This is a consequence of the use of An_0, which already corrects for the influence of γ.

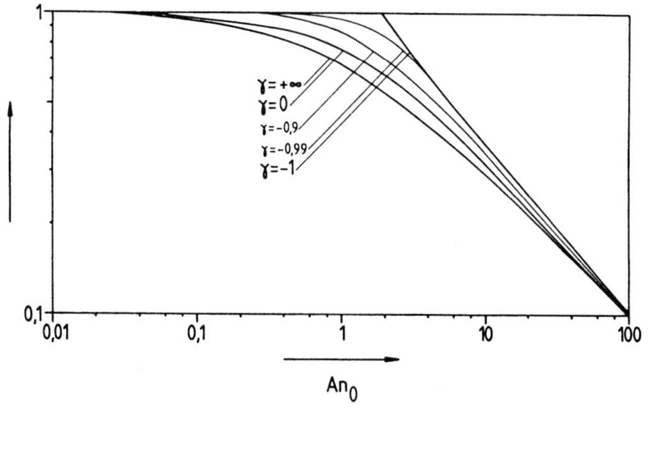

Figure 7.7 Effectiveness facto η versus zeroth Aris number An_0 for a simple, first-order reaction occurring in an infinite slab. The figure was drawn for nondiluted gases; lines for several values of γ are given.

7.3.4 Negligibility Criteria

To derive criteria which determine whether or not the effect of a gas being nondiluted can be neglected, we use the same approach as outlined before and introduce the following relative deviations:

- For the high η region

$$\delta_{\gamma,1} = \left| \frac{\eta|_{\gamma} - \eta|_{\gamma=0}}{1 - \eta|_{\gamma=0}} \right| \tag{7.87}$$

which can be written as

$$\delta_{\gamma,1} = \left| \frac{An_1|_{\gamma} - An_1|_{\gamma=0}}{An_1|_{\gamma=0}} \right| \tag{7.88}$$

Substitution of Equation 7.84 yields:

$$\delta_{\gamma,1} = |\gamma| \tag{7.89}$$

- For the low η region

$$\delta_{\gamma,0} = \frac{\left|\eta|_{\gamma} - \eta|_{\gamma=0}\right|}{\eta|_{\gamma=0}} \tag{7.90}$$

which can be written as

$$\delta_{\gamma,0} = \frac{\left| \dfrac{1}{\sqrt{An_0|_{\gamma}}} - \dfrac{1}{\sqrt{An_0|_{\gamma=0}}} \right|}{\dfrac{1}{\sqrt{An_0|_{\gamma=0}}}} \tag{7.91}$$

In this case δ_{γ} will depend on the value of γ and the reaction kinetics. For example, for first-order kinetics we find (see also Equation 7.85):

$$\delta_{\gamma,0} = \left| \frac{\sqrt{2\{\gamma - \ln(1+\gamma)\}}}{\gamma} - 1 \right| \tag{7.92}$$

Since we want to derive negligibility criteria, we can assume that the value of γ is small. Then for nth-order kinetics Equation 7.91 yields

$$\delta_{\gamma,0} = \left| \sqrt{1 - \frac{n+1}{n+2}\gamma} - 1 \right| \approx \frac{n+1}{2(n+2)} \times |\gamma| \tag{7.93}$$

The effect of the dependency of the diffusion coefficient on the concentration can be neglected if $\delta_{\gamma,1} < 10$ % and $\delta_{\gamma,0} < 10$ %, which yields, for the high η region,

$$|\gamma| < \frac{1}{10} \tag{7.94}$$

for the low η region,

$$|\gamma| < \frac{n+2}{5(n+1)} \tag{7.95}$$

Again, the criterion for the high η region is more stringent than the that for the low η region. With the generalized formulae, criteria for other types of kinetics can easily be obtained.

7.3.5 Bimolecular Reactions

The formulae derived for simple reactions can be extended to bimolecular reactions, although some of the equations will become more complex. We will only present the results and not give any calculations, since they are similar to those for simple reactions. The basis is again the dusty gas model, neglecting viscous flow. Consider the reaction

$$A + \nu_B B \rightarrow \nu_P P + \nu_Q Q$$

with the gas phase consisting of A, B, P, Q and m inert components D_i ($i = 1, 2, \ldots, m$). With the aid of the dusty gas model it can be shown that the internal pressure gradient follows from:

$$\begin{aligned}\Psi\kappa_{A,s}\left(1 - \frac{C_A}{C_{A,s}}\right) &= \frac{\sigma^2}{2Kn_A^*} + \sigma\left(1 + \frac{1}{Kn_A^*}\right) \\ &+ \left(1 + \Psi\kappa_{A,s} + \frac{1}{2Kn_A^*}\right) \times \left(1 - \exp\left\{-\frac{\Delta\nu}{\Psi Kn_A^*}\sigma\right\}\right)\end{aligned} \tag{7.96}$$

where σ is again the dimensionless pressure difference

$$\sigma = \frac{P - P_s}{P_s} \tag{7.66}$$

And Kn_A* a modified Knudsen number, now for bimolecular reactions:

$$Kn_A^* = \frac{1+\kappa_{A,s}\Delta\nu}{D_{Ak}} \times \left(\frac{\kappa_{B,s} - \nu_B \kappa_{A,s}}{D_{AB}(P_s)} + \frac{\kappa_{P,s} + \nu_P \kappa_{A,s}}{D_{AP}(P_s)} + \frac{\kappa_{Q,s} + \nu_Q \kappa_{A,s}}{D_{AQ}(P_s)} + \sum_{i=i}^{m} \frac{\kappa_{D_i,s}}{D_{AD_i}(P_s)} \right)^{-1}$$

(7.97)

Kn_A* does not depend on the pressure and the gas composition inside the catalyst pellet. Furthermore the following symbols are used in Equation 7.97:

$$\Delta\nu = \nu_P + \nu_Q - \nu_B - 1 \tag{7.98}$$

$$\Psi = \nu_P \sqrt{\frac{M_P}{M_A}} + \nu_Q \sqrt{\frac{M_Q}{M_A}} - \nu_B \sqrt{\frac{M_B}{M_A}} - 1 \tag{7.99}$$

Notice the similarity between Equations 7.65 and 7.96. For $\nu_B = 0$ the modified Knudsen number for bimolecular reactions Kn_A* simplifies to the modified Knudsen number for simple reactions Kn^*. Also, for $\nu_B = 0$ and $\nu_Q = 0$, Equation 7.96 merges into 7.65.

Again the maximum pressure difference inside a catalyst pellet will be obtained for very small pore radii (Kn_A*$\rightarrow \infty$) and very low effectiveness factors ($C_A \downarrow 0$):

$$\sigma_{max} = \Psi \kappa_{A,s} \tag{7.100}$$

Which is comparable to Equation 7-70.

Similar to Equation 7.65, 7-96 can be approximated by a simplified equation which, for bimolecular reactions, becomes

$$\sigma = \psi_\beta \Psi \kappa_{A,s} \left(1 - \frac{C_A}{C_{A,s}} \right) \tag{7.101}$$

with

$$\psi_\beta = \frac{\left(Kn_A^*\right)^2}{\left(Kn_A^*\right)^2 + \left(\kappa_{A,s}\Delta\nu + 1 + \frac{\Delta\nu}{\Psi} \right) Kn_A^* + \frac{\Delta\nu}{2\Psi}} \tag{7.102}$$

This approximation is similar to Equation 7.71.

In case of nondiluted gas, the effective diffusion coefficient again becomes dependent on the concentration C_A, as for simple reactions. We can derive

$$D_{e,A} = \frac{\varepsilon_P}{\gamma_P} D_{Ak} \left(1 + \frac{1+\kappa_A \Delta\nu}{Kn_A^*} \times \frac{P}{P_s} \right)^{-1} \tag{7.103}$$

Which is comparable to Equation 7.74. With the aid of approximation 7.101, Equation 7.103 is conveniently written as

$$\frac{D_{e,A}\big|_{C_A}}{D_{e,A}\big|_{C_A=0}} = \left(1+\gamma_\beta \frac{C_A}{C_{A,s}}\right)^{-1} \tag{7.104}$$

with

$$\gamma_\beta = \frac{\Delta\nu - \psi_\beta \Psi}{1+Kn_A^* + \psi_\beta \Psi \kappa_{A,s}} \times \kappa_{A,s} \tag{7.105}$$

which corresponds to Equations 7.77 and 7.78 for simple reactions.

The maximum value for γ_β will be obtained for very broad pore radii ($Kn_A^* \downarrow 0$, and, hence, according to Equation 7.102 also $\Psi_\beta \downarrow 0$) and equals

$$\gamma_\beta\big|_{Kn_A^* \downarrow 0} = \Delta\nu\ \kappa_{A,s} \tag{7.106}$$

Now that the effect of a gas being nondiluted is quantified, relationships for the Aris numbers can be derived. For the Aris numbers for diluted bimolecular reactions Equations 7.30 and 7.31 were used to relate the concentrations C_A and C_B inside the catalyst pellet. For nondiluted bimolecular reactions, two facts must be understood:

- The effective diffusion coefficient will be a function of the gas composition. This can be incorporated as was done for simple reactions (Equations 7.81 and 7.82).
- The number β will be a function of the gas composition, since both $D_{e,A}$ and $D_{e,B}$ depend on the composition. This means that Equation 7.31 cannot be used to express C_B in C_A.

For the second fact, and with the aid of the dusty gas model, we now calculate:

$$\beta = \nu_B \frac{D_{e,A}(C_A)C_{A,s}}{D_{e,B}(C_A)C_{B,s}} = \nu_B \frac{D_{Ak}\, C_{A,s}\, Kn_A^*}{D_{Bk}\, C_{B,s}\, Kn_B^*} \times \left\{1+\frac{Kn_B^* - Kn_A^*}{\left(1+Kn_A^* + \psi_\beta \Psi \kappa_{A,s}\right)\left(1+\gamma_\beta \frac{C_A}{C_{A,s}}\right)}\right\} \tag{7.107}$$

where Kn_B^* is the modified Knudsen number for component B, which follows from

$$Kn_B^* = \frac{1+\kappa_{A,s}\Delta\nu}{D_{Bk}} \times \left(\frac{\kappa_{A,s} - \frac{\kappa_{B,s}}{\nu_B}}{D_{AB}(P_s)} + \frac{\kappa_{P,s} + \nu_P \kappa_{B,s}}{D_{BP}(P_s)} + \frac{\kappa_{Q,s} + \frac{\nu_Q}{\nu_B}\kappa_{B,s}}{D_{BQ}(P_s)} + \sum_{i=1}^{m} \frac{\kappa_{D_i,s}}{D_{BD_i}(P_s)}\right)^{-1} \tag{7.108}$$

D_{Bk} is the Knudsen diffusion coefficient of B inside the catalyst pores, $D_{BP}(P_s)$ is the Maxwell diffusion coefficient of B in a binary mixture with P, the other diffusion coefficients are defined in a similar way. All other symbols are as defined before. Notice the following:

- If $\gamma_\beta = 0, \beta$ does not depend on the gas composition, since the effective diffusion coefficients of A and B do not depend on the gas composition,
- If $Kn_A{}^* = Kn_B{}^*$, β will not depend on the gas composition either. For this case the effective diffusion coefficients do depend on the gas composition, but the ratio $D_{e,A}(C_A)/D_{e,B}(C_A)$ does not.

Now that we have a relationship for the dependence of β on the gas composition, we can express the concentration C_B in C_A. Incorporating the dependence of β in a micro mass balance, instead of Equation 7.31 the following equation is obtained

$$C_B = C_{B,s}\left[1-\beta^*\left\{1-\frac{C_A}{C_{A,s}}+\Delta Kn^*\ln\left(\frac{1+\gamma_\beta}{1+\gamma_\beta\dfrac{C_A}{C_{A,s}}}\right)\right\}\right] \tag{7.109}$$

with

$$\beta^* = \nu_B\frac{D_{Ak}}{D_{Bk}}\frac{C_{A,s}}{C_{B,s}}\frac{Kn_A^*}{Kn_B^*} = \nu_B\sqrt{\frac{M_B}{M_A}}\frac{C_{A,s}}{C_{B,s}}\frac{Kn_A^*}{Kn_B^*} \tag{7.110}$$

and

$$\Delta Kn^* = \frac{Kn_A^* - Kn_B^*}{\left(\Delta\nu - \psi_\beta\Psi\right)\kappa_{A,s}} \tag{7.111}$$

Using the same line of reasoning as in Appendices A and B, it can be seen that for nondiluted bimolecular reactions the Aris numbers follow from

$$An_0 = \left(\frac{V_P}{A_P}\right)^2\frac{\left\{R\left(C_{A,s},C_{B,s}\right)\right\}^2}{2D_{e,A}\big|_{C_A=0}} \times\left(\int_0^{C_{A,s}}\frac{1}{1+\gamma_\beta\dfrac{C_A}{C_{A,s}}}\times R\left\langle C_A, C_{B,s}\left[1-\beta^*\left(1-\frac{C_A}{C_{A,s}}+\Delta Kn^*\ln\left\{\frac{1+\gamma_\beta}{1+\gamma_\beta\dfrac{C_A}{C_{A,s}}}\right\}\right)\right]\right\rangle dC_A\right) \tag{7.112}$$

$$An_1 = (1+\gamma_\beta)\left(\frac{V_P}{A_P}\right)^2 \frac{\Gamma}{D_{e,A}\big|_{C_A=0}} \times$$

$$\left.\frac{\partial R\left\langle C_A, C_{B,s}\left[1-\beta^*\left(1-\frac{C_A}{C_{A,s}}+\Delta Kn^* \ln\left\{\frac{1+\gamma_\beta}{1+\gamma_\beta \frac{C_A}{C_{A,s}}}\right\}\right)\right]\right\rangle}{\partial C_A}\right|_{C_A=C_{A,s}} \tag{7.113}$$

These formulae, together with the generalized approximation 6.59, enable the effectiveness factor for nondiluted bimolecular reactions to be calculated.

With the Equations 7.112 and 7.113 negligibility criteria can be derived in a similar way as before. Here only the results for (1,1) kinetics are given (Equation 7.40). A gas can be regarded as being diluted if the criterion

$$|\gamma_\beta| < \frac{1}{10} \tag{7.114}$$

is adhered to for the high η region. And, for the low η region,

$$|\gamma_\beta| < \frac{2(3-\beta^{**})}{5(4-\beta^{**})} \tag{7.115}$$

where

$$\beta^{**} = \beta^*(1+\gamma_\beta \Delta Kn^*) = \beta^* \frac{1+Kn_B^* - \psi_\beta \Psi \kappa_{A,s}}{1+Kn_A^* - \psi_\beta \Psi \kappa_{A,s}} \tag{7.116}$$

Since β^{**} will usually be small (between zero and one) criterion 7.115 also can be written as

$$|\gamma_\beta| < \frac{3}{10} \tag{7.117}$$

Again, the criterion for the high η region is more stringent than that for the low η region. With the generalized Equations 7.112 and 7.113, criteria for other forms of kinetics can also be obtained.

Example 9.21 illustrates the above criteria to determine whether gases can be regarded diluted. In this specific example this is in fact found to be a valid assumption. More generally, it can be stated that the dependence of the effective diffusion coefficient on the concentration will usually be of minor importance, if significant at all.

7.4 Anisotropic Catalyst Pellets

Cylindrical catalyst pellets, whether hollow or not, are often given shape by extrusion of a paste of wet catalyst (Chapter 3). During this process the diameter of pores in the radial direction will become smaller, whereas for pores in the longitudinal direction the length will decrease. This can result in a radial effective diffusivity being smaller than the longitudinal one. Thus cylindrical catalyst pellets can be anistropic.

This anisotropy can be accounted for in the Aris numbers. Considering a ring-shaped catalyst pellet, the material balance on a micro scale, for simple reactions, reads as (Appendix C)

$$\frac{D_{e,A}^{R}}{r}\frac{\partial}{\partial r}\left(r\frac{\partial C_A(r,h)}{\partial r}\right)+D_{e,A}^{H}\frac{\partial^2 C_A(r,h)}{\partial h^2}=R(C_A) \tag{7.118}$$

By substituting

$$r^{+}=r\sqrt{\frac{D_{e,A}^{H}}{D_{e,A}^{R}}} \tag{7.119}$$

Equation 7.118 can be written as:

$$D_{e,A}^{H}\left\{\frac{1}{r^{+}}\frac{\partial}{\partial r^{+}}\left(r^{+}\frac{\partial C_A(r^{+},h)}{\partial r^{+}}\right)+\frac{\partial^2 C_A(r^{+},h)}{\partial h^2}\right\}=R(C_A) \tag{7.120}$$

Differential equation 7.120 describes the concentration profile in an isotropic ring-shaped catalyst pellet with an effective diffusion coefficient $D_{e,A}{}^{H}$, a height H, and an inner radius

$$R_i^{+}=R_i\sqrt{\frac{D_{e,A}^{H}}{D_{e,A}^{R}}} \tag{7.121}$$

and an outer radius

$$R_u^{+}=R_u\sqrt{\frac{D_{e,A}^{H}}{D_{e,A}^{R}}} \tag{7.122}$$

Hence, anisotropy can be accounted for in the Aris numbers if the effective diffusion coefficient $D_{e,A}$ is replaced by $D_{e,A}{}^{H}$ and the characteristic dimension V_p/A_p by

$$\left(\frac{V_P}{A_P}\right)^{+}=\frac{1}{2}\left(\frac{1}{H}+\frac{1}{R_u^{+}-R_i^{+}}\right)^{-1}=\frac{1}{2}\left(\frac{1}{H}+\frac{1}{R_u-R_i}\sqrt{\frac{D_{e,A}^{R}}{D_{e,A}^{H}}}\right)^{-1} \tag{7.123}$$

Thus the following formulae for the Aris numbers for simple reactions are obtained:

$$An_0 = \left\{\left(\frac{V_P}{A_P}\right)^+\right\}^2 \frac{\{R(C_{A,s})\}^2}{2D_{e,A}^H}\left(\int_0^{C_{A,s}} R(C_A)dC_A\right)^{-1} \tag{7.124}$$

$$An_1 = \left\{\left(\frac{V_P}{A_P}\right)^+\right\}^2 \frac{\Gamma}{D_{e,A}^H}\left.\frac{\partial R(C_A)}{\partial C_A}\right|_{C_A=C_{A,s}} \tag{7.125}$$

Rather than accounting for anisotropy by modifying both the characteristic dimension and effective diffusion coefficient, this is achieved by modifying the effective diffusion coefficient only. Thus, for anisotropic catalyst pellets a modified effective diffusion coefficient $D_{e,A}^+$ is defined, which accounts for the anisotropy. Hence, for anisotropic catalyst pellets and simple reactions the Aris numbers can be calculated from

$$An_0 = \left(\frac{V_P}{A_P}\right)^2 \frac{\{(R(C_{A,s})\}^2}{2D_{e,A}^+}\left(\int_0^{C_{A,s}} R(C_A)dC_A\right)^{-1} \tag{7.126}$$

$$An_1 = \left(\frac{V_P}{A_P}\right)^2 \frac{\Gamma}{D_{e,A}^+}\left.\frac{\partial R(C_A)}{\partial C_A}\right|_{C_A=C_{A,s}} \tag{7.127}$$

Comparison of Equations 7.126 and 7.127 with 7.124 and 7.125 shows that for ring-shaped catalyst pellets the modified effective diffusion coefficient $D_{e,A}^+$ follows from

$$D_{e,A}^+ = \frac{\left(\sqrt{D_{e,A}^H} + \phi\sqrt{D_{e,A}^R}\right)^2}{(1+\phi)^2} \tag{7.128}$$

Where the number ϕ accounts for the catalyst geometry as

$$\phi = \frac{\lambda}{1-\gamma} = \frac{H}{R_u - R_i} \tag{7.129}$$

From this discussion it follows that all conclusions drawn for isotropic catalyst pellets hold for anisotropic catalyst pellets as well. In fact, with a single catalyst geometry, it is not possible to distinguish between isotropic and anisotropic pellets. The effect of anisotropy is lumped in with the effective diffusion coefficient. If the catalyst pellet is isotropic, then from Equation 7.128 it follows that we measure the effective diffusion coefficient $D_{e,A}{}^+ = D_{e,A}{}^H = D_{e,A}{}^R$. For anisotropic pellets, we measure $D_{e,A}{}^R$ for pellets with a large height (ϕ large) and $D_{e,A}{}^H$ for flat pellets (ϕ small). For intermediate values of ϕ the value of $D_{e,A}{}^+$ is between $D_{e,A}{}^R$ and $D_{e,A}{}^H$.

The only situation where anisotropy can give rise to errors, is when the catalyst geometry is changed (and thus ϕ) and if the value of the effective diffusion coefficient which was determined for another catalyst geometry is used. This is illustrated in Example 9.22.

The procedure given above can be extended for other catalyst geometries. For example, for an anisotropic parallelepiped with the dimensions width ⇔ length ⇔ height = W ⇔ L ⇔ H, the Aris numbers can still be calculated from the Equation 7.126 and 7.127. However, the modified effective diffusion coefficient $D_{e,A}{}^{+}$ now becomes:

$$D_{e,A}^{+} = \frac{\left(\dfrac{\sqrt{D_{e,A}^{H}}}{H} + \dfrac{\sqrt{D_{e,A}^{L}}}{L} + \dfrac{\sqrt{D_{e,A}^{W}}}{W}\right)^2}{\left(\dfrac{1}{H} + \dfrac{1}{L} + \dfrac{1}{W}\right)^2} \tag{7.130}$$

Modified effective diffusion coefficients for other catalyst geometries can easily be obtained.

Anisotropy has not been greatly studied, therefore it is difficult to obtain typical values of, for example, the ratio $D_{e,A}{}^{H}/D_{e,A}{}^{R}$. It is expected, however, that this ratio may be as high as 10, for example if, due to the manufacturing process, cracks occur in the longitudinal direction of the catalyst pellet. Those cracks will act as macro pores, which can increase the value of $D_{e,A}{}^{H}$ by some orders of magnitude. Consequently if the same overall or modified effective diffusion coefficient is used for different catalyst geometries, this may lead to serious errors. To avoid these errors the effective diffusion coefficient in each direction must be determined. For a ring-shaped catalyst pellet this can be done by measuring the overall or modified effective diffusion coefficient for two different catalyst geometries. With these data, $D_{e,A}{}^{H}$ and $D_{e,A}{}^{R}$ can be calculated from Equation 7.128. From the values of $D_{e,A}{}^{H}$ and $D_{e,A}{}^{R}$, $D_{e,A}{}^{+}$ can be calculated for arbitrary catalyst geometries. Thus, for a parallelepiped, $D_{e,A}{}^{+}$ must be determined for at least three catalyst geometries so as to be able to fully predict the effect of anisotropy.

Summarizing, anisotropy is unimportant if only one catalyst geometry is used. If more catalyst geometries are used, anisotropy must be investigated, since it can lead to serious deviations.

7.5 Summary Complex Situations

In Table 7.7 the most important dimensionless numbers introduced in this chapter are summarized (together with negligibility criteria). With these numbers, it is now possible to give generalized definitions for Aris numbers, for which two situations may be distinguished:

Diluted Gases

This is the most usual case. If it is assumed that the conversion rate depends on the concentrations C_A, C_B, C_C, ... and the temperature T, thus

$$R = R(C_A, C_B, C_C, \ldots, T) \tag{7.131}$$

Table 7.7 Summary of the characteristic numbers, their ranges and the negligibility criteria for complex situations

Effect	Characteristic number	Range	Negligibility criteria (*n*th-order and (n,m) kinetics)
Nonisothermal catalyst pellets	$\zeta = \frac{E_a}{RT_s} \times \frac{(-\Delta H) D_{e,A} C_{A,s}}{\lambda_p T_s}$	$\zeta \in]-\infty, +\infty[$	$\zeta < \frac{\lvert n \rvert}{10}$ (*high* η) $\zeta < \frac{n+2}{5}$ (*low* η)
bimolecular reactions	$\beta = \upsilon_B \frac{D_{e,A}}{D_{e,B}} \frac{C_{A,s}}{C_{B,s}}$	$\beta \in [0,1]$	$\beta < \frac{1}{10} \left\lvert \frac{n}{m} \right\rvert$ (*high* η) $\beta < \frac{1}{10} \left\lvert \frac{\sqrt{n+m+1}}{\sqrt{n+m+1} - \sqrt{n+1}} \right\rvert$ (*low* η)
Nondiluted gases	$\gamma = \frac{\sqrt{\upsilon_P} + 1 - \psi}{1 + Kn^* + \psi(\sqrt{\upsilon_P} - 1)\kappa_{A,s}} \times (\sqrt{\upsilon_P} - 1)\kappa_{A,s}$ $\psi = \frac{(Kn^*)^2}{(Kn^*)^2 + \{(\sqrt{\upsilon_P} - 1)\kappa_{A,s} + \sqrt{\upsilon_P} + 2\} Kn^* + \frac{\sqrt{\upsilon_P} + 1}{2}}$	$\gamma \in [-1, +\infty[$	$\gamma < \frac{\lvert n \rvert}{10}$ (*high* η) $\gamma < \frac{n+2}{5(n+1)}$ (*low* η)

and all the effects discussed in the definitions are incorporated, we obtain

$$An_1 = \left(\frac{V_P}{A_P}\right)^2 \frac{\Gamma}{D^+_{e,A}} \frac{\partial}{\partial C_A} R\left[C_A, C_{B,s}\left\{1-\beta_B\left(1-\frac{C_A}{C_{A,s}}\right)\right\}, C_{C,s}\left\{1-\beta_C\left(1-\frac{C_A}{C_{A,s}}\right)\right\}, \ldots\right.$$
$$\left.\ldots, T_s\left\{1+\alpha\left(1-\frac{C_A}{C_{A,s}}\right)\right\}\right]_{C_A=C_{A,s}} \tag{7.132}$$

$$An_0 = \left(\frac{V_P}{A_P}\right)^2 \frac{\{R(C_{A,s}, C_{B,s}, C_{C,s}, \ldots, T_s)\}^2}{2D^+_{e,A}} \times \left\langle \int_0^{C_{A,s}} R\left[C_A, C_{B,s}\left\{1-\beta_B\left(1-\frac{C_A}{C_{A,s}}\right)\right\},\right.\right.$$
$$\left.\left. C_{C,s}\left\{1-\beta_C\left(1-\frac{C_A}{C_{A,s}}\right)\right\}, \ldots, T_s\left\{1+\alpha\left(1-\frac{C_A}{C_{A,s}}\right)\right\}\right] dC_A \right\rangle^{-1} \tag{7.133}$$

where (Equation 7.60):

$$\beta_B = \nu_B \frac{D_{e,A}}{D_{e,B}} \frac{C_{A,s}}{C_{B,s}}$$

and

$$\beta_C = \nu_C \frac{D_{e,A}}{D_{e,C}} \frac{C_{A,s}}{C_{C,s}} \tag{7.134}$$

and

$$\alpha = \frac{(-\Delta H) D_{e,A} C_{A,s}}{\lambda_P T_s} \tag{7.135}$$

It is assumed that component A is the component not in excess, thus $\beta_B < 1, \beta_C < 1, \ldots$ and that the numbers $\beta_B, \beta_C, \ldots, \alpha$ do not depend on the temperature. They do not depend on the gas composition inside the catalyst pellet either, since the gas diluted. Notice the superscript '+' of the effective diffusion coefficient, which denotes that the catalyst pellet may be anisotropic.

Nondiluted Gases

For this case Equations 7.132 and 7.133 cannot be used, because the effective diffusion coefficients depend on the gas composition and, because of that, so do the numbers $\alpha, \beta_B, \beta_C, \ldots$ etc.

For simple reactions only the first point is important, and the following formulae (Equations 7.81 and 7.82) are found for the Aris numbers:

$$An_0 = \left(\frac{V_P}{A_P}\right)^2 \frac{\{R(C_{A,s})\}^2}{2} \left(\int_0^{C_{A,s}} D_{e,A}(C_A) R(C_A) dC_A\right)^{-1} \quad (7.136)$$

$$An_1 = \left(\frac{V_P}{A_P}\right)^2 \frac{\Gamma}{D_{e,A}(C_{A,s})} \left.\frac{\partial R(C_A)}{\partial C_A}\right|_{C_A = C_{A,s}} \quad (7.137)$$

for which $D_{e,A}$ as a function of C_A can be found with the aid of the dusty gas model, as discussed in the text.

For complex situations and nondiluted gasses both the above-mentioned points play a role. This has been illustrated for bimolecular reactions. With the aid of the dustygas model (neglecting viscous flow), formulae can also be found for the Aris numbers.

Many complex situations have not been addressed, such as simultaneous intraparticle temperature and pressure gradients and nondiluted gases with anisotropic catalyst pellets. Calculations for these and other complex situations proceed along the same line as demonstrated for bimolecular reactions and nondiluted gases. A framework that can be used to investigate the effect of complex situations on the effectiveness factor is given. Also presented are criteria that can be used for a quick estimate as to whether or not certain phenomena are important.

References

1. Hlavacek, V., Kubicek, M.A. (1970) *Chem. Eng. Sci.*, **25**, 1537.
2. Weisz, P.B., Hicks, J.S. (1962) *Chem. Eng. Sci.*, **17**, 265.
3. Mears, D.A. (1971) *Ind. Eng. Chem., Process Des. Dev.*, **10**, 541.
4. Doraiswamy, L.K., Sharma, M.M. (1984), Heterogeneous Reactions: Analysis, Examples, and Reactor Design, Vol. 1, New York: John Wiley & Sons.
5. Mason, E.A. Malinauskas, E.A.P. (1983) Gas Transport in porous Media: the Dusty-Gas Model, Amsterdam: Elsevier.
6. Kehoe, J.P.G., Aris, R. (1973) *Chem. Eng. Sci.*, **28**, 2094.
7. Haynes, H.W., Jr. (1978) *Can. J. Chem. Eng.*, **56**, 582.
8. Weekman, V.W., Gorring, R.L. (1965) *J. Catal.*, **4**, 260.

8 Design of Catalyst Pellets

The reaction rate per unit volume of catalyst as well as its selectivity depend on both the specific catalytic activity and the surface area of the active component per unit catalyst volume, as well as on its pore structure. These characteristics are determined by the conditions of catalyst preparation. Therefore, when developing a new catalyst, it is extremely important to be able to determine in advance the required internal surface area and the most suitable pore structure of the catalyst for the given reaction.

The more active a catalyst is, the more difficult it is to obtain benefits, due to an increased influence of transport phenomena on the conversion rate for fast chemical reactions. For some types of chemical reactions, such as consecutive reactions with the intermediate as the desired product, an increase of catalytic activity may lead to undesired effects if transport phenomena inside and outside the catalyst pellet play a role.

This chapter is concerned with the improvement of catalyst performance through a better pellet design. This design relates to the physical properties of the catalyst pellets for given kinetics and does not involve the chemical composition of the catalyst. Examples are given to illustrate the influence of structural parameters on catalyst performance.

8.1 Porous Structure and Observed Reaction Rate

8.1.1 Porous Structure and Catalyst Activity

The nature and arrangement of the pores determine transport within the interior porous structure of the catalyst pellet. To evaluate pore size and pore size distributions providing the maximum activity per unit volume, simple reactions are considered for which the concept of the effectiveness factor is applicable. This means that reaction rates can be presented as a function of the key component A only, hence $R_A(C_A)$. Various systems belonging to this category have been discussed in Chapters 6 and 7. The focus is on gaseous systems, assuming the resistance for mass transfer from fluid to outer catalyst surface can be neglected and the effectiveness factor does not exceed unity. The mean reaction rate per unit particle volume can be rewritten as

$$\langle R_A \rangle = \frac{1}{V_p} \iiint_{V_p} R_A dV = \eta R_A(C_{A,s}) = \eta S \tilde{R}_A(C_{A,s}) \tag{8.1}$$

where S is the pore surface area per unit pellet volume and R_A is the reaction rate per unit surface area.

To examine the dependence of the mean reaction rate on the porous structure we should describe the relationships between all parameters of Equation 8.1 and the structural parameters: mean pore radius, void fraction ε_p, and surface area per unit particle volume S.

The reaction rate per unit surface area $\tilde{R}_A$ does not depend on the porous structure and is determined only by the chemical composition of the active components.

The specific surface area S can be related to the particle porosity ε_p and the average pore radius r_e as:

$$S = \frac{2\varepsilon_p}{r_e} \tag{8.2}$$

As seen in Chapter 6, the effectiveness factor can be approximated by

$$\tilde{\eta} = \frac{1}{\sqrt{1+\tilde{\eta}An_1 + (1-\tilde{\eta})An_0}} \tag{8.3}$$

The Aris numbers An_0 and An_1 are

$$An_0 = g_0\phi^2 \tag{8.4}$$

$$An_1 = g_1\phi^2 \tag{8.5}$$

where the dimensionless coefficients g_0 and g_1 are found to be

$$g_0 = \frac{C_{A,s}R_A(C_{A,s})}{2\int_0^{C_{A,s}} R_A(C_A)dC_A} = \frac{C_{A,s}\tilde{R}_A(C_{A,s})}{2\int_0^{C_{A,s}} \tilde{R}_A(C_A)dC_A} \tag{8.6}$$

$$g_1 = \Gamma\frac{C_{A,s}}{R_A(C_{A,s})}\left.\frac{dR_A(C_A)}{dC_A}\right|_{C_{A,s}} = \Gamma\frac{C_{A,s}}{\tilde{R}_A(C_{A,s})}\left.\frac{d\tilde{R}_A(C_A)}{dC_A}\right|_{C_{A,s}} \tag{8.7}$$

Here, Γ is the geometry factor (described in Chapter 6), which depends only on the shape of the catalyst pellet, and ϕ is the Thiele modulus defined by

$$\phi^2 = \frac{V_p^2}{A_p^2}\frac{R_A(C_{A,s})}{D_{e,A}C_{A,s}} = \frac{V_p^2}{A_p^2}\frac{S\tilde{R}_A(C_{A,s})}{D_{e,A}C_{A,s}} \tag{8.8}$$

Substitution of Equations 8.4 and 8.5 into 8.3 yields

$$\tilde{\eta} = \frac{1}{\sqrt{1+[\tilde{\eta}g_1 + (1-\tilde{\eta})g_0]\phi^2}} \tag{8.9}$$

The coefficients g_0 and g_1 do not depend on the porous structure as Equations 8.6 and 8.7 indicate. From calculations of g_0 and g_1 for different systems it follows their values are, as a rule, of the order of unity. For example, for a reaction rate described by power low kinetics $R_A = k_n C_A^n$,

$$g_0 = \Gamma n$$

$$g_1 = \frac{n+1}{2}$$

The effective diffusivity for gases, which is incorporated in expression 8.8 for ϕ^2 can be approximated by the equation (see Equation 3.27)

$$D_{e,A} = \frac{\varepsilon_p}{\gamma_p} \frac{D_{Am} D_{Ak}}{D_{Am} + D_{Ak}} \tag{8.10}$$

in which D_{Am} and D_{Ak} are the molecular diffusivity of component A and its Knudsen diffusivity in a straight cylindrical pore. Molecular diffusivity is independent of the pore size and, according to kinetic gas theory, can be found as

$$D_{Am} = \frac{\lambda \bar{v}}{3}$$

Where λ is the mean free path and $\bar{v}$ is the mean molecular speed. The Knudsen diffusivity is proportional to the mean pore radius r_e:

$$D_{Ak} = \frac{2\bar{v} r_e}{3}$$

Thus,

$$D_{Ak} = D_{Am} \frac{2r_e}{\lambda} \tag{8.11}$$

Substitution of Equation 8.11 into 8.10 gives the dependence of the effective diffusivity on pore radius for gases as

$$D_{e,A} = \frac{\varepsilon_p}{\gamma_p} \frac{D_{Am}}{1+Kn} \tag{8.12}$$

where $Kn = \lambda/(2r_e) = \lambda/d_e$ is the Knudsen number. Combination of Equations 8.2, 8.8 and 8.12 yields

$$\phi^2 = \phi_\lambda^2 \frac{(1+Kn)Kn}{2}$$

Here,

$$\phi_\lambda^2 = \frac{V_p^2}{A_p^2}\frac{\tilde{R}_A(C_{A,s})}{C_{A,s}}\frac{8\gamma_p}{\lambda D_{Am}} \tag{8.13}$$

is the Thiele modulus for a catalyst with an average pore diameter d_e equal to the molecule mean free path λ.

Equations 8.1, 8.9 and 8.13 describe the dependence of the reaction rate on the pore size, particle porosity and tortuosity.

First to be considered are the limiting forms of this dependence. In the kinetic regime, that is without any diffusion limitations ($\phi^2 << 1$, and hence usually $An_0 << 1$ and $An_1 << 1$), the effectiveness factor approaches unity and the mean reaction rate according to Equation 8.1 is proportional to the specific surface area:

$$\langle R_A \rangle \sim S \sim \frac{\varepsilon_p}{r_e}$$

If diffusion limitations are significant ($\phi^2 >>1$), the effectiveness factor is much less than unity ($\eta g_1 << (1-\eta)g_0 \approx g_0$) and, from Equation 8.9,

$$\eta = \frac{1}{\phi\sqrt{g_0}} \tag{8.14}$$

From Equations 8.1, 8.8 and 8.14

$$\langle R_A \rangle = \frac{S\tilde{R}_A(C_{A,s})}{\phi\sqrt{k_0}} = \frac{A_p}{V_p}\sqrt{\frac{SC_{A,s}\tilde{R}_A(C_{A,s})D_{e,A}}{g_0}} \sim \sqrt{SD_{e,A}} \tag{8.15}$$

The dependence of the effective diffusivity on the pore size is determined by the ratio of the molecule mean free path λ and the pore diameter (see Equation 8.12). For $\lambda << d_e$ ($Kn<<1$) ordinary diffusion is the predominant transport mechanism in the pores, the effective diffusivity does not depend on the pore size and

$$\langle R_A \rangle \sim \sqrt{SD_{e,A}} \sim \sqrt{\frac{S\varepsilon_p}{\gamma_p}} \sim \frac{\varepsilon_p}{\sqrt{\gamma_p r_e}} \tag{8.16}$$

For Knudsen diffusion ($\lambda >> d_e$) the effective diffusivity is proportional to r_e and:

$$\langle R_A \rangle \sim \sqrt{SD_{e,A}} \sim \sqrt{\frac{S\varepsilon_p r_e}{\gamma_p}} \sim \frac{\varepsilon_p}{\sqrt{\gamma_p}} \tag{8.17}$$

The dependence of the reaction rate on the pore size as calculated from Equations 8.1, 8.9 and 8.12 for arbitrary reaction regimes is shown in Figure 8.1 for typical values of the reference Thiele modulus ϕ_λ. The results are presented as a ratio of the reaction rate $\langle R_A \rangle$ to its maximum value $\langle R_{A0} \rangle$ observed in the limit of infinitely small pore diam-

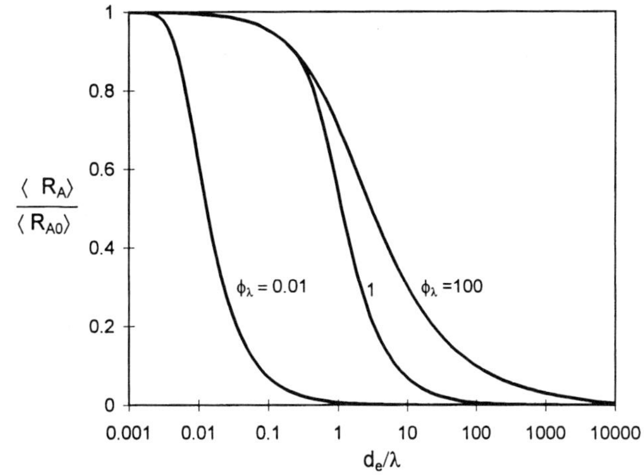

Figure 8.1 Dependence of the reaction rate per unit catalyst volume on pore diameter.

eter. As can be seen, in spite of an increase of diffusion limitations the catalytic activity increases with decreasing pore size. Figure 8.1 shows also that for reactions with $\phi_\lambda > 1$ the catalyst activity approaches a maximum value in the region of relatively large pore diameters of about $(0.1 - 1)\lambda$. At atmospheric pressure $\lambda \approx 50$ nm, thus the reaction rate close to maximum is reached when $d_e \approx 5 - 50$ nm. This means that a mean pore diameter is still about 10-100 times larger than the minimum pore size we encounter in nature. In this situation diffusion control inhibits the increase of the reaction rate by increasing the specific surface area.

Experimental investigations reporting the influence of pore structure on catalyst effectiveness are quite scarce. Topchieva et al. [1] have studied the kinetics of cumene cracking over aluminosilicate catalysts of the same chemical composition, but widely varying pore radii (1.2-11.5 nm). In the diffusion limited region, as found from an Arrhenius plot, the authors observed the reaction rate is proportional to the pore radius. This observation at first sight contradicts Equations 8.16 or 8.17, but can be explained as follows. Due to the small pore size compared to the mean free path and diffusion limitation Equation 8.17 should be considered. For the catalyst used in the experiments the pore size had been altered approximately 10-fold, whereas the surface area changed over relatively narrow range. For constant area $\varepsilon_p \sim r_e$ (see Equation 8.2) and $\langle R_A \rangle \sim r_e$ as Equation 8.17 predicts.

A significant increase in catalytic activity as compared to the limiting values, shown in Figure 8.1, can be achieved by the use of bidisperse porous structures. Such catalyst pellets are formed by compressing, extruding or in some other manner compacting finely powdered microporous material into a pellet. Ideally the micropores are due to the porosity in the individual microparticles of catalyst. The macropores result from voids between the microparticles, after pelletization or extrusion. In such catalysts, most of the catalytic surface is contained in the micropores, since $S \sim 1/r_e$. The bidisperse structure is illustrated in Figure 8.2 compared to monodisperse particle.

A single effective diffusion coefficient cannot adequately characterize the mass transfer within a bidisperse-structured catalyst when the influence of the two individual systems is equally important. In a realistic model the separate identity of the macropore and micropore structures must be maintained, and the diffusion must be described in

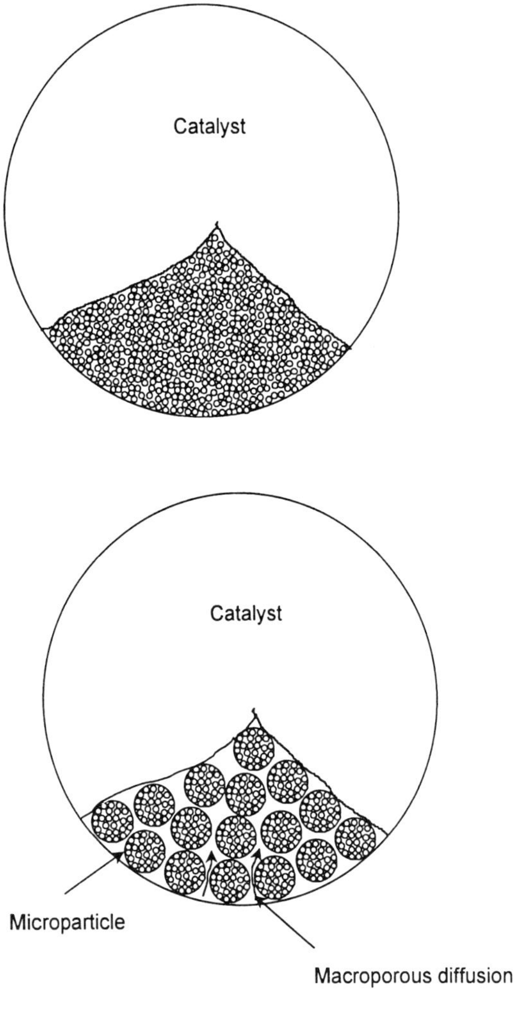

Figure 8.2 Schematic of (a) monodisperse and (b) bidisperse catalyst.

terms of two coefficients, an effective macropore diffusivity $D_{e,m}$ and an effective micropore diffusivity, $D_{e,\mu}$ [2]. Other workers have employed such a macro-micropore model in their evaluation of effectiveness factors in bidisperse structured catalysts [3-5]. It has been shown that, for such systems, the pellet effectiveness is equal to the product of the micro- and macropore effectiveness factors. This model might be expected to apply when the diffusivity in the micropores or microparticles $D_{e,\mu}$ is considerably smaller than the macropore diffusivity $D_{e,m}$, for only in such circumstances can the microparticle contribution to the radial flux in the catalyst particle be neglected.

A bidisperse structure can be advantageous because the effectiveness factor in the microparticles is often close to unity (their size being three to four orders of magnitude smaller than the usual size of the industrial catalysts). It is of interest to estimate the ratio of the conversion rates with mono- and bidisperse structures, having the same size, when the porous structure of the microparticles is identical to that of a monodisperse pellet. This ratio can easily be found when the micropore effectiveness factor is close to unity, as in the case for many industrial systems. Since the external surface area of the microparticle can be neglected compared to its internal surface, the pore surface per unit volume of the bidisperse catalyst is

$$(1-\varepsilon_m)S$$

Thus the reaction rate per unit volume of the bidisperse pellet is

$$R_{A,bd}(C_A) = (1-\varepsilon_m)S\tilde{R}_A(C_A)$$

where the subscript m designates the macropores (and μ – the micropores), and S is the specific pore surface of the microparticle. The corresponding reaction rate for monodisperse catalyst is

$$R_{A,md}(C_A) = S\tilde{R}_A(C_A)$$

Conventional procedures can be used to find the reaction effectiveness factors of the mono- and bidisperse pellets. If diffusion limitations in the macropores are significant the average reaction rates are

$$\langle R_A \rangle_{bd} = \frac{R_{A,bd}(C_{A,s})}{\sqrt{An_{0,bd}}}$$

$$\langle R_A \rangle_{md} = \frac{R_{A,md}(C_{A,s})}{\sqrt{An_{0,md}}}$$

where

$$An_{0,bd} = g_0 \frac{V_p}{A_p}\sqrt{\frac{R_{A,bd}(C_{A,s})}{D_{e,m}C_{A,s}}}$$

$$An_{0,md} = g_0 \frac{V_p}{A_p}\sqrt{\frac{R_{A,md}(C_{A,s})}{D_{e,\mu}C_{A,s}}}$$

are the Aris numbers the bi- and monodisperse structures.

The above equations give directly a ratio of the reaction rates:

$$\frac{\langle R_A \rangle_{bd}}{\langle R_A \rangle_{md}} = \left(\frac{D_{e,m}R_{A,bd}(C_{A,s})}{D_{e,\mu}R_{A,md}(C_{A,s})}\right)^{0.5} = \left(\frac{D_{e,m}(1-\varepsilon_m)}{D_{e,\mu}}\right)^{0.5}$$

The magnitude of the ratio of effective gas diffusivities in the macro- and micropore regions is in the range 10 – 100 for most industrially important catalysts [6]. This ratio depends upon the pore size distributions of the pellets. Hashimoto and Smith [7] have determined this ratio as 80 at 30° C for the diffusion of *n*-butane in alumina pellets with macro and micro mean pore radii of 120 and 1.7 nm, respectively. A typical value of macropore porosity is 0.5. Thus, upon transition to a bidisperse structure, the reaction rate may increase 2 – 7 times.

In the above considerations the total pressure should also be taken into account. With increasing pressure, the efficiency of a bidisperse catalyst decreases because diffusion in micropores turns from Knudsen to ordinary and the difference between $D_{e,m}$ and $D_{e,\mu}$ disappears. At a pressure of 1-10 MPa, a uniform porous structure with a pore size close to the mean free path is the most favorable [6].

8.1.2 Influence of Mass Transport on Selectivity

A deficient design of a catalyst pellet for a single reaction leads to a poor utilization of the often expensive catalyst mass. Fortunately, we can usually adjust some operating conditions, such as the temperature or residence time, or install additional catalyst material to compensate for a design error. This section discusses the influence of diffusion limitations on the yield of the desired product in a multipath reaction. Here, when the diffusion intrusions are not properly accounted for, it is often impossible to counterbalance the impact of the design error by a change in the operating conditions.

Wheeler [8] presented the first theoretical analysis of the influence of diffusion on the yield of a desired product. Many additional investigations have been reported since then, and an excellent survey is available [9]. Only one simple case is described here, to illustrates the influence of diffusion limitations on the selectivity. Consider a first-order isothermal catalytic sequence

$$A \xrightarrow{k_1} B \xrightarrow{k_2} C$$

in which B is the desired product. Mass transfer inside the porous pellet is assumed to obey Fickian-type diffusion. The concentration profiles for a plate geometry are described by the equations

$$D_{e,A}\frac{d^2C_A}{dx^2} = k_1 C_A \tag{8.18}$$

$$D_{e,B}\frac{d^2C_B}{dx^2} = k_2 C_B - k_1 C_A \tag{8.19}$$

with boundary conditions

$$x = 0 \qquad \frac{dC_A}{dx} = 0, \qquad \frac{dC_B}{dx} = 0 \tag{8.20}$$

$$x = R \qquad C_A = C_{A,s}, \qquad C_B = C_{B,s} \tag{8.21}$$

where x is the distance across the pellet and R the half-width of the plate. Solution of these equations gives the concentration profiles inside the pellet as

$$C_A = C_{A,s}\frac{\cosh(\phi_A x/R)}{\cosh(\phi_A)} \tag{8.22}$$

$$\begin{aligned} C_B = & \left(C_{B,s} + C_{A,s}\frac{D_{e,A}}{D_{e,B}}\frac{\phi_A^2}{(\phi_A^2-\phi_B^2)}\right)\frac{\cosh(\phi_B x/R)}{\cosh(\phi_B)} \\ & - C_{A,s}\frac{D_{e,A}}{D_{e,B}}\frac{\phi_A^2}{(\phi_A^2-\phi_B^2)}\frac{\cosh(\phi_A x/R)}{\cosh(\phi_A)} \end{aligned} \tag{8.23}$$

where

$$\phi_A = R\sqrt{\frac{k_1}{D_{e,A}}}$$

$$\phi_B = R\sqrt{\frac{k_2}{D_{e,B}}}$$

The total amounts of A and B converted inside the catalyst pellet per unit surface area and per unit time can be found as the fluxes of A and B to the pellet:

$$R_A'' = D_{e,A}\left.\frac{dC_A}{dX}\right|_{X=R} = Rk_1 C_{A,s}\eta_A \tag{8.24}$$

$$R_B'' = D_{e,B}\left.\frac{dC_B}{dX}\right|_{X=R} = Rk_2 C_{B,s}\eta_B - Rk_1 C_{A,s}\eta_A \Psi_{AB} \tag{8.25}$$

where

$$\eta_A = \frac{\tanh(\phi_A)}{\phi_A}$$

$$\eta_B = \frac{\tanh(\phi_B)}{\phi_B}$$

$$\Psi_{AB} = \frac{\phi_A^2\eta_A - \phi_B^2\eta_B}{\phi_A^2 - \phi_B^2}$$

The selectivity σ_B (the ratio between the amount of B obtained and the amount of reactant A converted) can be computed from

$$\sigma_B = -\frac{R''_B}{R''_A} = \Psi_{AB} - \frac{C_{B,s}}{C_{A,s}}\frac{k_2}{k_1}\frac{\eta_B}{\eta_A} \tag{8.26}$$

This selectivity may be called a point selectivity, because it depends on the concentrations in the fluid phase, which may be different at different points in the reactor. If $C_{B,s} = 0$, Equation 8.26 predicts

$$\sigma_B = \Psi_{AB} = \frac{\phi_A^2\eta_A - \phi_B^2\eta_B}{\phi_A^2 - \phi_B^2} \tag{8.27}$$

When the diffusion resistances are negligible, $\eta_A \to 1$ and $\eta_B \to 1$ and Equation 8.27 yields $\sigma_B = 1$; as expected the selectivity is not affected by the diffusion. When the diffusion limitations are large $\eta_A \to 1/\phi_A$ and $\eta_B \to 1/\phi_B$ and the selectivity asymptotically approaches the value

$$\sigma_B = \frac{1}{1+\phi_B/\phi_A} \tag{8.28}$$

Equation 8.28 therefore predicts that diffusion limitations reduce the selectivity below the value attained without limitations. The dependence of the point selectivity, determined by Equation 8.27, on ϕ_B/ϕ_A for different values of ϕ_A is presented in Figure 8.3. The selectivity decreases both with ϕ_B/ϕ_A and ϕ_A.

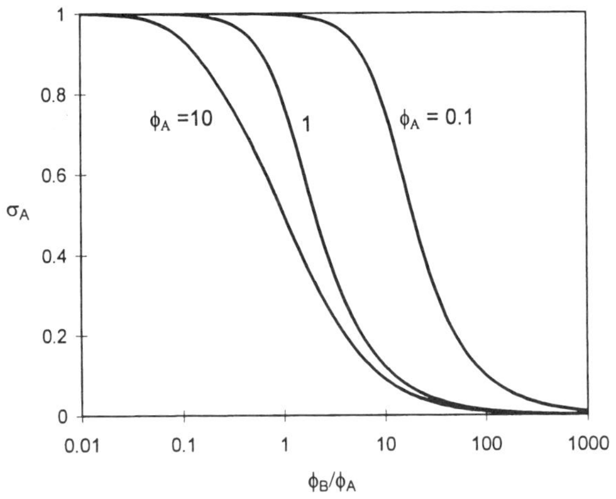

Figure 8.3 Effect of the diffusion limitations on catalyst selectivity.

Equation 8.26 shows that diffusion-limited selectivity depends on the surface conditions. Therefore, when a reacting mixture flows through a packed bed reactor the local selectivity is dependent on the position. The total yield (the ratio of desired product B

obtained with respect to the amount of reactant A fed) and total selectivity can be found for reactor which can be described by simplest plug flow model [10]. In this case the distributions of the reactant and product over the reactor length are described by the equations

$$u_s \frac{dC_{A,s}}{dz} = -A_{pv} R''_A \tag{8.29}$$

$$u_s \frac{dC_{B,s}}{dz} = -A_{pv} R''_B \tag{8.30}$$

with reactor inlet boundary conditions of

$$z=0 \quad C_{A,s}=C_{A,0}, \quad C_{B,s}=C_{B,0} \tag{8.31}$$

where z is the axial coordinate, u_s is the superficial velocity (this is the average linear velocity of the fluid if no packing were present in the column), R''_A and R''_B are determined by Equations 8.24 and 8.25 and A_{pv} is the outer surface area of the particles per unit reactor volume. Equations 8.29, 8.30 and 8.26 provide an equation relating the concentrations of A and B in the fluid:

$$\frac{dC_{B,s}}{dC_{A,s}} = \frac{R''_B}{R''_A} = -\Psi_{AB} + \frac{C_{B,s}}{C_{A,s}} \frac{k_2 \eta_B}{k_1 \eta_A} \tag{8.32}$$

Integrating Equation 8.32 and accounting for the inlet conditions of Equation 8.31, gives the following relationship between the concentrations:

$$\frac{C_{B,s}}{C_{A,0}} = \frac{\Psi_{AB}}{1-p}\left[\left(\frac{C_{A,s}}{C_{A,0}}\right)^p - \frac{C_{A,s}}{C_{A,0}}\right] + \frac{C_{B,0}}{C_{A,0}}\left(\frac{C_{A,s}}{C_{A,0}}\right)^p \tag{8.33}$$

where

$$p = \frac{k_2 \eta_B}{k_1 \eta_A} \tag{8.34}$$

Equation 8.33 is the most useful with which to determine the conversion level for maximum yield of the desired product and for computing the impact of diffusion on the yield. To demonstrate the influence of diffusion limitation on the yield, determined by Equation 8.33, consider the particular case of $C_{B,0} = 0$ and $D_{e,A} = D_{e,B}$. In this case

$$\frac{C_{B,s}}{C_{A,0}} = \frac{k_1}{k_1 - k_2}\left[\left(\frac{C_{A,s}}{C_{A,0}}\right)^p - \frac{C_{A,s}}{C_{A,0}}\right] \tag{8.35}$$

For the diffusion limitation free condition $\eta_A \approx \eta_B \approx 1$ and the yield is

$$\frac{C_{B,s}}{C_{A,0}} = \frac{k_1}{k_1 - k_2}\left[\left(\frac{C_{A,s}}{C_{A,0}}\right)^{k_2/k_1} - \frac{C_{A,s}}{C_{A,0}}\right] \tag{8.36}$$

For strong diffusion limitations $\eta_B/\eta_A \approx \phi_A/\phi_B = (k_1/k_2)^{1/2}$, and the corresponding equation is

$$\frac{C_{B,s}}{C_{A,0}} = \frac{k_1}{k_1 - k_2}\left[\left(\frac{C_{A,s}}{C_{A,0}}\right)^{\sqrt{k_2/k_1}} - \frac{C_{A,s}}{C_{A,0}}\right] \tag{8.37}$$

Thus, the internal mass transport causes the system to behave as though selectivity were not governed by k_1/k_2 but by less favorable ratio $(k_1/k_2)^{1/2}$. When the reaction rates are equal ($k_1 = k_2$) Equations 8.36 and 8.37 transform to

$$\frac{C_{B,s}}{C_{A,0}} = \frac{C_{A,s}}{C_{A,0}}\ln\left(\frac{C_{A,s}}{C_{A,0}}\right)$$

$$\frac{C_{B,s}}{C_{A,0}} = \frac{1}{2}\frac{C_{A,s}}{C_{A,0}}\ln\left(\frac{C_{A,s}}{C_{A,0}}\right)$$

Thus diffusion limitations decrease the yield twofold. These results may be generalized to include interface and intraparticle diffusion for bidisperse catalysts [5]. The effect of diffusion limitation on the concentration distribution over the reactor length can be calculated from Equations 8.24, 8.29 and 8.33. An example of such calculations is shown in Figure 8.4 for the case $C_{B,0} = 0$, $D_{e,A} = D_{e,B}$, and $k_2/k_1 = 0.1$. Although the rate of

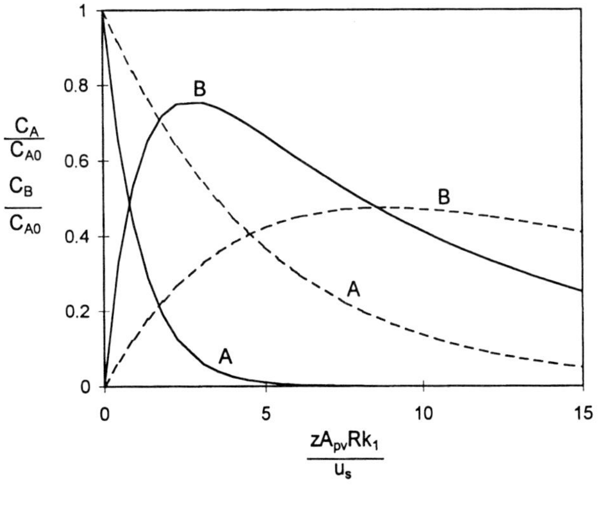

Figure 8.4 Effect of diffusional limitations on the concentration distributions of A and B over the reactor length for $k_2/k_1 = 0.1$, $\phi_A = 0.5$ (solid lines) and $\phi_A = 5$ (dashed lines).

the second reaction is small compared to the rate of the first, its influence on the yield and selectivity is significant.

8.2 Optimal Particle Shape and Size

The problem of the optimal particle shape and size is crucial for packed bed reactor design. Generally, the larger the particle diameter, the cheaper the catalyst. This is not usually a significant factor in process design – more important are the internal and external diffusion effects, the pressure drop, the heat transfer to the reactor walls and a uniform fluid flow.

The most commonly found shape of catalyst particle today is the hollow cylinder. One reason is the convenience of manufacture. In addition there are often a number of distinct process advantages in the use of ring-shaped particles, the most important being enhancement of the chemical reaction under conditions of diffusion control, the larger transverse mixing in packed bed reactors, and the possible significant reduction in pressure drop. It is remarkable (as discussed later) that the last advantage may even take the form of reduced pressure losses and an increased chemical reaction rate per unit reactor volume [11].

8.2.1 Pressure Drop

In many applications the pressure drop ΔP over the reactor bed is a governing factor in reactor design. The pressure drop (although seldom larger than 10% of the total pressure, and thus not a major factor in changing the chemical reaction rate) may be a very important factor in the assessment of the energy costs, especially when a recycle is required. Reduction of ΔP in such situations can produce considerable savings, especially for large units [12]. Several equations for pressure-drop calculations have been proposed, of which the Ergun equation [13] has been preferred for years. It accounts for the pressure drop caused by simultaneous inertial and viscous losses:

$$\frac{\Delta P}{L} = 150 \frac{(1-\varepsilon_b)^2}{\varepsilon_b^3} \frac{\mu}{d_p^2} u_s + 1.75 \frac{(1-\varepsilon_b)}{\varepsilon_b^3} \frac{\rho_f}{d_p} u_s^2 \tag{8.38}$$

where L is the bed length, ε_b is the void fraction of the bed, μ is the fluid viscosity, ρ_f is the fluid density, u_s is the superficial velocity and d_p is the equivalent particle diameter.

The Ergun equation can be applied to gases by using the density and viscosity of the gas at the arithmetic average of the end conditions. For large pressure drops and temperature changes, it seems better to use Equation 8.38 with the pressure gradient in differential form: $\Delta P/L$ should be replaced by $-dP/dz$, where z is the distance in the direction of flow.

The superficial velocity can be found if the mass or volume flow rates (G or Φ_v) are known:

$$u_s = \frac{4G}{\pi d_t^2 \rho_f} = \frac{4\Phi_v}{\pi d_t^2}$$

The equivalent particle diameter is defined as

$$d_p = \frac{6V_p}{S_p} \tag{8.39}$$

With this definition, for spheres, the use of Equation 8.39 gives just the diameter of sphere. Expressions of equivalent diameters for different particle shapes as used in packed bed reactors are presented in Table 8.1.

Table 8.1 Equivalent diameters for particles of different shape

Shape	Equivalent diameter (d_p)
Sphere, diameter d_s	d_s
Cylinder with length H equal to diameter $d_c = 2R_u$	d_c
Long extrudates with radius R_u	$3R_u$
Ring with length H, outside radius R_u, and inside diameter R_i	$\dfrac{3(R_u + R_i)H}{R_u - R_i + H}$

The bed porosity depends on many factors, including the particle shape, ratio of tube to particle diameter, roughness of the particle surface and the charging method. Void fractions in packed tubes for variously shaped particles and at different particle to tube diameter ratios have been given [14]. Any values from 0 to 1 can be obtained for the porosity. For example, for a regular packing of cylinders of equal size the porosity may reach a value of 0.096. Low porosity values can also be obtained with mixtures of the particles of different sizes. Using rings with a thin wall a porosity of close to 1 can be achieved. Normally the porosity ranges from 0.3 to 0.6, and is consistently higher for some reactors than for others.

Equation 8.38 may be rewritten in terms of dimensionless groups [15]

$$\left(\frac{\Delta P}{\rho_f u_s^2}\right)\left(\frac{d_p}{L}\right)\left(\frac{\varepsilon_b^3}{1-\varepsilon_b}\right) = 150\frac{1-\varepsilon_b}{Re} + 1.75 \tag{8.40}$$

where $Re = \rho_f u_s d_p/\mu = G d_p/\mu$ is the Reynolds number.

Equation 8.40 shows that the two terms in the right-hand side are equal at $Re = 85.7(1 - \varepsilon_b)$. So, for typical bed void fractions of 0.4 the viscous and inertial resistance's are approximately equal at $Re \approx 50$.

The void fraction is the most significant factor determining the loss of pressure during fluid flow through a packed bed. At low flow rates ($Re << 50$) the pressure drop is proportional to

$$\Delta P \propto \frac{(1-\varepsilon_b)^2}{\varepsilon_b^3}$$

and, at high flow rates ($Re >> 50$), to

$$\Delta P \propto \frac{1-\varepsilon_b}{\varepsilon_b^3}$$

The dependence of the relative pressure drop on the porosity at otherwise equal conditions in the viscous and inertial flow regimes are shown in Figure 8.5, where $\Delta P_{0.4}$ is the pressure drop at $\varepsilon_b = 0.4$. All other variables in the equation remaining constant; a change in the void fraction from 0.4 to 0.5 reduces the pressure drop more than 2.8 times in viscous flow regime and more than 2.3 times in the inertial regime.

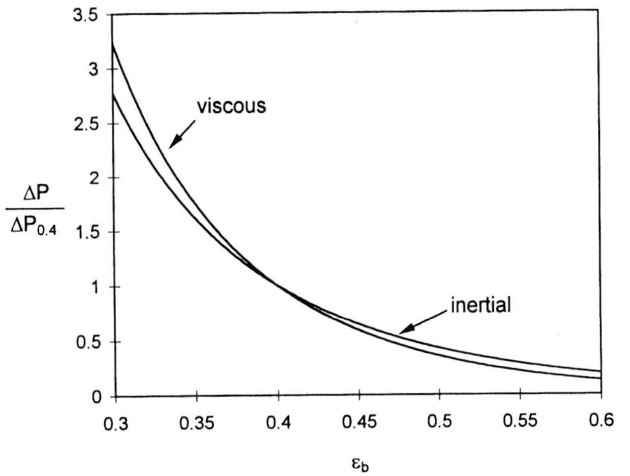

Figure 8.5 Effect of the bed porosity on pressure drop for viscous resistance and inertial resistance.

The standard way to decrease the pressure drop is to use a catalyst in the form of hollow cylinders or rings. The expected gain in void fraction, when full cylinders with a height H and radius R_u are replaced by the hollow cylinders of the same external sizes and an internal radius R_i, can be calculated on the basis of some geometrical considerations. The void fractions of beds packed with the cylinders and rings are:

$$\varepsilon_{b,cyl} = \frac{V_{tube} - V_{cyl} n_{cyl}}{V_{tube}} \tag{8.41}$$

and

$$\varepsilon_{b,ring} = \frac{V_{tube} - V_{ring} n_{ring}}{V_{tube}} \tag{8.42}$$

where V_{cyl} and V_{ring} are the volumes of one single cylinder or ring, n_{cyl} and n_{ring} the numbers of cylinders or rings in the column, and V_{tube} is the volume of the tube filed with the catalyst. England and Gunn [11] have found experimentally that for cylinders of the same diameter and length:diameter ratio the number of particles in the unit volume of the packed bed was independent of the size of thickness of wall of the cylinders in the bed, so $n_{ring} = n_{cyl}$. Eliminating n_{cyl}/V_{tube} from Equations 8.41 and 8.42 gives

$$\varepsilon_{b,ring} = 1-(1-\varepsilon_{b,cyl})\frac{V_{ring}}{V_{cyl}} \tag{8.43}$$

Substitution of the expressions of V_{cyl} and V_{ring} into Equation 8.43 yields

$$\varepsilon_{b,ring} = 1-(1-\varepsilon_{b,cyl})(1-\iota^2) \tag{8.44}$$

where $\iota = R_i/R_u$. The bed porosity as a function of the inner to outer radius is shown in Figure 8.6 for $\varepsilon_{b,cyl} = 0.35$ as also found by England and Gunn [11]. For $\iota = 0.5$, Equation 8.44 gives $\varepsilon_{b,ring} = 0.5125$. Thus a significant decrease in pressure drop can be expected.

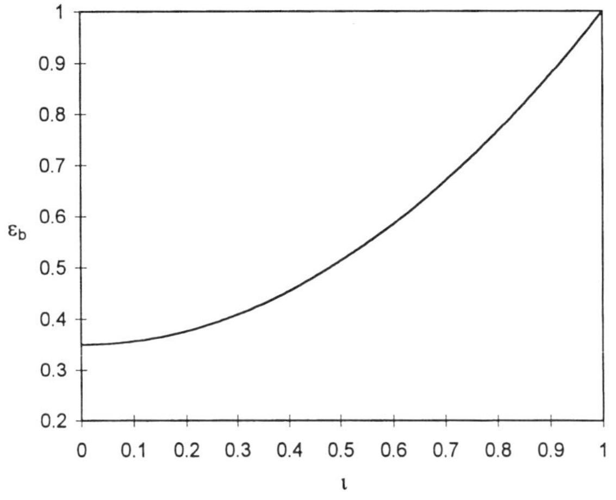

Figure 8.6 Porosity of packed beds as a function of ratio (inner radius of cylinder:outer radius of cylinder).

It is interesting to see how the particle shape can influence the pressure drop. England and Gunn [11] have compared experimentally four types of cylindrical particles. The outer diameter and the length of all particles were the same and equal to 6.35 mm. The ratio of inside to outside radius and also the bed porosity's are given in the Table 8.2

Table 8.2 Characteristics of catalyst pellets [11]

Type	R_i/R_u	ε_b
Grade 1 (solid)	0	0.35
Grade 2 (PVC)	0.5	0.53
Grade 3 (glass)	0.752	0.70
Grade 4 (metal)	0.855	0.81

The measured pressure drops as a function of the Reynolds number $Re' = 2\rho_f u_s R_u/\mu$, are presented in Figure 8.7. It is clear there is a significant reduction in pressure drop as

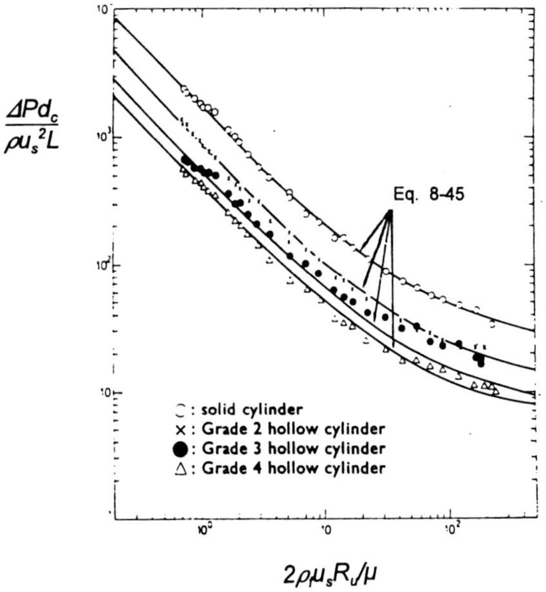

Figure 8.7 Dimensionless pressure drop, $2R_u\Delta P/(\rho u_s 2L)$, as a function of Reynolds number for the hollow and solid cylinders (after England and Gunn [11]).
Reproduced with permission of the Institution of Chemical Engineers.

the ratio of outside to the inside diameter decreases. According to the authors their experimental data are better described by the equation

$$\frac{\Delta P d_c}{\rho u_s^2 L} = 1000\left(\frac{3}{Re'^2} + \frac{0.15}{Re'} + 6\times10^{-4}\right)^{1/2}\left(\frac{\varepsilon_{b,cyl}}{\varepsilon_b}\right)^{1.65} \tag{8.45}$$

where ε_b is the bed porosity, and $\varepsilon_{b,cyl}$ is the bed porosity when packed with solid cylinders.

It is also important [11] that the magnitude of axial and radial mixing is significantly greater in beds of hollow particles than in beds with solid particles. The increase in radial mixing is an additional reason to use a ring-shaped catalyst for highly endo- and exothermic processes, such as methane conversion or ethene oxidation, where heat removal through the reactor wall is a decisive factor in the reactor stability. Other advantages of the ring-shaped catalyst are described in the next section.

8.2.2 Diffusion Limitations

Equation 8.38 shows that the particle diameter (as does the bed porosity) also has a strong influence on the pressure drop. It appears that the larger the particles the better. This is true if diffusion resistance inside a porous particle is not a problem. If it is, larger particles lead to lower conversions, so some compromise may be necessary. An attractive particle shape for such a compromise is the hollow cylinder.

Gunn [16] and England and Gunn [11] have investigated the effect of diffusion and geometry in hollow cylinders on the effectiveness factor. The results of their computations for a first-order reaction demonstrate that the adverse effect of diffusion upon chemical reaction, can be minimized in design. Numerical investigations of the optimal geometrical parameters of ring-shape catalysts have been made by Evenchik and Sokolinskii [17] for the water-gas shift reaction and for SO_2 oxidation: they confirm the trends found for first-order reactions.

If a catalyst in a form of solid cylinders is replaced by hollow cylinders with an optimal internal radius, the reactor productivity will increase although the total mass of catalyst decreases. For example, the use of hollow cylinders with an internal radius of $0.35R_u$ instead of solid cylinders of the same external size for the water-gas shift reaction at a pressure of 3 MPA and a temperature of 417 ^{0}C may increase the effectiveness of the unit reactor volume by 18%. This is particularly important for processes under elevated pressures, because of the strong dependence of the reactor cost on its volume [17].

The use of the generalized approach described in previous chapters demonstrates the significance of the catalyst shape in a wide variety of situations. To this end, as an example, ring-shaped particles are also considered.

For comparison of solid and hollow cylinders it is acceptable, as in the previous investigations, to use a modified effectiveness factor, defined as the ratio of the apparent reaction rate to that in a solid cylinder with the same height and radius and in the absence of diffusion limitations. This modified effectiveness factor may be expressed through the usually used effectiveness factor as

$$\eta_m = \eta \frac{V_{ring}}{V_{cyl}}$$

where V_{ring} is the volume of hollow cylinder and V_{cyl} is the volume of the solid cylinder of the same external size. The last equation can be rewritten as

$$\eta_m = \eta(1-\iota^2) \tag{8.46}$$

with $\iota = R_i/R_u$. The value of η_m characterizes the effectiveness of the utilization of the volume of a solid cylinder rather than the effectiveness of the catalyst volume.

A hole in the cylinder leads to the improvement of the use of the internal catalyst surface and this improvements increases with increasing ι, whereas the catalyst volume decreases. Therefore the hole in a cylinder has a twofold influence on the value of η_1. Consider the function of $\eta_m(\iota)$. $\eta_m \rightarrow 0$ as $\iota \rightarrow 1$. From Equation 8.47:

$$\frac{\partial \eta_m}{\partial \iota} = \frac{\partial \eta}{\partial \iota}(1-\iota^2) + \eta(-2\iota)$$

and $\partial\eta_1/\partial\iota = \partial\eta/\partial\iota > 0$ at $\iota \rightarrow 0$. Therefore η_m as a function of ι must have a maximum when α varies from 0 to 1. To quantify this effect calculations of η_m are required. In case of ring-shaped particles they can be simplified through an approximate calculation of the geometry factor, necessary for the calculation of the first Aris number.

The geometrical factor for solid and hollow cylinders, can be calculated with a high accuracy as

$$\Gamma = \frac{2}{3} \frac{I(\iota)\left[\lambda^2 + 4/3\lambda(1-\iota) + (1-\iota)^2\right]}{\left(I(\iota) + \lambda^2/3\right)\left(1-\iota^2\right)} \tag{8.47}$$

where $\lambda = H/R_u$ and

$$I(\iota) = \iota^2 + \frac{1-\iota^2}{2}\left(1 + \frac{1}{\ln(\iota)}\right)$$

If $\iota > 0.1$, that is for all practically interesting situations,

$$I(\iota) \approx \frac{1-\iota^2}{3}$$

and Equation 8.47 can be replaced by

$$\Gamma = \frac{2}{3}\frac{\lambda^2 + 4/3\lambda(1-\iota) + (1-\iota)^2}{(1-\iota^2+\lambda^2)}$$

For a solid cylinder $\iota = 0$, and

$$\Gamma = \frac{\lambda^2 + 4/3\lambda + 1}{3/2 + \lambda^2}$$

The dependence of the modified effectiveness factor on ι for a first-order isothermal irreversible reaction ($R_A = k_A C_A$) is presented in Figure 8.8 for different values of Thiele modulus $\phi = R_u\sqrt{k_A / D_{e,A}}$. All curves in this case are characterized by a gentle maximum, which becomes more pronounced when the ratio of the cylinder height to its external radius $\lambda = H/R_u$ increases. The value of ι providing the maximum also depends on λ: with increasing λ the maximum position shifts towards higher values of ι. The maximum relative increase of the modified effectiveness factor

$$\Xi = \frac{\eta_{m,\max} - \eta_m(0)}{\eta_m(0)}$$

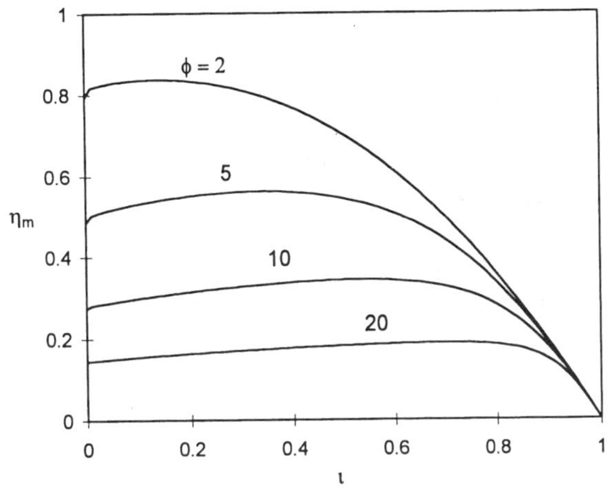

Figure 8.8 Modified effectiveness factor versus ratio of inner and outer diameters of the hollow cylinder for a first-order isothermal reaction.

is achieved at $\lambda = 4$ and is equal to 0.2. At $\lambda = 2$ and 1 the values of Ξ are 14% and 4%, respectively. In the more extreme cases the productivity of unit reactor volume may be increased by as much as a factor two when solid cylinders are replaced by hollow ones [11]. These calculations can guide the choice of particle.

The advantages of the ring-shaped particles are also found for other type of reactions. To demonstrate this, consider an adiabatic plug flow reactor assuming that the external mass and heat transfer limitations are negligible. Equations for fluid-phase concentration and temperature (which are equal to the concentration and temperature at the surface of the pellet) are

$$u_s \frac{dC_{A,s}}{dz} = -(1-\varepsilon_b)\eta(C_{A,s},T_s)R_A(C_{A,s},T_s) \tag{8.48}$$

$$u_s\rho_f c_p \frac{dT_s}{dz} = (-\Delta H)(1-\varepsilon_b)\eta(C_{A,s},T_s)R_A(C_{A,s},T_s) \tag{8.49}$$

with boundary conditions at the reactor inlet:

$$z=0 \quad C_{A,s}=C_{A,0}, \quad T_s=T_0 \tag{8.50}$$

From Equations 8.48 and 8.49 it follows that

$$\frac{d}{dz}\left(C_{A,s}+\frac{\rho_f c_p}{(-\Delta H)}T_s\right)=0 \tag{8.51}$$

Integrating Equation 8.51 with the boundary conditions of Equation 8.50 shows that the fluid phase temperature can be related to the fluid phase concentration of component A via:

$$T_s = T_0 + \frac{(-\Delta H)}{\rho_f c_p}(C_0 - C_{A,s}) \tag{8.52}$$

Therefore, Equations 8.48 and 8.49 can be combined into one equation for concentration only. The effectiveness factor for the case considered can be calculated with the technique described in Chapter 7. It is important to stress that the effectiveness factor changes along the reactor because parameters of the reaction rate expression, Equation 6.18, ε and α depend on the surface concentration and temperature. The calculated modified effectiveness factors for nonisothermal first-order reaction at different conversions $\zeta = (1- C_{A,s}/C_0)$ are shown in Figure 8.9 versus the ratio of inner and outer diameters of the hollow cylinder. The parameters chosen for the calculation are:

$$\varepsilon\big|_{z=0} = \frac{E_a}{RT_0} = 10, \quad \alpha\big|_{z=0} = \frac{(-\Delta H)D_{e,A}C_0}{\lambda_p T_0} = 0.01$$

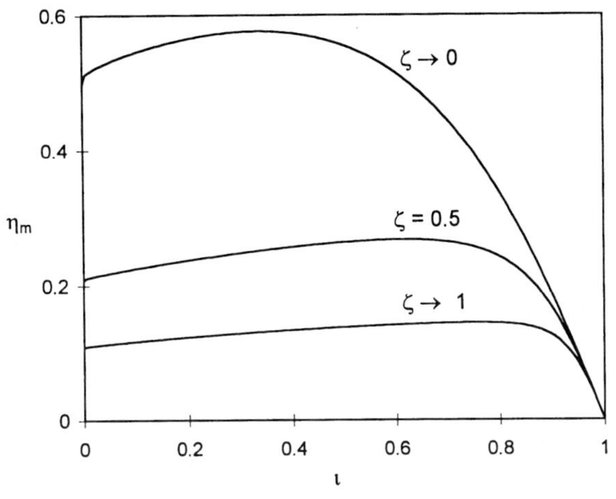

Figure 8.9 Modified effectiveness factor versus the ratio of inner and outer diameters of the hollow cylinder for a first-order nonisothermal reaction.

$$\frac{\Delta T_{ad}}{T_0} = \frac{(-\Delta H)C_0}{\rho_f c_p T_0} = 0.5, \qquad \lambda = \frac{H}{R_u} = 2$$

Qualitatively the results are similar to those found for isothermal reaction, although maximums in the dependence of η_m on ι are more pronounced.

The replacement of solid cylinders by hollow cylinders may significantly influence the performance of a packed bed reactor as a whole. The dependence of the reactor length, required for a conversion of 90%, on the relative hole radius ι is presented in Figure 8.10 for the same parameters as in Figure 8.9. A decrease in the reactor volume of 20 % is possible. It is important that this volume reduction is accompanied by an increase in porosity from 0.35 to 0.5. Use of Equation 8.45 shows that the pressure drop can be decreased 2.2 times.

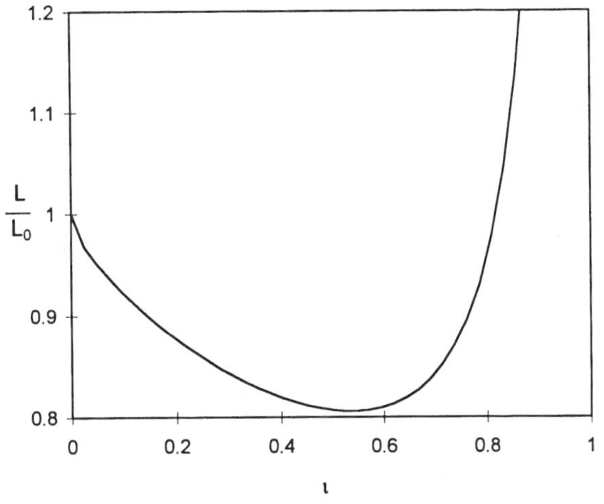

Figure 8.10 Adiabatic reactor length, required for a conversion of 90%.

8.3 Distribution of Catalytic Material in Pellets

Catalysts, prepared by impregnation of the porous support, in many cases exhibit intrapellet activity gradients, which are traditionally thought to be detrimental to catalyst performance. The effect of a deliberate nonuniform distribution of the catalytic material in the support, on the performance of a catalyst pellet received attention as early as the late 1960s [18,19]. These, as well as later studies, both experimental and theoretical, demonstrated that nonuniformly distributed catalysts can offer superior conversion, selectivity, durability, and thermal sensibility characteristics over those wherein the activity is uniform.

The performance indexes, which define an optimal catalyst distribution, include effectiveness, selectivity, yield and deactivation rate. The key parameters, affecting the choice of the optimal catalyst profile, are the reaction kinetics, the transport resistances, and the production cost of the catalyst. An extensive review of the theoretical and experimental developments in this area is available [20]. Two typical examples to demonstrate the importance of an appropriate distribution of the active components are now described.

8.3.1 Dehydrogenation of Butene

The following simple example by Wohlfahrt [21] which demonstrates the effect on selectivity for an industrial reaction; it is based on an experimental investigation by Voge and Morgan [22].

The dehydrogenation of *n*-butenes, diluted with steam, on a Shell 205 catalyst can be described approximately by two first-order reactions in series (A → B → C):

$$\text{butene} \xrightarrow{k_1} \text{butadiene} + \text{hydrogen}$$
$$\text{butadiene carbon} \xrightarrow{k_2} \text{dioxide} + \text{cracked products.}$$

The rate constants can be determined from the experimental data as [22]

$$k_1 = 1.97\times10^8 \exp(-16125/T),\ \mathrm{s^{-1}}$$

$$k_2 = 0.9k_1$$

Because of the large pores of the catalysts and the dilution with steam in a water:butene ratio of 12, constant molecular diffusion coefficients of 9.69×10^{-5} m^{2}s^{-1} (butene/water) and 9.95×10^{-5} m^2 s^{-1} (butadiene/water) at a temperature of 933 K can be assumed. The porous structure of the catalyst is represented by a ε_p/γ_p value of 0.1. The concentration profiles are given by Equations 8.22 and 8.23 for plate geometry. The results for R = 10 mm and R = 2 mm are shown in Figure 8.11 with $C_{B,s} = 0$.

The fraction of butadiene beyond the maximal concentration diffuses to the interior of the catalyst pellet and is lost. Thus the selectivity, given by Equation 8.27 amounts to 52% only. By decreasing the active layer of the catalyst to one-fifth of the original value, transport of butadiene from the point of maximum concentration to the pellet interior can be stopped, and selectivity is raised to 85 %.

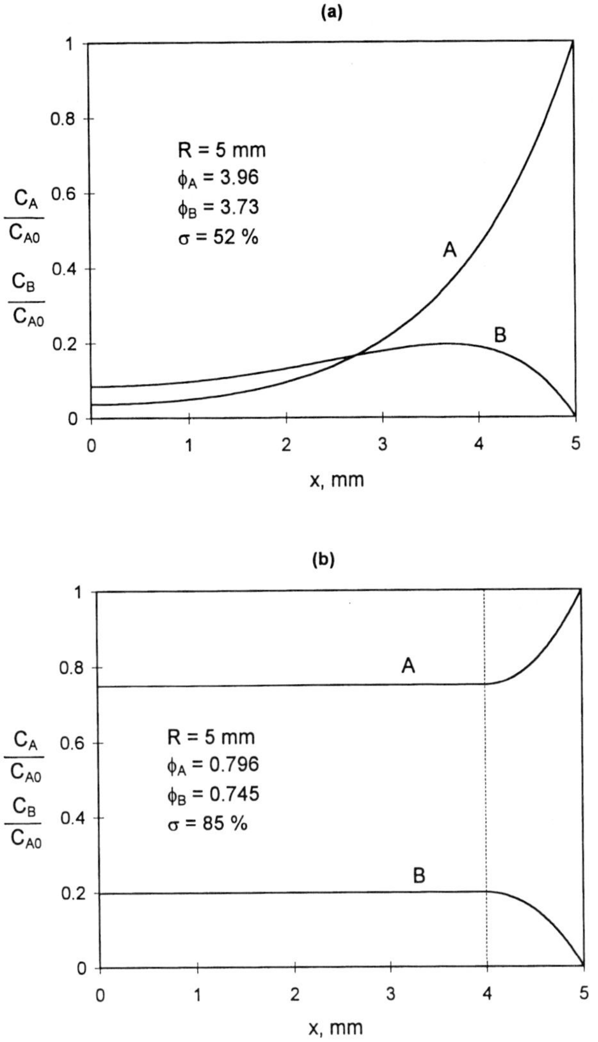

Figure 8.11 Concentration profiles of butene dehydrogenation: (a) uniformly impregnated pellet; (b) external layer of only 2 mm is impregnated (adapted from Wohlfahrt [21]). Reprinted from Chemical Engineering Science, **37**, K. Wohlfahrt, The design of catalyst pellets, 283-290, 1982, with kind permission from Elsevier Science Ltd, The Boulevard, Langford Lane, Kidlington OX5 1GB, UK.

8.3.2 Poison-resistant Catalyst for Automotive Emission Control

In automotive emission control, a catalyst is required which simultaneously catalyzes three reactions (oxidation of carbon monoxide and residual hydrocarbons, as well as reduction of nitrogen oxides) over a range of air/fuel ratio values centered around the stoichiometric point. This is a rather ambitious goal that requires simultaneous operation of the various active elements to perform different functions. Also, under normal, operating conditions, the catalyst is at a sufficiently high temperature that the reactions are diffusion controlled. For this reason, in catalyst manufacture the active components are deposited only close to the external surface of the pellet. An additional complication arises from the occurrence of two deactivating processes: poisoning due to the presence of trace quantities of lead and phosphorus in the feed, and thermal sintering [20].

Platinum and palladium, the noble metals that are currently employed in oxidative emission control, are unable to effectively control emissions of nitrogen oxides. For this reason rhodium is added to the catalyst, which catalyzes the reduction of nitrogen oxides, even in the presence of traces of oxygen [23]. However, the high cost of rhodium as compared to that of platinum and palladium and its strong sensitivity to phosphorus and lead poisoning motivates the use of small amounts of rhodium in the pellet and protecting it from poisoning.

The catalyst that satisfies these requirements was designed and tested by Hegedus et al. [23]. On average, the catalyst pellets contained 1.12 g of Pt, 0.44 g of Pd, and 0.06 g of Rh per car. These are located in three separate bands, as shown in Figure 8.12. The

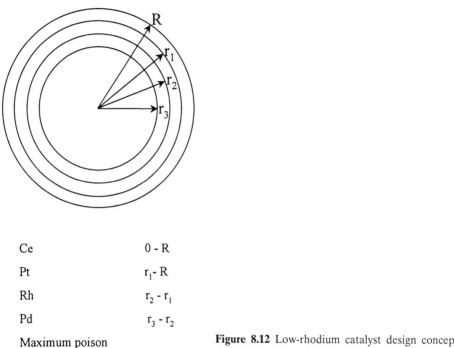

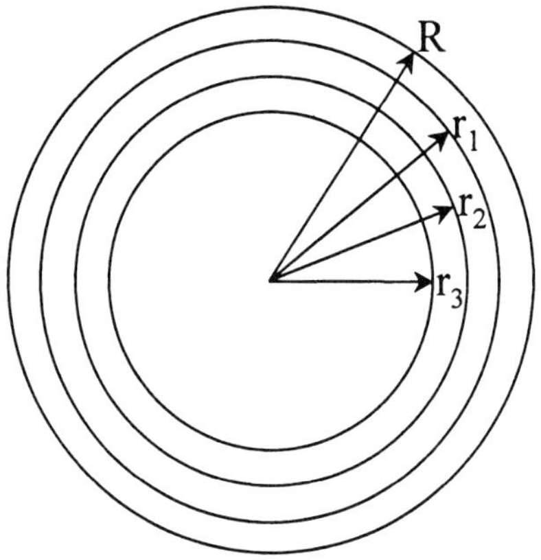

Ce	0 - R
Pt	r_1 - R
Rh	$r_2 - r_1$
Pd	$r_3 - r_2$
Maximum poison penetration	r_1 - R

Figure 8.12 Low-rhodium catalyst design concept (from Hegedus et al [23]).
Permission granted by Academic Press, INC.

first, Pt band, has a depth determined so as to entirely contain the expected penetration depth of the poison during the expected catalyst lifetime. Next is the Rh band, so that diffusion resistances in the nitric oxide reduction process are minimized and protection from poisoning is given. Finally the Pd band, at the deepest location, provides reasonable activity even at relatively low temperatures, typically encountered during engine startup.

This catalyst design indeed constitutes a remarkable example, being based only on experimental observations rather than quantitative models [20].

References

1. Topchieva, K.V., Antipina, T.V., Li He-Syan. (1960) *Kinetics and Katalysis*, **1**, 438.
2. Haynes, H.W., Sarma, P.N. (1973) *AIChE J.*, **19**, 1043.
3. Mingle, J.O. Smith, J.M. (1961) *AIChE J.*, **7**, 243.
4. Carberry, J.J. (1962a) *AIChE J.*, **8**, 557.
5. Carberry, J.J. (1962b) *Chem. Eng. Sci.*, **17**, 675.
6. Boreskov, G.K. (1972) in *The Porous Structure of Catalysts and Transport Processes in Heterogeneous Catalysis*: G.K. Boreskov (ed.). Budapest: Akadémiai Kiadó, pp. 1-20.
7. Hashimoto, N., Smith, J.M. (1974) *Ind. Eng. Chem., Fundam.*, **13**, 115.
8. Wheeler, A. (1951) *Advances in Catalysis*, **3**, 249.
9. Aris, R. (1975) *The Mathematical Theory of Diffusion and Reaction in Permeable Catalyst*, Vol. 1 Oxford: Clarendon Press.
10. Westerterp, K.R., Van Swaaij, W.P.M., Beenackers, A.A.C.M. (1987) *Chemical Reactor Design and Operation*. Chichester: Wiley.
11. England, R., Gunn, D.J. (1970) *Trans. Instn Chem. Engrs*, **48**, T265.
12. Rase, H.F. (1990) *Fixed-Bed Reactor Design and Diagnostics. Gas-Phase Reactions.* Boston: Butterworths.
13. Ergun, S. (1952) *Chem. Eng. Progr.*, **48**, 89.
14. Leva, M., Grummer, M. (1947) *Chem. Eng. Progr.*, **43**, 713.
15. Bird, R.B., Stewart W.E., Lightfoot, E.N. (1960) *Transport Phenomena*. New York: Wiley.
16. Gunn, D.J. (1967) *Chem. Eng. Sci.*, **22**, 1439.
17. Evenchik, N.S., Sokolinskii, Yu.A. (1977)*Teor. Found. Chem. Eng.*, **11**, 561.
18. Kasaoka, S., Sakata, Y. (1968) *J. Chem. Eng. Japan*, **1**, 138.
19. Minhas, S., Carberry, J.J. (1969) *J. Catalysis*, **14**, 270.
20. Gavriilidis, A., Varma, A. (1993) *Catal. Rev.-Sci. Eng.*, **35**(3), 399.
21. Wohlfahrt, K. (1982) *Chem Eng. Sci.*, **37**, 283.
22. Voge, H.H., Morgan, C.Z. (1972) *Ind. Eng. Chem. Process Des. Develop.*, **11**, 454.
23. Hedegus, L.L., Summers, J.C., Schlatter, J.C., Baron,K. (1979) *J. Catalysis*, **56**, 321.

9 Examples

9.1 Analysis of Rate Equations

Janssen et al. studied the kinetics of the hydrogenation of 2,4-dinitrotoluene (DNT) dissolved in methanol [1,2]. They distinguished a reaction scheme:

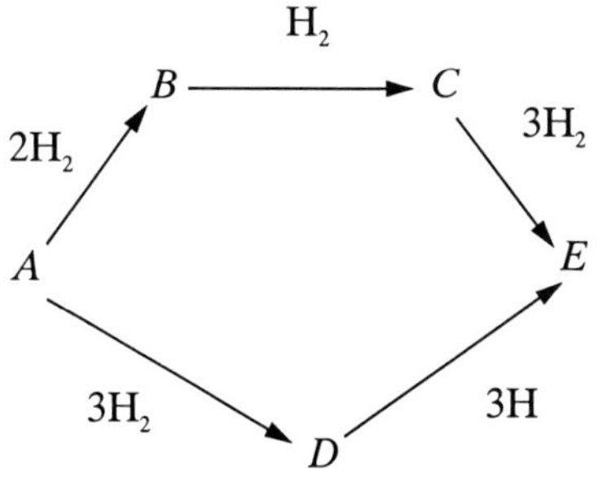

in which A = DNT, B = 4-hydroxylamino-2-nitrotoluene (4HA-2NT), C = 4-amino-2 nitrotoluene (4A-2NT), D = 2-amino-4 nitrotoluene (2A-4NT) and E = 2,4-diaminotoluene (DAT). They determined for all five reactions the reaction rate equations for a 5% Pd/C catalyst with an average particle size of 25 μm. Their rate equations are of the Langmuir-Hinshelwood type

$$R_{ij} = k_{AB} q_{ij} C_{cat} \theta_i \theta_H$$

where q_{ij}=k_{ij} /k_{AB} , C_{cat} is the catalyst concentration in $kg_{cat} m^{-3}$ liquid Further, k_{AB} = $4.0\times10^8\times\exp$ (-6850/T) mol DNT kg_{cat}^{-1} s^{-1}. θ_H is given by

$$\theta_H = \frac{K_H \sqrt{P_H / RT}}{1 + K_H \sqrt{P_H / RT}}$$

where p_H is the partial pressure of hydrogen in the vapor phase and $K_H = 4.3\times10^{-7}\times\exp$ (3630/T) $m^{3/2}$ $mol^{-1/2}$. Finally, θ_i is given by

$$\theta_i = \frac{Q_i c_i}{c_A + Q_B c_B + Q_c c_c + Q_D c_D + Q_E c_D}$$

in which c_i=C_i/C_0, C_0 is the initial DNT concentration in moles per m^3 liquid, and Q_i=K_i/K_A, where K_i are the chemisorption constants for the components $A - E$. The values for the various rate and chemisorption constants ratios is given in Table 9.1.

Table 9.1 Reaction rate and chemisorption constants for the hydrogenation of DNT

q_{AB}	= 1	Q_A	= 1
q_{BC}	= 290 exp(-1730/*T*)	Q_B	= 645 exp (-2800/*T*)
q_{CE}	= 0.013 exp (1160/*T*)	Q_C	= 0.13
q_{AD}	= 2.3 exp (-770/*T*)	Q_D	= 0.06
q_{DE}	= 1.1	Q_E	= 1.9×10^{-14} exp (8854/*T*)

From the rate equations follows a two-site mechanism: on one site hydrogen is adsorbed, and on the other DNT or its consecutive products. In the expression for the hydrogen concentration the partial pressure in the vapor phase is taken, so that K_H is a combination constant for the chemisorption and the solubility of hydrogen in the liquid phase.

Janssen et al. [1,2] tested the rate equation over the following ranges: C_{cat} from 0.5 to 2.0 kg m^{-3}, C_0 from 100 to 450 mol DNT per m^3 liquid, *T* from 308 to 357 K, P_H from 0.21 to 3.72 MPa and P_H/RT from 78 to 1437 mol H_2 per m^3 vapor phase

The above rate equations are now analyzed, especially the influence of pressure, temperature and the degrees of occupation. First we have to determine what the adiabatic temperature rise is in order to know how dangerous the reactions may be. The data of Janssen et al. show that for the reaction of DNT → DAT, for maximum hydrogen consumption, $(-\Delta H) = 1180\times10^6$ J $kmol^{-1}$, and $\rho c_p = 2.08\times10^6$ J $m^{-3}K^{-1}$, so that for a concentration of 0.45 kmol DNT m^{-3}

$$\Delta T_{ad} = 256\,K$$

This temperature increase is so high, that care must be taken with these reactions.

In Figures 9.1 and 9.2. the initial reaction rate R_{AB}=$k_{AB}\theta_H$ and θ_H are plotted as a function of the hydrogen partial pressure and the reaction temperature. Further, C_{cat} = 2 kgm^{-3}, C_0=400 $DNTm^{-3}$ and c_A=1 in these plots. At low temperatures of say 25 °C, it hardly makes sense to increase the hydrogen pressure above 2.0 MPa and, at higher temperatures of say 100 °C, not above 4.0 MPa. It can be observed that the degree of occupation of the catalyst with hydrogen at low temperature is relatively high – between 0.45 and 0.70 in the appropriate pressure range – but at higher temperatures has decreased considerably to between 0.06 and 0.2. In the range studied the value of k_{AB} still increases more rapidly than the value of θ_H decreases with temperature.

In Figure 9.3 $k_{AB}\theta_H$ is plotted as a function of 1/*T* as well as the initial selectivity R_{AB}/R_{AD}. An ever-decreasing slope for increasing temperature of the initial rate R_{AB} is observed, which is caused by the decreasing degrees of occupation of the catalyst by hydrogen. Also, from 25 °C to 100 °C the ratio of the initial formation of *B* over that of *D* decreases from 5.75 to 3.45. Therefore to make component *B* as selectively as possible, we have to operate at low temperatures.

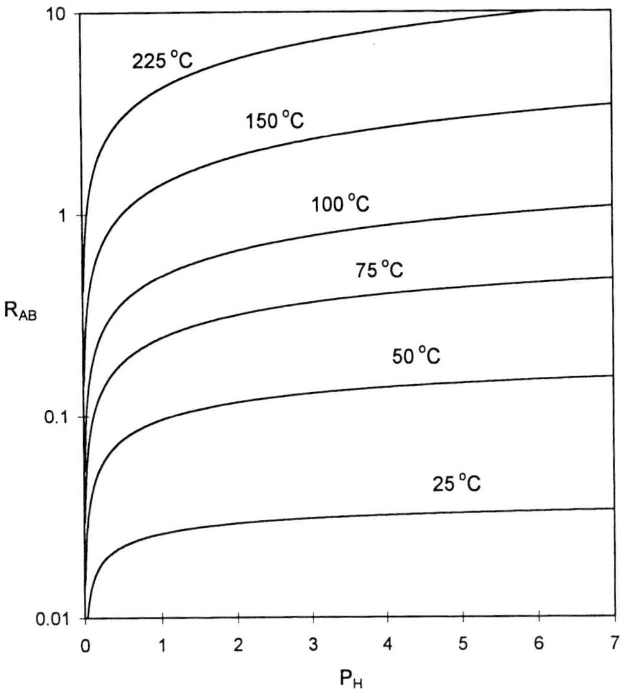

Figure 9.1 Initial reaction rate as a function of the hydrogen partial pressure and temperature.

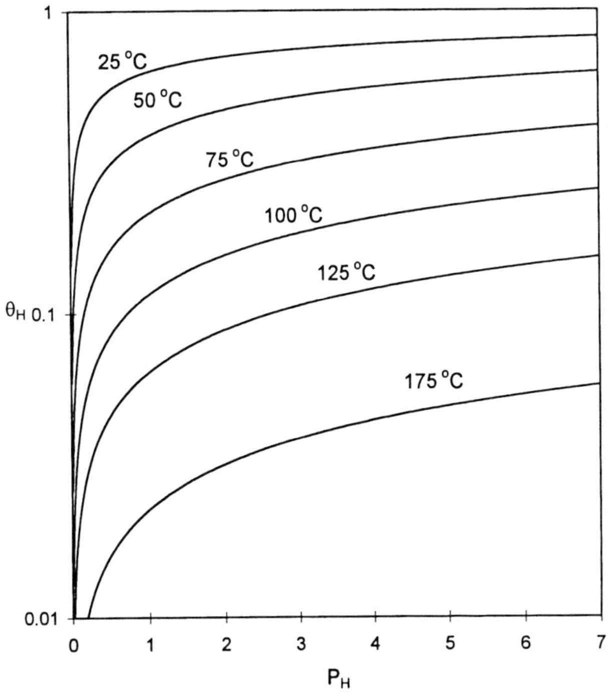

Figure 9.2 Degree of occupation for hydrogen as a function of the hydrogen partial pressure and temperature.

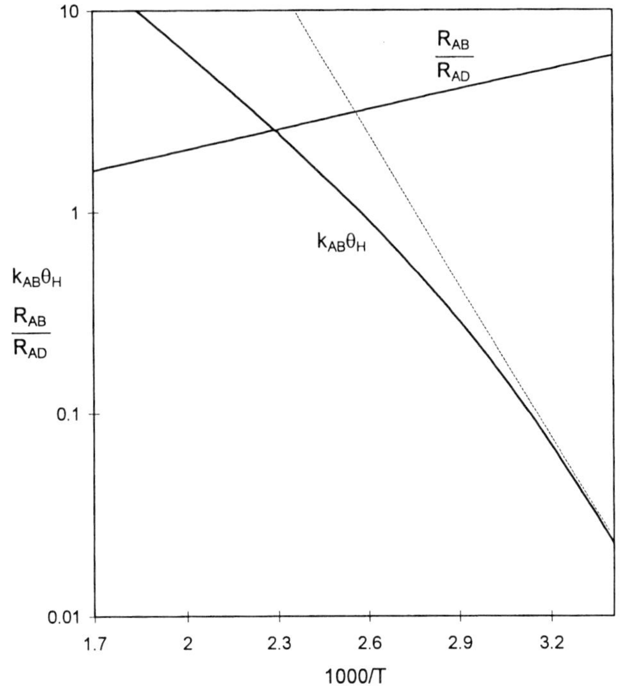

Figure 9.3 Initial reaction rate and selectivity as functions of temperature.

In Figure 9.4 the degrees of occupation have been plotted under certain conditions for components A, B and D as a function of the conversion of A. Over almost the whole conversion range the catalyst is almost exclusively occupied by component A. For a conversion below 73% θ_B is below 0.10 and for values of below 97% θ_D remains below 0.10. This is known as molecular queuing: as the conversion proceeds, the catalyst must first be cleaned of the first component, before the next one can be adsorbed and converted. To this end Figure 9.5 shows for the same conditions R_{BC}/R_{AB} and R_{DE}/R_{AD} as a function of the conversion of A. It is seen that R_{DE} becomes equal to R_{AD} only after 92% of the available A has been converted, and R_{BC} becomes equal to R_{AB} only after a conversion $\zeta_A = 0.96$.

9.2 Comparison of Published Rate Equations

In literature many rate equations can be found. It is often difficult to make a choice between them. To this end an analysis has to be made of the experimental conditions and the fits of the data by rate equations. As a general rule use of rate equations derived for a different catalyst or based on data outside the range of application should be avoided. A number of rate equations derived by different methods but for the same catalyst are compared.

Table 2.1 gives results for the two best fitting rate equations for the synthesis of methanol from CO and H_2 as obtained by two different researchers. Kuczynski et al. [3] de-

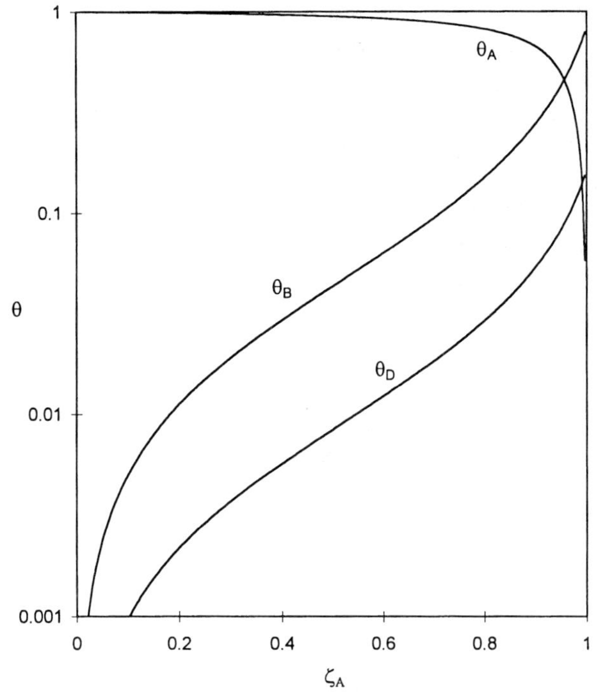

Figure 9.4. Degrees of occupation as a function of the conversion of A.

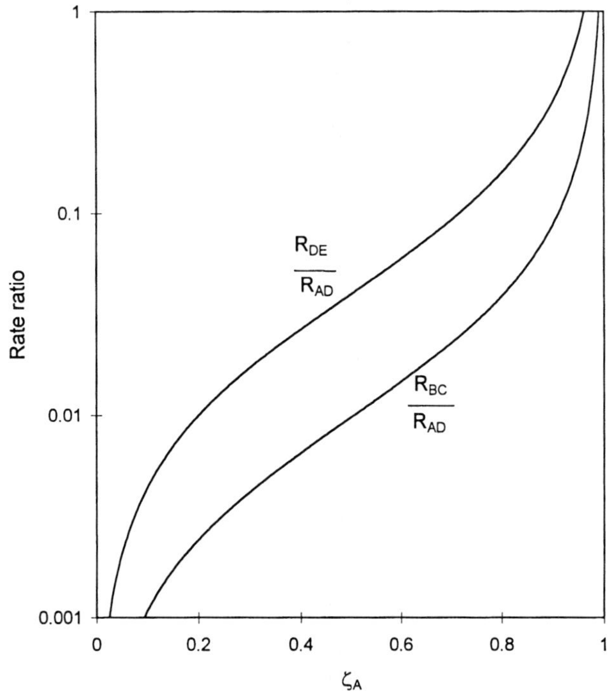

Figure 9.5. Ratio of reaction rates as a function of the conversion of A.

termined the reaction rates in a small integral reactor with a tube diameter of 8 mm, for CO fractions of 0.10 – 0.54, pressures of 3 – 9 MPa and temperatures of 210 – 272 °C and for the BASF S3.85 low-pressure Cu catalyst. Later, Bos et al. [4] repeated the experiments with the same catalyst in a Berty reactor and obtained more accurate results over a narrower range of conditions. In Figure 9.6 a plot is given of the initial reaction rate for the synthesis of methanol over the temperature range studied. The results of Kuczynski et al. for rate equation 1 deviate strongly from the other three data fits. The discrepancy is so acute that it is reasonable to conclude that some of the parameters reported must be wrong, most probably the activation temperature. The three remaining correlations exhibit a similar slope, with the two of Bos et al. not deviating by more than 30% of each other. The correlation of Kuczynski et al. for rate equation 2, especially in the lower temperature region, produces lower rates. This may be caused by the experimental method: two travelling thermocouples were used, located in sheaths. Heat conduction in the sheaths will considerably smooth out the measured temperature profile. If the measured temperature profiles had been under reaction conditions around 10-12 °C higher, the data of Kuczynski et al. and of Bos et al. would more or less coincide. The last question is which correlation it is preferable to use. The above discussion suggests a preference for those of Bos et al. leaving a choice between the correlations according to equations 1 and 2. The degrees of occupation of the Cu on ZnO catalyst by CO and H_2 could be investigated. Over the temperature range considered (not shown here, but easily calculable) for θ_{CO} a range of 0.50 – 0.57 and for θ_{H_2} of 0.19 – 0.30 is found for rate equation 2 of Bos et al., whereas for equation 2 θ_{CO} varies from 0.80 to 0.025 and θ_{H_2} varies from 0.0006 to 0.011. In particular, the θ_{H_2} values for equation 2 do not seem very realistic. They point to an Eley-Rideal mechanism. This suggests a slight preference for

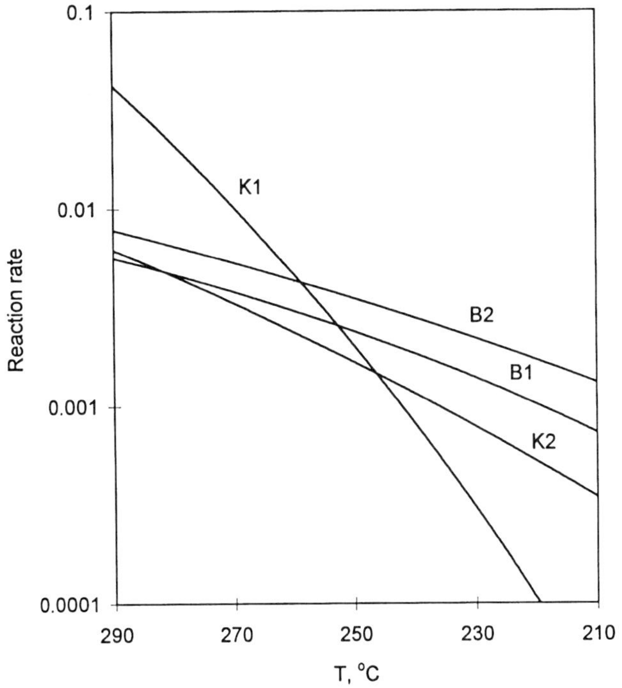

Figure 9.6 Initial reaction rate for the synthesis of methanol over the temperature range studied: curves K1 and K2 are equations 1 and 2 derived by Kuczynski et al. [3]; curves B1 and B2 are equations 1 and 2 derived by Bos et al. [4].

the correlation equation 1 of Bos et al. for the methanol synthesis rates. (Note, however, that the analysis has been made for the initial rates only.)

9.3 Permeation of Gases through a Porous Slug

Chapter 3 described a new model for transport through porous media, developed recently by Kerkhof [5] and called the binary friction model (BFM). It is of interest to see how this model can be applied to the description of available experiments and to compare the results with those of the dusty gas model (DGM). Kerkhof [5] took the experimental data of Evans et al. [6,7] for the permeation of He and Ar through a low-permeability porous graphite septum. The experimental set-up, similar to the Wicke-Kallenbach diffusion cell, is sketched in Figure 9.7. Of interest are the steady

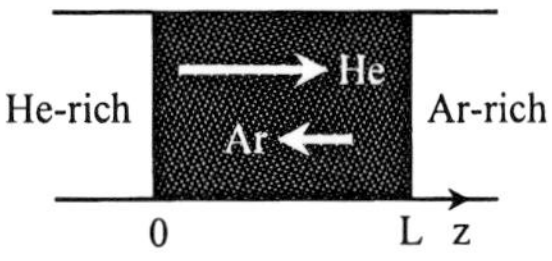

$\Delta P = P(L) - P(0)$

Figure 9.7 Schematic set-up of experiments by Evans et al. [6,7].

state fluxes through a porous septum, at both sides of which the partial pressures of the components are set constant. For a binary mixture of A and B, the flux equations are

$$-\frac{1}{RT}\frac{dP_A}{dz} = \frac{P_B J_A - P_A J_B}{PD^e_{AB}} + f_A J_A \tag{9.1}$$

$$-\frac{1}{RT}\frac{dP_A}{dz} = \frac{P_B J_A - P_A J_B}{PD^e_{AB}} + f_A J_A \tag{9.2}$$

where:

$$f_A = \left(D_{AK} + \frac{B_0 P}{\mu_{A,f}}\right)^{-1} \; ; \; f_B = \left(D_{BK} + \frac{B_0 P}{\mu_{B,f}}\right)^{-1} \tag{9.3}$$

Equations 9.1 and 9.2 contain five unknown variables: the partial pressures and fluxes of the components and the total pressure. To find the fluxes five additional equations are necessary. One obvious equation is

$$P = P_A + P_B \tag{9.4}$$

The boundary conditions at the opposite surfaces of the septum give the other four. Assuming there is no mass transfer resistance other than inside the septum, these boundary conditions can be written as

$$z = 0 \quad P_A = P_{A,0}, \quad P_B = P_{B,0} \tag{9.5}$$

$$z = L \quad P_A = P_{A,L}, \quad P_B = P_{B,L} \tag{9.6}$$

Thus there are two first-order differential equations (Equations 9.1 and 9.2) and five algebraic equations (Equations 9.4 – 9.6) with which to determine the two integration constants and the five variables. Different numerical techniques can be used to solve the problem. One way is to linearize Equations 9.1 and 9.2 and apply the iteration procedure described by Kerkhof [5]. An equation describing the variation of the total pressure inside the septum,

$$-\frac{dP}{dz} = RT(f_A J_A + f_B J_B) \tag{9.7}$$

is obtained by adding Equations 9.1 and 9.2. In one of their experiments, Evans et al. maintained a zero pressure difference over the septum, so that the total pressures at the surfaces were equal:

$$P_{A,0} + P_{B,0} = P_{A,L} + P_{B,L} \tag{9.8}$$

For this specific situation, it may be assumed that $dP/dz = 0$. Now, the flux of component B can be eliminated from Equation 9.7, which yields

$$J_B = -J_A \frac{f_A}{f_B} \tag{9.9}$$

$$\frac{dP_A}{dz} = RT\left[\frac{1}{PD^e_{AB}}\left(P_A\left(1-\frac{f_A}{f_B}\right) - P_A\right) - f_A\right] J_A \tag{9.10}$$

The fluxes can be obtained by direct numerical integration of Equation 9.10:

$$J_A = \frac{1}{RTL}\int_{P_{A,0}}^{P_{A,L}} \frac{dP_A}{\frac{1}{PD^e_{AB}}\left(P_A\left(1-\frac{f_A}{f_B}\right) - P_A\right) - f_A} \tag{9.11}$$

$$J_B = -\frac{1}{RTL}\int_{P_{A,0}}^{P_{A,L}} \frac{f_A}{f_B} \frac{dP_A}{\frac{1}{PD^e_{AB}}\left(P_A\left(1-\frac{f_A}{f_B}\right)-P_A\right)-f_A} \tag{9.12}$$

Kerkhof used the following values of parameters required to calculate the fluxes of He and Ar: molecular diffusivity $D_{AB} = 7.29\times10^{-5}$ m^2s^{-1} at 1 bar (as given by Reid et al., [8]), permeability $B_0 = 1.66\times10^{-14}$ m^2, porosity- tortuosity ratio $\varepsilon_p/\gamma_p = 1.28\times10^{-4}$ and the Knudsen diffisivities for He and Ar $D_{He,k} = 3.07\times10^{-4}$ m^2s^{-1} and $D_{Ar,k} = 9.72\times10^{-5}$ m^2s^{-1}, respectively. For the fractional viscosities Equation 3.25 was used. The experimental data of Evans et al. compared to the predictions of Equations 9.11 and 9.12 are shown in Figure 9.8. The coverage is good, except at higher pressures in which the He flow as predicted by the model is lower than the experimental values. The Ar flux is predicted well over the whole range. Similar agreements and discrepancies were found for the DGM.

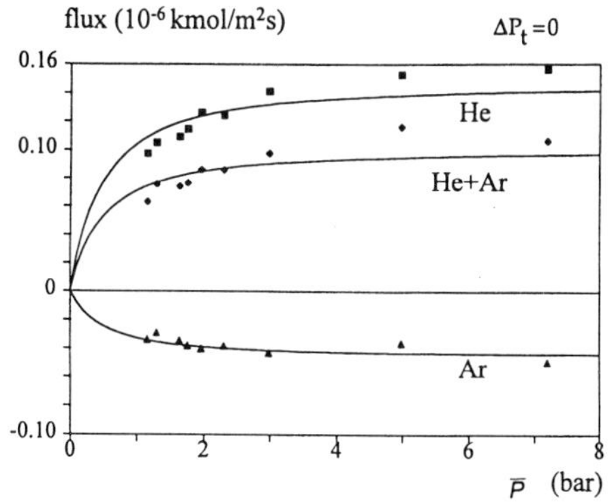

Figure 9.8 Fluxes of He and Ar through a low permeability graphite septum in the case of zero pressure difference over the septum at variable average pressure. Composition difference $\Delta x_{He} = 0.963$. Symbols show the experimental data of Evans et al. [6,7]; lines show results according to the binary friction model (from Kerkhof [5]). Reprinted from Chemical Engineering Journal, **64**, P.J.A.M. Kerkhof, A modified Maxwell-Stefan model for transport through inert membranes: the binary friction model, 319-344, 1996, with kind permission from Elsevier Science S.A., P.O. Box 564, 1001 Lausanne, Switzerland.

In other experiments, Evans et al. varied the pressure drop over the septum at various levels of average total pressure and for constant compositions at both sides. Some experimental data compared to the numerical solution of Kerkhof [5] are shown in Figure 9.9. For the average pressure

$$\overline{P} = (P(0) + P(L))/2$$

of 1.49 bar, the BFM describes the data well and only shows minor deviations for the Ar flux at large pressure differences. At the 4.93 bar level, there is an excellent correspondence. Mason and Malinauskas [9] and Krishna [10] found approximately the same results for the DGM.

One remark should be made. According to all the available theories the total pressure inside the septum should be uniform if the total pressures at its surfaces are equal. This is not the case for the BFM. However, the numerical solution of the problem shows that this total pressure variation can be neglected in the calculation of the fluxes and the assumption of a constant pressure over the septum made by Kerkhof [5] is correct.

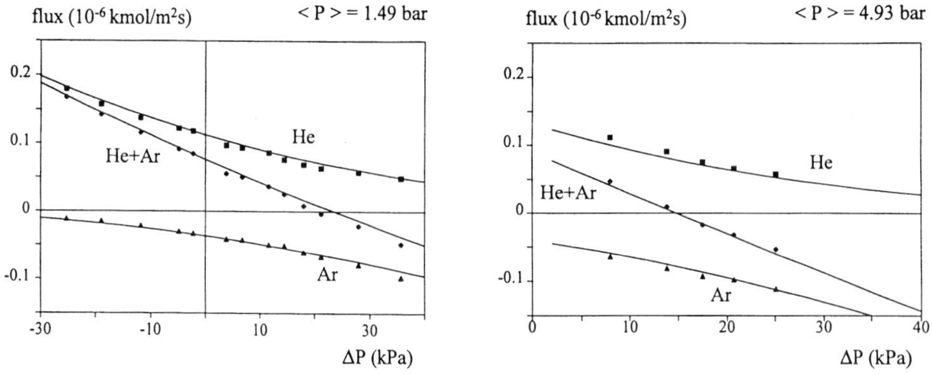

Figure 9.9 Fluxes of He and Ar through a low permeability graphite septum in the case of constant average pressures of (a) 1.49 bar and (b) 4.93 bar at varying pressure difference over the septum. Composition difference $\Delta x_{He} = 0.963$. Symbols show the experimental data of Evans et al. [6,7] (from Kerkhof [5]).
Reprinted from Chemical Engineering Journal, **64**, P.J.A.M. Kerkhof, A modified Maxwell-Stefan model for transport through inert membranes: the binary friction model, 319-344, 1996, with kind permission from Elsevier Science S.A., P.O. Box 564, 1001 Lausanne, Switzerland.

9.4 Gaseous Counter-diffusion in a Capillary

Some of the results of BFM experiments by Waldmann and Schmitt [11] cannot be explained with the DGM. These experiments are similar to the classic experiments by Kramers and Kistemaker [12] on gas counter-diffusion in a capillary, but Waldmann and Schmitt conducted experiments for a broader range of gas pairs. The experimental set-up is shown schematically in Figure 9.10. It basically consists of two chambers of 64.2

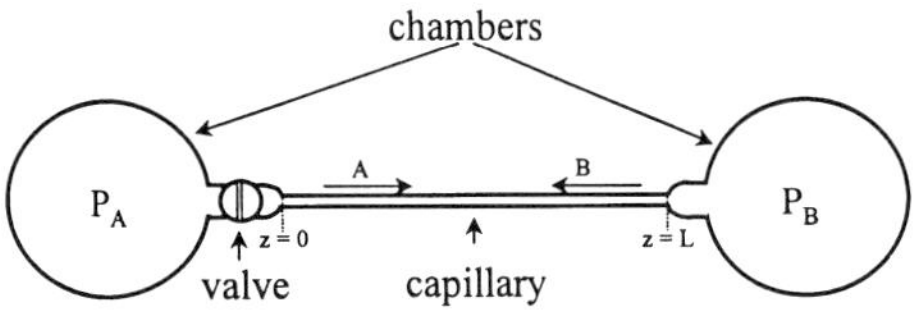

Figure 9.10 Schematic set-up of Waldmann and Schmitt [11].

and 32.1 cm^3 connected by a capillary of 0.048 m length and 0.198 mm diameter. Initially the left chamber is filled with a pure gas A to the desired pressure and the chamber and capillary on the right are filled with another pure gas B at the same pressure. On opening the valve the gases interdiffuse through the capillary. The pressure difference across the capillary during the counter diffusion were measured as a function of time. All measurements were made at 20°C. During these experiments the pressure difference builds up, reaches a maximum, and after that the system goes slowly to equilibrium. This means that, in spite of the uniform pressure, there was a net molar flux. The reason for this is a difference of the mean molecular speeds, which are inversely proportional to the square root of the molecular mass, see Chapter 3.

The two gas pairs – $N_2 - C_2H_4$ and $Ar - CO_2$ – are of interest, because the experimental data for these systems contradict the DGM. The sum of Equations 3.21 and 3.22 of the DGM gives

$$-\frac{1}{RT}\left[1+\frac{B_o}{\mu}\left(\frac{P_A}{D_{Ak}}+\frac{P_B}{D_{Bk}}\right)\right]\frac{dP}{dz}=\frac{J_A}{D_{Ak}}+\frac{J_B}{D_{Bk}} \tag{9.13}$$

Initially the total pressure is uniform, so dP/dz = 0 and the ratio of the fluxes is

$$\frac{J_A}{J_B}=-\frac{D_{Ak}}{D_{Bk}} \tag{9.14}$$

Substituting the Knudsen diffusivities from Equation 3.6 into Equation 9.14 gives the famous and debatable Graham's relationship

$$\frac{J_A}{J_B}=-\sqrt{\frac{M_B}{M_A}} \tag{9.15}$$

The molecular weights of nitrogen and ethylene are almost equal: M_{N_2} = 28.013, $M_{C_2H_4}$= 28.054. Therefore, approximately equal molar fluxes of the components might be expected and, as a consequence, there should be no pressure difference between the chambers. Similarly for Ar and CO_2 we have M_{Ar} = 39.948 and N_{CO_2} = 44.010 and, according to Equation 9.15 the flux of Ar should be larger then for CO_2. Therefore, a higher pressure is expected in the chamber filled originally with CO_2. Both these expectations are confirmed by numerical solutions of the DGM equations. However, these results are in contrast with the experimental findings of Waldmann and Schmitt [11]. For the system $N_2 - C_2H_4$ they observed a pressure difference with the highest pressure on the nitrogen side. For the system $Ar - CO_2$ Waldmann and Schmitt made experiments over a wide range of average pressures and observed that, at low average pressures, the maximum pressure occurs at the CO_2 side, as expected from Equation 9.15, but at higher pressures this effect is reversed: the pressure increased at the Ar side.

The BFM predicts the observed qualitative behavior correctly. The calculations made by Kerkhof [5] for the two systems are presented in Figures 9.11 and 9.12. Mathematically the problem is solved in a similar way as described in the previous example, only the partial pressures in the chambers are not constant. The variations of the partial pressures in the chambers can be calculated assuming ideal mixing in the chambers. Physically, the observed behavior can be explained by the difference in the component viscosities, which indeed is taken into account in the BFM. The flux of the less viscous component is larger at the same partial pressure difference. More explanations of this interesting problem are given in a second paper by Kerkhof [13].

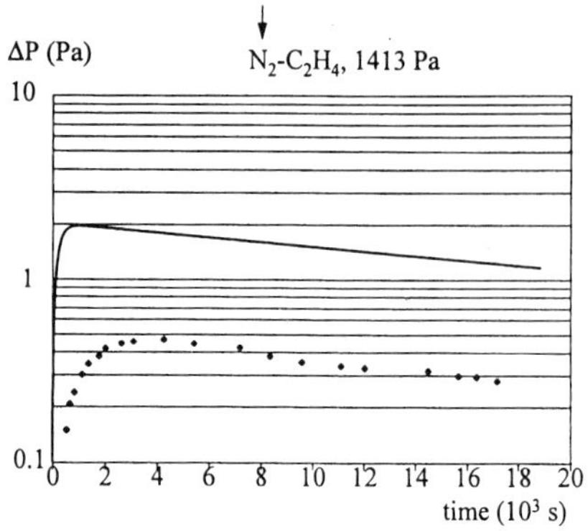

Figure 9.11 Net pressure difference in the counterdiffusion of N_2 and C_2H_4. Symbols show the experimental data of Waldmann and Schmitt [11]; drawn line show simulation with the binary friction model. Highest pressure on the nitrogen side. The DGM predicts no pressure difference (from Kerkhof [5]). Reprinted from Chemical Engineering Journal, **64**, P.J.A.M. Kerkhof, A modified Maxwell-Stefan model for transport through inert membranes: the binary friction model, 319-344, 1996, with kind permission from Elsevier Science S.A., P.O. Box 564, 1001 Lausanne, Switzerland.

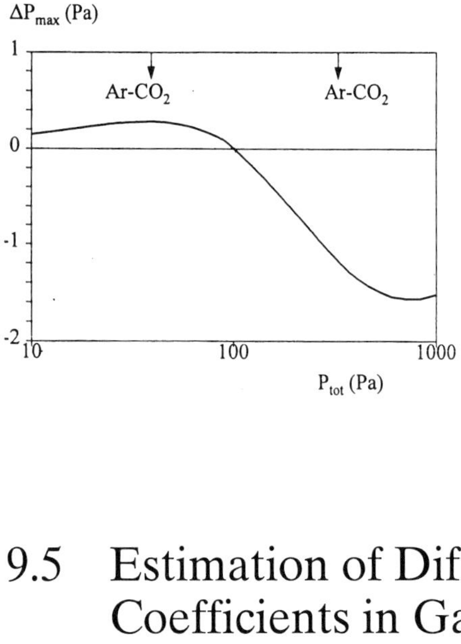

Figure 9.12 Reversal of the maximum pressure difference in the counterdiffusion of Ar and CO_2 as simulated with the binary friction model for the experimental situation of Waldmann and Schmitt [11]. At low average pressure, highest pressure on the CO_2 side (from Kerkhof [5]). Reprinted from Chemical Engineering Journal, **64**, P.J.A.M. Kerkhof, A modified Maxwell-Stefan model for transport through inert membranes: the binary friction model, 319-344, 1996, with kind permission from Elsevier Science S.A., P.O. Box 564, 1001 Lausanne, Switzerland.

9.5 Estimation of Diffusion Coefficients in Gases

Estimate the diffusion coefficient of allyl chloride (AC) in air at 298 K and 1 bar. The experimental value is 0.098 cm²s⁻¹. P = 1 bar; with M_{AC} = 76.5 and M_{air} = 29.0, M_{AB} = $2\times[(1/76.5) + (1/29.0)]^{-1}$ = 42.0; and T = 298 K. For air, $\sum_v$ = 19.7, and for allyl chloride (C_3H_5Cl) using Table 3.6, $\sum_v = 3\times15.9 + 5\times2.31 + 21.0 = 80.25$. Substitution in Equation 3.37 gives

$$D_{AB}=\frac{0.00143\times(298)^{1.75}}{1\times(42.0)^{1/2}\left[(19.7)^{1/3}+(80.25)^{1/3}\right]^2}=0.096\ \text{cm}^2\text{s}^{-1}$$

The error is -2%.

At low to moderate pressures, binary diffusion coefficients vary inversely with pressure as can be seen in Equation 3.37. At high pressures, the product $D \times P$ is no longer constant but decreases with increasing P. Also, as indicated earlier, at low pressures, the binary diffusion coefficient is independent of composition. At high pressures, where the gas phase may deviate significantly from an ideal gas, small effects of composition have been noted.

Although Equation 3.37 shows that D_{AB} is proportional to $T^{1.75}$, the power to which T should be raised depends on the actual temperature. Thus D_{AB} varies as $T^{3/2}$ to T^2 [8]. The value of 1.75 represents an average and may be not appropriate if wide temperature ranges are to be considered.

9.6 Estimation of Diffusion Coefficients in Liquids

Estimate D_{AB} for a dilute solution of TNT (2,4,6-trinitrotoluene) in benzene at 15° C [14].

The required data are: μ = 0.705 cP (for solution considered as pure benzene); V_b = 140 cm^3 mol^{-1}; X = 1 for benzene; M = 78.11 for benzene. Substitution in Equation 3.38 gives

$$D_L = 7.4\times 10^{-8}\,\frac{(1.0\times 78.11)^{0.5}(273+15)}{0.705(140)^{0.6}} = 1.38\times 10^{-5}\ \text{cm}^2\text{s}^{-1}$$

The measured value is 1.39×10^{-5} cm^{2}s^{-1}.

9.7 Measurement of Reaction Kinetics and Effective Pellet Diffusivity in a Single-pellet Reactor

The isomerization of n-hexane at 453° C and a total pressure of 1.6 bars has been investigated by Christoffel and Schuler [15]. They used a single pellet reactor with a pellet thickness L of 6.22×10^{-3} m. The porous pellet material contained 0.35 wt % Pt on Al_2O_3. The flow rate along the pellet was so high that the observed conversion rate was independent of the flow rate. The partial pressure of n-hexane in the feed was varied between 15 and 160 mbar. The ratio of the n-hexane concentration in the space below the pellet $C_{A,b}$ and in the gas mixture above the pellet $C_{A,s}$ was found to be independent of $C_{A,s}$: $C_{A,b}/C_{A,s} = 0.191$. The apparent conversion rate was found to be proportional to the n-hexane concentration in the feed and to be represented by $R_A = 0.184\times C_{A,s}$, kmol m^{-3} s^{-1}. From these results the following information can be obtained:

- the reaction order
- the value of the kinetic rate constant
- the effective diffusivity in the pellet.

Because the conversion rate is independent of the external flow rate, external diffusion limitations can be neglected. Both from the proportionality of the conversion rate with the feed concentration and the independence of $C_{A,b}/C_{A,s}$ from $C_{A,s}$ it follows that the reaction is first order, so that $C_{A,b}/C_{A,s} = 1/\cosh\phi$, $R_A = k_1 C_{A,s}(\tanh\phi)/\phi$ and $D_{e,A} = L^2 k_1/\phi^2$. From the data given, $\phi = 2.34$, $k_1 = 0.439\ \mathrm{s^{-1}}$ and $D_{e,A} = 3.1\times10^{-6}\ \mathrm{m^2\,s^{-1}}$.

9.8 Langmuir-Hinshelwood Kinetics in a Ring-shaped Catalyst

A ring-shaped catalyst pellet has the dimensions $R_i \times R_u \times H = 2 \times 4 \times 8$ mm. In the pellet a simple reaction is carried out with Langmuir-Hinshelwood kinetics:

$$R(C_A) = \frac{kC_A}{1+kC_A}$$

Further data are: $k = 0.8\ \mathrm{s^{-1}}$, $KC_{A,s} = 1$ and $D_e = 2\times10^{-7}\ \mathrm{m^2\,s^{-1}}$. To calculate the effectiveness factor, Equation 6.29 we gives

$$An_0 = \left(\frac{V_P}{A_P}\right)^2 \frac{k}{D_e}\left(\frac{KC_{A,s}}{1+KC_{A,s}}\right)^2 \frac{1}{2\{KC_{A,s} - \ln(1+KC_{A,s})\}} = 1.043$$

Since ι = ½ and λ = 2, from Figure 6.10 the geometry factor Γ equals 1.12. Then, from Equation 6.38

$$An_1 = \left(\frac{V_P}{A_P}\right)^2 \Gamma \frac{k}{D_e} \frac{1}{(1+KC_{A,s})^2} = 0.712$$

According to Equation 6.57

$$\tilde{\eta} = \frac{1}{\sqrt{1+An_1}} = 0.763$$

and Equation 6.58 yields

$$\tilde{\eta} = \frac{1}{\sqrt{1+An_0}} = 0.700$$

Apparently we operate in the high η region and thus expect Equation 6.57 to be more accurate than Equation 6.58. This is confirmed when using the much more accurate approximation 6.59 to calculate the effectiveness factor:

$$\tilde{\eta} = \frac{1}{\sqrt{1 + \tilde{\eta} An_1 + (1 - \tilde{\eta}) An_0}} = 0.746$$

From the data given in Table 6.8 and Figure 6.17 we estimate that Δ_{max} must be a few per cent only. This means that the relative error Δ must be smaller than a few per cent, let us assume that it is 2 %. In the high η region, Δ is the relative error in 1-η and hence the absolute error in the calculated value of η must be roughly

$(1-\eta) \times \Delta = (1-0.746) \times 0.02 = 0.005$

Hence, $\eta = 0.746 \pm 0.005$

Thus the value of η obtained will be more than accurate enough for all practical cases.

9.9 Nonisothermal Catalyst Pellets and First-order Reactions

For first-order reactions the formulae 7.13 and 7.14 become:

$$An_0 = \left(\frac{V_P}{A_P}\right)^2 \frac{k}{D_e} \frac{\zeta^2}{2\left(e^{\zeta} - 1 - \zeta\right)}$$

$$An_1 = \left(\frac{V_P}{A_P}\right)^2 \Gamma \frac{k}{D_e} (1 - \zeta)$$

9.10 Effectiveness Factor for Nonisothermal Catalyst Pellets

A reaction of the order ½ is carried out in a spherical catalyst pellet. For the given surface temperature and concentration the product $k_S C_{A,s}^{-½}$ equals 0.2 s^{-1}. The effective diffusion coefficient has been determined as 2×10^{-7} m^2 s^{-1}. The diameter of the sphere is 6 mm. Fur-

thermore, the following data have been estimated: E_a = 40 kJ mole^{-1}, $(-\Delta H)$ = 400 kJ mole^{-1}, $C_{A,s}$ = 20 mole m^{-3}, T_s = 550 K, λ_p = 0.5 W m^{-1} K^{-1}.
The effectiveness factor must be calculated, with and without neglecting the heat effects. According to Equations 7.13 and 7.14, for the Aris numbers

$$An_0 = \left(\frac{V_P}{A_P}\right)^2 \frac{kC_{A,s}^{-\frac{1}{2}}}{D_e} \frac{\zeta}{2}\left(\frac{1}{2}\sqrt{\frac{\pi}{\zeta}}\, e^{\zeta}\, erf\sqrt{\zeta} - 1\right)^{-1}$$

$$An_1 = \left(\frac{V_P}{A_P}\right)^2 \frac{\Gamma}{D_e} kC_{A,s}^{-\frac{1}{2}}\left(\frac{1}{2} - \zeta\right)$$

For a sphere the geometry factor Γ equals 6/5 (Table 6.3). Furthermore, V_p/A_p = 1 mm and ζ = 0.051. Thus, with heat effects (ζ = 0.051)

$$An_0 = 0.733$$

$$An_1 = 0.540$$

Hence

$$\tilde{\eta} = \frac{1}{\sqrt{1 + \tilde{\eta}An_1 + (1 - \tilde{\eta})An_0}} = 0.796$$

Without heat effects (ζ = 0)

$$An_0 = 0.750$$

$$An_1 = 0.600$$

Hence

$$\tilde{\eta} = \frac{1}{\sqrt{1 + \tilde{\eta}An_1 + (1 - \tilde{\eta})An_0}} = 0.783$$

The intraparticle temperature gradients result in an increase in the effectiveness factor. This is obvious since the reaction is strongly exothermic. The increase, however, is only 2 % relative. Thus in this case intraparticle temperature gradients can be neglected.

9.11 Exothermic, Zeroth-order Reaction in a Ring-shaped Catalyst

A zeroth-order reaction is carried out in a ring-shaped catalyst. The following numbers have been evaluated from measurements:

$$\phi^2 = \left(\frac{V_P}{A_P}\right)^2 \frac{k_s}{D_e C_{A,s}} = 0.500$$

$$\zeta = \frac{E_a}{RT_s} \frac{(-\Delta H) D_e C_{A,s}}{\lambda_P T_p} = 0.500$$

Hence, since $\zeta > 0$, the reaction is exothermic. The geometry of the catalyst is given by $\lambda = 1$ and $\iota = ¼$ (Equations 6.46 and 6.47). Then from Figure 6.10 the geometry factor Γ equals 1.11. The problem is to calculate the effectiveness factor.

For this situation the Aris numbers follow from (Equations 7.13 and 7.14)

$$An_0 = \phi^2 \frac{\zeta}{2(e^\zeta - 1)} = 0.193$$

$$An_1 = -\Gamma \phi^2 \zeta = -0.278$$

Since $An_1 < 0$, approximation 6.59 cannot be used. To calculate the effectiveness factor exactly involves solving partial differential equations, which is very time consuming. The effectiveness factor is therefore estimated as follows: construct an infinite slab in such a way, that for an exothermic zeroth-order reaction, it has the same Aris numbers as given above. Since the Aris numbers are generalized the hollow cylinder under consideration and the constructed slab will have almost the same effectiveness factor. Calculation of the effectiveness factor for a slab is relatively easy. Hence an estimate for the effectiveness factor for the hollow cylinder is obtained relatively easily.

First, construct the slab. Changing the geometry from a hollow cylinder to a slab changes the geometry factor from $\Gamma = 1.11$ to $\Gamma = 2/3$. To arrive at the same Aris numbers we must change the values of ϕ^2 en ζ. For the slab those values follow from:

$$An_0 = \phi^2 \frac{\zeta}{2(e^\zeta - 1)} = 0.193$$

$$An_1 = -\Gamma \phi^2 \zeta = -\frac{2}{3} \phi^2 \zeta = -0.278$$

Which yields $\phi^2 = 0.568$ and $\zeta = 0.732$ (compare with $\phi^2 = 0.500$ and $\zeta = 0.500$ for the hollow cylinder).

For an infinite slab, analytical solutions can be derived for the effectiveness factor. If η is plotted versus An_0 two branches occur (Figure 7.3). For the low An_0 branch, concentration profiles inside the catalyst pellet are as given in Figure 6.12a. For this branch the effectiveness factor can be calculated from

$$An_0 = \frac{1}{e^{\zeta} - 1}\left(\frac{\ln\left(x + \sqrt{x^2 - 1}\right)}{x}\right)^2$$

$$\eta = \frac{x\sqrt{x^2 - 1}}{\ln\left(x + \sqrt{x^2 - 1}\right)}$$

$$x \in \left[1, e^{\frac{1}{2}\zeta}\right] \qquad and \qquad \zeta > 0$$

For the high An_0 branch the concentration profiles are of the form as in Figure 6.12c. For this branch the effectiveness factor follows from

$$\eta = \frac{1}{\sqrt{An_0}}$$

When the two branches meet (concentration profile as in Figure 6.12b), the η-An_0 graph is nondifferentiable. For this specific point the zeroth Aris number has a critical value:

$$An_{o,cr} = \frac{\left\{\ln\left(\sqrt{e^{\zeta}} + \sqrt{e^{\zeta} - 1}\right)\right\}^2}{e^{\zeta}\sqrt{e^{\zeta} - 1}}$$

Notice that multiplicity can occur. Multiplicity can never occur, however, if

$$An_0 < An_{0,cr}$$

This condition is sufficient but not necessary to rule out multiplicity.

Returning now to the slab, it can be calculated (since $\zeta = 0.732$) that

$$An_{0,cr} = 0.606$$

Thus, $An_0 < An_{0,cr}$ (since $An_0 = 0.193$) and two conclusions can be drawn: we operate in the low An_0 branch and there is no multiplicity. The formulae for the low An_0 branch which yields $\eta = 1.19$ (and the dummy variable $x = 1.138$). Then, for the hollow cylinder given in the beginning of the example, there will be no multiplicity and the effectiveness factor must be roughly equal to 1.19.

9.12 Negligibility of Intraparticle Temperature Gradients

Apply the three criteria 7.23, 7.24 and 7.28 to the data given in example 9.10.

The reaction order $n = ½$ and hence the criteria for isothermal operation are

- in the high η region $|\zeta| < 0.05$
- in the low η region $|\zeta| < 0.5$
- of Mears $|\zeta| < 0.025$.

The actual value of ζ was 0.051. Thus according to criteria 7.23 and 7.28 the catalyst can be regarded isothermal. Even the high η criterion is as good as fulfilled. Since we operate in the intermediate η region, the relative error introduced by assuming isothermal operation must be much smaller than 10 %. Actually it is 2 %, as was calculated in Example 9.10.
According to Mears' criterion the catalyst cannot be regarded as being isothermal because $\zeta = 0.051 > 0.025$. This illustrates that Mears' criterion is too stringent.

9.13 Effectiveness Factors Larger than one for Langmuir-Hinshelwood Kinetics

Consider the following type of Langmuir-Hinshelwood kinetics:

$$R(C_A, C_B) = \frac{kC_A C_B}{(1 + K_A C_A + K_B C_B)^2} \tag{9.16}$$

and indicate when effectiveness factors larger than one can be expected.
By matter of definition this is the case when the high η Aris number An_1 becomes negative. An_1 can be calculated from Equation 7.35, which yields

$$An_1 = \left(\frac{V_P}{A_P}\right)^2 \frac{\Gamma}{D_{e,A}} kC_{B,s} \frac{(1+\beta) + (1-\beta)(K_B C_{B,s} - K_A C_{A,s})}{(1 + K_A C_{A,s} + K_B C_{B,s})^3} \tag{9.17}$$

It therefore follows that effectiveness factors larger than one can occur if

$$K_A C_{A,s} - K_B C_{B,s} > \frac{1+\beta}{1-\beta}$$

that is, if there is a large excess of one of the components and if the component not in excess is absorbed much more strongly than the other.

9.14 Effectiveness Factor Bimolecular Langmuir-Hinshelwood Kinetics

The following reaction is carried out in a cylindrical catalyst pellet:

$$A + 2B \rightarrow P$$

The kinetic expression for this reaction is given by Equation 9.16. Further data are: k = 1 m^3 mole^{-1} s^{-1}; K_A = 0.1 m^3 mole^{-1}; K_B = 1 m^3 mole^{-1}; $C_{A,s}$ = 2 mole m^{-3}; $C_{B,s}$ = 5 mole m^{-3}. The pellet dimensions are 4 mm height by 4 mm diameter. The effective diffusion coefficients of A and B are 2×10^{-7} m^2 s^{-1} and 3×10^{-7} m^2 s^{-1}, respectively. We wish to calculate the effectiveness factor.

The characteristic pellet dimension V_p/A_p is 2/3 mm. For the Aris numbers we find:

- For An_1. The geometry factor Γ equals 1.04, since $\lambda = 2$ and $\iota = 0$ (Figure 6.10). The number β is given by

$$\beta = \nu_B \frac{D_{e,A}}{D_{e,B}} \frac{C_{A,s}}{C_{B,s}} = 0.53$$

- An_1 can be calculated from Equation 9.17 as

$$An_1 = \left(\frac{V_P}{A_P}\right)^2 \frac{\Gamma}{D_{e,A}} kC_{B,s} \frac{(1+\beta)+(1-\beta)(K_B C_{B,s} - K_A C_{A,s})}{(1+K_A C_{A,s} + K_B C_{B,s})^3} = 6.59$$

- An_0 can be calculated from Equation 7.34, which yields

$$An_0 = \left(\frac{V_P}{A_P}\right)^2 \times \left(\frac{K_A C_{A,s} + \beta K_B C_{B,s}}{(1+K_A C_{A,s} + K_B C_{B,s})^2}\right)^2 \left(\frac{2\beta\mu + 2\beta - 1}{1+\mu} - (2\beta\mu + \beta - 1)\ \mathrm{ln}\frac{1+\mu}{\mu}\right)^{-1} \quad (9.18)$$

With β defined as in Equation 7.30 and μ as

$$\mu = \frac{1 + (1-\beta)K_B C_{B,s}}{K_A C_{A,s} + \beta K_B C_{B,s}} \tag{9.19}$$

from the available data we calculate $\mu = 1.18$ and $An_0 = 8.96$.

Now the effectiveness factor η can be calculated from

$$\tilde{\eta} = \frac{1}{\sqrt{1 + \tilde{\eta} An_1 + (1-\tilde{\eta}) An_0}} = 0.330$$

The relative error in the value obtained for $\tilde{\eta}$ is estimated to be roughly 1 % and thus

$$\tilde{\eta} = 0.330 \pm 0.003$$

9.15 Langmuir-Hinshelwood Kinetics and Intraparticle Temperature Gradients

Give a criterion for isothermal operation of a catalyst pellet for the Langmuir-Hinshelwood kinetics of Equation 9.16 in the high η region. Criteria for the high η region are more restrictive than for the low η region, so they will be more than adequately stringent for all values of η.

By combining the Equations 7.14 and 7.35 it can be seen that, for this case, An_1 is given by

$$An_1 = \left(\frac{V_P}{A_P}\right)^2 \frac{\Gamma}{D_{e,A}} \times \left. \frac{\partial \left(\dfrac{k C_A C_{B,s} \left\{1 - \beta\left(1 - \dfrac{C_A}{C_{A,s}}\right)\right\}}{\left(1 + K_A C_A + K_B C_{B,s}\left\{1 - \beta\left(1 - \dfrac{C_A}{C_{A,s}}\right)\right\}\right)^2} e^{\xi\left(1 - \frac{C_A}{C_{A,s}}\right)} \right)}{\partial C_A} \right|_{C_A = C_{A,s}} \tag{9.20}$$

Heat effects are negligible if the number

$$\delta_{\zeta,1} \approx \left| \frac{An_1|_{\zeta} - An_1|_{\zeta=0}}{An_1|_{\zeta=0}} \right| \tag{7.19}$$

is small enough (Equation 7.19). In this equation $An_1|_{\zeta}$ is the real Aris number and $An_1|_{\zeta=0}$ is the number which would have been found if the catalyst pellet were to operate isothermally. It can be calculated by putting ζ in Equation 9.20 equal to zero. Substitution of Equation 7.19 yields

$$\delta_{\zeta,1} = \left| \frac{1 + K_A C_{A,s} + K_B C_{B,s}}{(1+\beta) + (1-\beta)(K_B C_{B,s} - K_A C_{A,s})} \right| \times |\zeta|$$

Assuming that heat effects are negligible if $\delta_{\zeta,1}$ is smaller than 10%, the criterion for isothermal operation becomes

$$|\zeta| < \frac{1}{10} \times \left| \frac{(1+\beta) + (1-\beta)(K_B C_{B,s} - K_A C_{A,s})}{1 + K_A C_{A,s} + K_B C_{B,s}} \right| \tag{9.21}$$

From the discussion about simple reactions it is readily seen that a general criterion for isothermal operation in the high η region is:

$$|\zeta| < \frac{1}{10} \times \frac{C_{A,s}}{R(C_{A,s})} \left| \frac{\partial R(C_A)}{\partial C_A} \right|_{C_A = C_{A,s}} \tag{9.22}$$

Since all conclusions drawn for simple reactions also hold for bimolecular reactions, the general criterion for bimolecular reactions becomes:

$$|\zeta| < \frac{1}{10} \times \frac{C_{A,s}}{R(C_{A,s}, C_{B,s})} \times \left| \frac{\partial R\left(C_A, C_{B,s}\left\{1 - \beta\left(1 - \frac{C_A}{C_{A,s}}\right)\right\}\right)}{\partial C_A} \right|_{C_A = C_{A,s}} \tag{9.23}$$

Upon assuming Langmuir-Hinshelwood kinetics this formula directly yields Equation 9.21.

9.16 Approximation of the Effectiveness Factor for Bimolecular Reactions

Calculate the effectiveness factor for the data given in Example 9.14, using the approximation 7.36.

- For $\beta = 1$ the kinetic expression reduces to:

$$R(C_A) = \frac{\left(k\frac{C_{B,s}}{C_{A,s}}\right)\times C_A^{\ 2}}{\left\{1+\left(K_A + K_B\frac{C_{B,s}}{C_{A,s}}\right)C_A\right\}^2} = \frac{k^* C_A^{\ 2}}{(1+K^*C_A)^2}$$

From Equations 6.29 and 6.38 the Aris numbers are calculated as

$$An_1 = \left(\frac{V_P}{A_P}\right)^2 \frac{\Gamma}{D_{e,A}} \frac{2k^* C_{A,s}}{(1+K^*C_{A,s})^3} = 5.917$$

$$An_0 = \left(\frac{V_P}{A_P}\right)^2 \frac{k^* C_{A,s}}{2D_{e,A}} \frac{(K^*C_{A,s})^2}{(1+K^*C_{A,s})^4} \left(\frac{2+K^*C_{A,s}}{1+K^*C_{A,s}} - 2\frac{\ln(1+K^*C_{A,s})}{K^*C_{A,s}}\right)^{-1} = 7.964$$

The effectiveness factor follows from

$$\tilde{\eta} = \frac{1}{\sqrt{1+\tilde{\eta}An_1 + (1-\tilde{\eta})An_0}} = 0.348 = \tilde{\eta}\big|_{\beta=1}$$

- For $\beta = 0$ the kinetic expression reduces to:

$$R(C_A) = \frac{(kC_{B,s})\cdot C_A}{\{(1+K_B C_{B,s}) + K_A C_A\}^2} = \frac{k^{**} C_A}{(1+K^{**}C_A)^2}$$

Then the Aris numbers follow from

$$An_1 = \left(\frac{V_P}{A_P}\right)^2 \frac{\Gamma}{D_{e,A}} k^{**} \frac{1-K^{**}C_{A,s}}{(1+K^{**}C_{A,s})^3} = 10.124$$

$$An_0 = \left(\frac{V_P}{A_P}\right)^2 \frac{k^{**}}{2D_{e,a}} \frac{K^{**}C_{A,s}}{\left(1+K^{**}C_{A,s}\right)^3} \left(\frac{1+K^{**}C_{A,s}}{K^{**}C_{A,s}} \ln\left(1+K^{**}C_{A,s}\right)-1\right)^{-1} = 10.181$$

The effectiveness factor follows from:

$$\tilde{\eta} = \frac{1}{\sqrt{1 + \tilde{\eta}An_1 + (1 - \tilde{\eta})An_0}} = 0.299 = \tilde{\eta}\big|_{\beta=0}$$

From Example 9.14 $\beta = 0.533$, hence the estimate for the effectiveness factor is

$$\tilde{\tilde{\eta}}\big|_\beta = \tilde{\eta}\big|_{\beta=0} + \beta\left(\tilde{\eta}\big|_{\beta=1} - \tilde{\eta}\big|_{\beta=0}\right) = 0.326$$

If the relative deviation between χ and β is of the order of 10 %, the relative error in $\tilde{\tilde{\eta}}\,|_{\mathrm{b}}$ will be, from Equation 7.44:

$$\frac{\left|\tilde{\tilde{\eta}}\big|_\beta - \tilde{\eta}\big|_\beta\right|}{\tilde{\eta}\big|_\beta} \approx \frac{\left|\tilde{\tilde{\eta}}\big|_\beta - \tilde{\eta}\big|_{\beta=0}\right|}{\tilde{\eta}\big|_\beta} \times \left|\frac{\beta - \chi}{\chi}\right| = 1\ \%$$

This is in agreement with Example 9.14 where it was found $\tilde{\eta}\,|_{\mathrm{b}} = 0.330$ and thus:

$$\frac{\left|\tilde{\tilde{\eta}}\big|_\beta - \tilde{\eta}\big|_\beta\right|}{\tilde{\eta}\big|_\beta} \approx 1.2\ \%$$

Therefore, we may conclude that approximation 7.36 is in fact very accurate.

One final remark: the formulae given in this example for An_1 and An_0 could also have been obtained from Equations 9.17 and 9.18 by putting $\beta = 1$ and $\beta = 0$, respectively.

9.17 Negligibility Criteria for Langmuir-Hinshelwood Kinetics

We want to know whether $C_{B,s}$ could have been substituted for C_B in Example 9.14. The criteria 7.55 and 7.56 yield:

- for the high η region: $\beta < 0.066$.
- for the low η region: $\beta < 0.77$ (since $\mu_0 = 1.033$ and $\mu_1 = 6.2$).

Again, the criterion for the high η region is more stringent than that for the low η region. Since we operate in the low η region, the latter criterion of $\beta < 0.77$ must be used. From Example 9.14 $\beta = 0.53$, and hence we can substitute $C_{B,s}$ for C_B. This is in agreement with what was found in example 9.16, where it was calculated that, by substituting $C_{B,s}$ for C_B a relative error is introduced of

$$\delta_{\beta,0} = \frac{\left|\eta|_{\beta} - \eta|_{\beta=0}\right|}{\eta|_{\beta=0}} = \frac{|0.326-0.299|}{0.299} = 9\%$$

Viewed in retrospect, a lot of work can be saved in Example 9.14 by using criterion 7-56.

9.18 Calculation of $\kappa_D|_{\kappa_A=0}$

For a reactor feed composition of 10 mole % A, 30 mole % P and 60 mole % D and for one molecule of A converted into three molecules of P, for every place inside the reactor $\kappa_D|_{\kappa_A=0}$ equals:

$$\kappa_D|_{\kappa_A=0} = \frac{\textit{total moles of } D\ (\textit{if all } A \textit{ is converted})}{\textit{total moles of } A+P+D \textit{ if all } A \textit{ is converted}}$$

$$= \frac{0.6}{3\times 0.1 + 0.3 + 0.6} = 0.5$$

9.19 Maximum Pressure Difference in a Porous Pellet

What is the maximum pressure difference inside a catalyst for the reactor feed given in Example 9.18? The reactor pressure is 20 bar.

Since $\nu_P = 3$ and $\kappa_{A,s} = 0.1$ we calculate $\sigma_{\max} = 0.073$. From the definition of σ it then follows that

$$(P-P_s)_{\max} = \sigma_{\max} \times P_s = 1.5\, bar$$

Hence the maximum obtainable pressure inside the catalyst pellet is 21.5 bar, irrespective the size and shape of the catalyst pellet and the reaction kinetics.

9.20 Can a Gas be Regarded as Being Diluted

Determine whether intraparticle pressure gradients can be neglected for the data given in Example 9.16.
The value of γ will be highest at the reactor inlet, where $\kappa_{A,s}$ is the highest. Furthermore, high values of γ can only be obtained for low values of the Knudsen number Kn^*. This can be seen from Equation 7.79 where the maximum value for γ is given as

$$\gamma\Big|_{Kn^*\downarrow 0}=(\nu_P-1)\kappa_{A,s} \tag{7.79}$$

Thus it follows that γ must always be smaller than

$$\gamma<(\nu_P-1)\kappa_{A,s}\Big|_{reactor\ inlet}=0.2$$

Comparison with criterion 7.95 shows that if we operate in the low η region, neglecting intraparticle pressure gradients and the dependence of the effective diffusion coefficient on the gas composition is always allowed. According to criterion 7.94 care must be taken when operating in the high η region. For low values of Kn^*, that is for broad catalyst pores, an error up to 20 % may be introduced by neglecting pressure gradients and gas composition influences.

9.21 Are Gases Diluted for the Oxidation of Ethylene

Consider the oxidation of ethylene and assume only the following reaction occurs:

$$C_2H_4+\frac{1}{2}O_2\rightarrow C_2H_4O$$

Oxygen is present in excess. We calculate $\Delta\nu$ = -½ and Ψ = -0.28 for $M_{C_2H_4}$ = 28 g mole^{-1} , M_{O_2} = 32 g mole^{-1} , $M_{C_2H_4O}$ = 44 g mole^{-1} . The feed consists of 10 % C_2H_4, 40 % O_2 and 50 % inert on a mol basis. The reactor pressure is 20 bar.

The largest pressure gradient inside the pellet will be obtained for very small catalyst pores ($Kn_A^*\rightarrow\infty$) and at the reactor inlet:

$$\sigma<\Psi\kappa_{A,s}\Big|_{reactor\ inlet}=-0.028$$

This means that the pressure difference inside a catalyst pellet, no matter the pellet size, will always be smaller than 0.028 × 20 = 0.56 bar. Hence, in general, internal pressure gradients can be neglected for the oxidation of ethylene.

Can the gas be considered as nondiluted? The maximum value of γ_β will be obtained for very broad catalyst pores ($Kn_A^* \downarrow 0$) and, again, at the reactor inlet:

$$\gamma_\beta < \Delta \nu \, \kappa_{A,s}\Big|_{reactor\ inlet} = -0.05$$

Therefore, from criteria 7.114 and 7.117 it follows that for the oxidation of ethylene the gas may be regarded as a diluted gas.

9.22 Anisotropy of a Ringshaped Catalyst Pellet

Reconsider the ring-shaped catalyst pellet given in Example 9.8, for which

$$\phi = \frac{H}{R_u - R_i} = \frac{8}{4-2} = 4$$

Assume the catalyst pellet to be anisotropic. The radial effective diffusion coefficient is $D_{e,A}{}^R = 5\times10^{-8}$ m^2 s^{-1}; the longitudinal effective diffusion coefficient equals $D_{e,A}{}^H = 1.8\times10^{-6}$ m^2 s^{-1}. Then, if the effective diffusion coefficient is determined from measurements of the effectiveness factor, this gives

$$D^+_{e,A} = \frac{\left(\sqrt{D^H_{e,A}} + \phi\sqrt{D^R_{e,A}}\right)^2}{(1+\phi)^2} = 2\times10^{-7} \quad \mathrm{m^2 s^{-1}}$$

which is in agreement with the value given in Example 9.8. So far nothing is wrong.

Now change the catalyst geometry and increase the height twofold. What will be the consequence if the anisotropy is overlooked and what will be the correct value for the effectiveness factor.

Changing the catalyst geometry leads to changes of:

- The characteristic dimension V_p/A_p, which was 8/10 mm and now becomes

$$\frac{V_P}{A_P} = \frac{1}{2}\left(\frac{1}{H} + \frac{1}{R_u - R_i}\right)^{-1} = \frac{8}{9} \quad \mathrm{mm}$$

- The geometry factor Γ; its value was 1.12. In the new situation $\lambda = 4$ and $\iota = ½$ and, thus according to Figure 6.10, the geometry factor is now $\Gamma = 0.80$.

- The effective diffusion coefficient, which is modified for anisotropy. Its value was 2×10^{-7} m^2 s^{-1}; in the new situation it becomes

$$D_{e,A}^{+}=\frac{\left(\sqrt{D_{e,A}^{H}}+\phi\sqrt{D_{e,A}^{R}}\right)^2}{(1+\phi)^2}=1.08\times10^{-7}\quad \mathrm{m^2s^{-1}}$$

Since $\phi = 8$.

Therefore, the following effectiveness factors are calculated:

- If anisotropy of the pellet is overlooked (thus $V_p/A_p = 8/9$ mm, $\Gamma = 0.80$ and $D_{e,A}^{+} = 2\times10^{-7}$ m^{2}s^{-1}), this gives $An_0 = 1.030$ and $An_1 = 0.785$, thus

$$\tilde{\eta}=\frac{1}{\sqrt{1+\tilde{\eta}An_1+(1-\tilde{\eta})An_0}}=0.714\pm0.005$$

(this is for a height of 16 mm; for a height of 8 mm as in Example 9.8 we calculated $\tilde{\eta} = 0.764$).

- If anisotropy of the pellet is accounted for correctly (thus $V_p/A_p = 8/9$ mm, $\Gamma = 0.80$ and $D_{e,A}^{+} = 1.08\times10^{-7}$ m^2 s^{-1}, this gives $An_0 = 1.908$ and $An_1 = 1.454$, thus

$$\tilde{\eta}=\frac{1}{\sqrt{1+\tilde{\eta}An_1+(1-\tilde{\eta})An_0}}=0.605\pm0.005$$

to be compared with $\tilde{\eta} = 0.714 \pm 0.005$ when neglecting anisotropy.
Therefore, if we do not take into account anisotropy of a catalyst pellet and we change the catalyst geometry, we can introduce serious errors.

References

1. Janssen, H.J., Kruithof, A.J., Steghuis, G.J. Westerterp, K.R. (1990) *Ind. Eng. Chem. Res.*, **29**, 754.
2. Janssen, H.J., Kruithof, A.J., Steghuis, G.J. Westerterp, K.R. (1990) *Ind. Eng. Chem. Res.*, **29**, 1822.
3. Kuczynski, M., Browne, W.I., Fontein H.J., Westerterp, K.R. (1987), *Chem. Eng. Process.*, **21**, 179.
4. Bos, A.N.R., Borman, P.C., Kuczynski, M., Westerterp, K.R., (1989) *Chem. Eng. Sci.*, **44**, 2435.
5. Kerkhof, P.J.A.M. (1996) *Chem. Eng. J.*, **64**, 319.
6. Evans, R.B., Watson, G.M., Truitt, J. (1962) *J. Appl. Phys.*, **33**, 2682.
7. Evans, R.B., Watson, G.M., Truitt, J. (1963) *J. Appl. Phys.*, **34**, 2020.
8. Reid, R.C., Prausnitz, J.M., Poling, B.E. (1987) *The Properties of Gases and Liquids.* New York: McGraw-Hill.

9. Mason, E.A., Malinauskas, A.P. (1983), *Gas Transport in Porous Media: The Dusty-Gas Model.* Amsterdam: Elsevier.
10. Krishna, R. (1987) *Chem. Eng. J.*, **35**, 75.
11. Waldmann Von, L., Schmitt, K.H. (1961) *Z. Naturforschg*, **16a**, 1343.
12. Kramers, H.A., Kistemaker, J. (1943) *Physica*, **10**, 699.
13. Kerkhof, P.J.A.M. (1997) *Ind. Eng. Chem. Res.*, **36**, 915.
14. Bird, R.B., Stewart W.E., Lightfoot, E.N. (1960) *Transport Phenomena*. New York: Wiley.
15. Christoffel, E.G., Schuler, J.C. (1980) *Chem.-Ing.-Tech.*, **52**, 844.

Appendix A

Derivation of a Formula for the Zeroth Aris Number An_0 for Simple Reactions

In Figure A.1 an arbitrary catalyst geometry is shown. On the external surface area of the catalyst an infinitesimally small surface element dA is chosen; the curvature of the element can be neglected. Perpendicular to the dA plane an X-coordinate is chosen, which has a positive value when directed inwards.

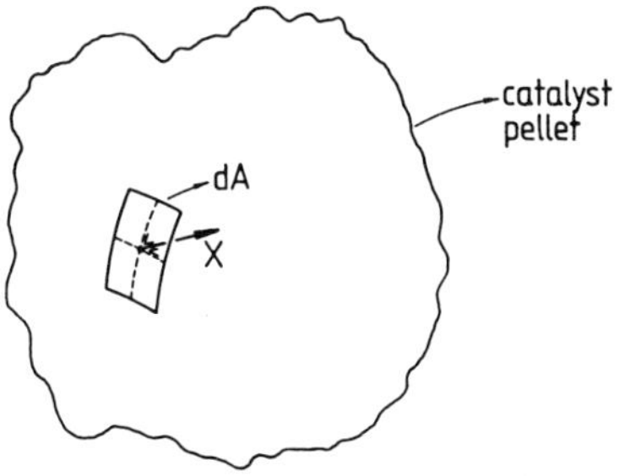

Figure A.1 Arbitrary catalyst geometry with an external surface element dA.

For very low values of the effectiveness factor the penetration depth of a reactant in the solid catalyst matrix is very small. Since the curvature of the surface element can be neglected, the Laplace operator reduces to

$$\nabla^2 C_A = \frac{d^2 C_A}{dX^2} \tag{A.1}$$

Then, for simple reaction schemes, the concentration profile can be calculated from

$$D_{e,A}\frac{d^2 C_A}{dX^2} = R(C_A) \tag{A.2}$$

After multiplying both sides by dC_A/dX and integrating from $X = 0$ to $X = X$ we find

$$\left(\frac{dC_A}{dX}\bigg|_{X=X}\right)^2 - \left(\frac{dC_A}{dX}\bigg|_{X=0}\right)^2 = \int_{C_A|_{X=0}}^{C_A|_{X=X}} R(C_A)dC_A \tag{A.3}$$

Since

$$C_A\big|_{X=0} = C_{A,s} \tag{A.4}$$

$(dC_A/dX)|_{X=0}$ can be calculated from Equation A.3 if both dC_A/dX and C_A are known for one and the same value of X.

Three basically different concentration profiles can occur inside the catalyst (Figure A.2):

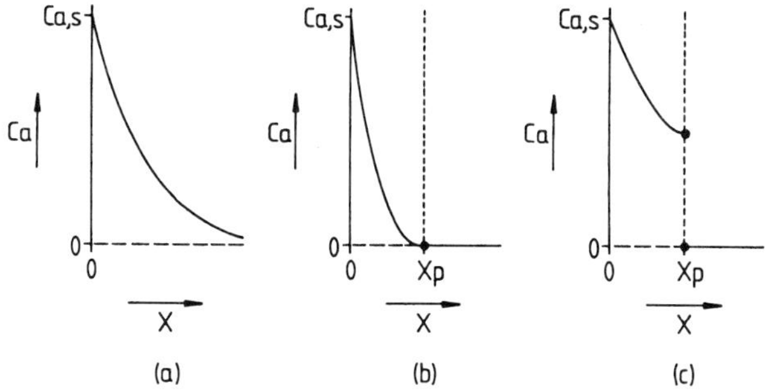

Figure A.2 Sketch of the concentration profiles that can occur inside a catalyst. The concentration C_A is plotted versus the inwardly directed coordinate X.

1. For the limit of $C_A \downarrow 0$, $R(C_A) = 0$, C_A decreases to zero as X increases, see Figure A.2a. For very large values of X (or for a single value of X for very low effectiveness factor) both C_A and dC_A/dX will become zero. Hence, by substituting a sufficiently large value of X (or a small enough value of the effectiveness factor), Equation A.3 gives the following limiting case

$$\left.\frac{dC_A}{dX}\right|_{X=0} = -\sqrt{\frac{2}{D_{e,A}}\int_0^{C_{A,s}} R(C_A)dC_A} \tag{A.5}$$

2. If in the limit of $C_A \downarrow 0$, $\{C_A\, R(C_A)\} = 0$ but $R(C_A) \neq 0$, the concentration C_A will become zero for a certain penetration depth X_p (Figure A.2b). For values of X larger than X_p, Equation A.2. yields a solution without any physicochemical significance. Since no reactant is converted on the right-hand side of the plane $X = X_p$, no reactant will penetrate beyond this plane. Therefore, there can be no concentration gradient:

$$\left.\frac{dC_A}{dX}\right|_{X=X_P} = 0 \tag{A.6}$$

Also, since

$$C_A\big|_{X=X_P} = 0 \tag{A.7}$$

substitution of $X = X_p$ in Equation A.3 yields Equation A.5 again.

3. For the limit of $C_A \downarrow 0$, $\{C_A\, R(C_A)\} \neq 0$ the concentration profile will again stop for a certain penetration depth X_p. As before, no reactant will penetrate beyond the plane $X = X_p$ and the derivative of C_A with respect to X must therefore equal zero for $X = X_p$.

The concentration C_A, however, in this case does not have to be zero for $X = X_p$. Because of that, a simple general formula cannot be obtained from Equation A.3 if for the limit of $C_A \downarrow 0$, $\{C_A R(C_A)\} \neq 0$. This point is elaborated further at the end of this appendix.

From Equation A.5 it can be calculated that for low values of the effectiveness factor η and $\{C_A R(C_A)\} = 0$ for the limit of $C_A \downarrow 0$, the flux of component A through the surface element dA must be equal to

$$J_A\big|_\eta = -D_{e,A} \left.\frac{dC_a}{dX}\right|_{X=0} = +\sqrt{2D_{e,A}\int_0^{C_{A,s}} R(C_A)dC_A} \tag{A.8}$$

Hence the total amount of A which is converted inside the catalyst pellet N_A is given by

$$N_A\big|_\eta = \oint\!\!\oint_{A_P} J_A\big|_\eta dA = A_P\sqrt{2D_{e,A}\int_0^{C_{A,s}} R(C_A)dC_A} \tag{A.9}$$

and the effectiveness factor follows from

$$\eta = \frac{N_A\big|_\eta}{N_A\big|_{\eta=1}} = \frac{A_P}{V_P R(C_{A,s})}\sqrt{2D_{e,A}\int_0^{C_{A,s}} R(C_A)dC_A} \tag{A.10}$$

Since the zeroth Aris number An_0 is defined as the number which becomes equivalent to $1/\eta^2$ for low values of the effectiveness factor η, it follows that An_0 can be calculated from

$$An_0 = \left(\frac{V_P}{A_P}\right)^2 \times \frac{\{R(C_{a,s})\}^2}{2D_{e,A}} \times \left(\int_0^{C_{A,s}} R(C_A)dC_A\right)^{-1} \tag{A.11}$$

which is the formula given in Table 6.4.

To arrive at this formula it was necessary to assume that for the limit of $C_A \downarrow 0$, $\{C_A R(C_A)\} = 0$. This is confirmed by the above equation for An_0: if for the limit $C_A \downarrow 0$, $\{C_A R(C_A)\} \neq 0$, the integral in Equation A.11 is nonexistent. The physicochemical meaning of this will be discussed in the text, where we will study Figure 6.11. No low η region exists for reaction orders of $n \geq -1$ in this figure or, more generally, for $\{C_A R(C_A)\} \neq 0$ when $C_A \downarrow 0$. In this situation, for low effectiveness factors, the effectiveness factor cannot be obtained from a stationary mass balance because it does not yield a mathematically unique solution. If such is the case, obviously, by matter of definition there is no sense in talking about a zeroth Aris number An_0. A unique number An_0 can be found if and only if the following criterion is adhered to:

$$\lim_{C_A \downarrow 0}\{C_A R(C_A)\} = 0 \tag{A.12}$$

Appendix B

Derivation of a Formula for the First Aris Number An_1 for Simple Reactions

The concentration profile inside a catalyst pellet for simple reactions is given by:

$$D_{e,A}\,\nabla^2 C_A = R(C_A) \tag{A.13}$$

In the high η region the concentration C_A will be close to $C_{A,s}$. Hence $R(C_A)$ can be estimated by making a first-order Taylor approximation around $C_A = C_{A,s}$:

$$\begin{aligned} R(C_A) &\approx R(C_{A,s}) + (C_A - C_{A,s})\left.\frac{\partial R(C_A)}{\partial C_A}\right|_{C_A=C_{A,s}} \\ &= \left.\frac{\partial R(C_A)}{\partial C_A}\right|_{C_A=C_{A,s}} \times \left\{ C_A - C_{A,s} + \frac{R(C_{A,s})}{\left.\frac{\partial R(C_A)}{\partial C_A}\right|_{C_A=C_{A,s}}} \right\} \end{aligned} \tag{A.14}$$

Thus, by substituting

$$C_A^{\phi} = C_A - C_{A,s} + \frac{R(C_{A,s})}{\left.\frac{\partial R(C_A)}{\partial C_A}\right|_{C_A=C_{A,s}}} \tag{A.15}$$

Equation A.13 can be written as:

$$D_{e,A}\nabla^2 C_A^{\phi} = \left\{ \left.\frac{\partial R(C_A)}{\partial C_A}\right|_{C_A=C_{A,s}} \right\} C_A^{\phi} \tag{A.16}$$

This is equivalent to the differential equation describing the concentration profile inside a catalyst pellet for first-order reactions with reaction rate

constant $\left.\frac{\partial R(C_A)}{\partial C_A}\right|_{C_A=C_{A,s}}$ and surface concentration $\frac{R(C_{A,s})}{\left.\frac{\partial R(C_A)}{\partial C_A}\right|_{C_A=C_{A,s}}}$

Therefore, if the effectiveness factor for first-order reactions η_1 is known as a function of the zeroth Aris number An_0, the effectiveness factor in the high η region can be estimated from:

$$\eta \approx \eta_1\left(An_0^{\phi}\right) \tag{A.17}$$

where the modified zeroth Aris number An_0^{ϕ} may be calculated from

$$An_0^{\phi} = \left(\frac{V_P}{A_P}\right)^2 \times \frac{1}{D_{E,A}} \times \left\{\left.\frac{\partial R(C_A)}{\partial C_A}\right|_{C_A=C_{A,s}}\right\} \tag{A.18}$$

This last number follows from the zeroth Aris number An_0 for first-order reactions, by replacing the reaction rate constant k with (see the formulae in Table 6.4) $\left.\frac{\partial R(C_A)}{\partial C_A}\right|_{C_A=C_{A,s}}$. The estimate A.17 holds for arbitrary reaction kinetics, and only for the high η region.

The function η_1 still depends on the catalyst geometry; it can also become very complex for certain catalyst geometries, such as for hollow cylinders (see Appendix C). A simpler estimate for the effectiveness factor is obtained by making a second Taylor approximation, now for the function η_1. This approximation must hold for high values of the effectiveness factor η, thus the Taylor approximation must be made around $An_0^{\phi} = 0$:

$$\eta \approx \left.\eta_1\left(An_0^{\phi}\right)\right|_{An_0^{\phi}=0} + \left.\frac{\partial \eta_1\left(An_0^{\phi}\right)}{\partial An_0^{\phi}}\right|_{An_0^{\phi}=0} \times An_0^{\phi} \tag{A.19}$$

Obviously;

$$\left.\eta_1\left(An_0^{\phi}\right)\right|_{An_0^{\phi}=0} = 1 \tag{A.20}$$

It therefore follows that the effectiveness factor in the high η region can be estimated from

$$\eta \approx 1 - An_0^{\phi} \times \left\{-\left.\frac{\partial \eta_1\left(An_0^{\phi}\right)}{\partial An_0^{\phi}}\right|_{An_0^{\phi}=0}\right\} \tag{A.21}$$

It is convenient at this stage to define a geometry factor Γ as

$$\Gamma = -2\frac{\partial\eta_1\left(An_0^{\phi}\right)}{\partial An_0^{\phi}}\Bigg|_{An_0^{\phi}=0} \tag{A.22}$$

alternatively

$$\Gamma = -2\frac{\partial\eta_1\left(An_0\right)}{\partial An_0}\Bigg|_{An_0=0} \tag{A.23}$$

Since the function η_1 only depends on the catalyst geometry, the geometry factor Γ will only depend on the catalyst geometry and not on the pellet size or reaction kinetics. Substitution of Equation A.22 in A.21 yields

$$\eta \approx 1-\frac{1}{2}\times\Gamma An_0^{\phi} \tag{A.24}$$

or

$$\Gamma An_0^{\phi} \approx (1-\eta)\times 2 \tag{A.25}$$

Since for values of the effectiveness factor close to one, $2 \approx 1+\eta$, Equation A.25 may also be written as

$$\Gamma An_0^{\phi} \approx 1-\eta^2 \tag{A.26}$$

Since the first Aris number An_1 is defined as the number which becomes equivalent to $1-\eta^2$ for values of η close to one, it follows that An_1 can be calculated from

$$An_1 \approx \Gamma An_0^{\phi} \tag{A.27}$$

After substitution of Equation A.18, this yields

$$An_1 = \left(\frac{V_P}{A_P}\right)^2 \frac{\Gamma}{D_{e,A}} \frac{\partial R(C_A)}{\partial C_A}\Bigg|_{C_A=C_{A,s}} \tag{A.28}$$

which is the generalized formula for An_1 for simple reactions.

It is emphasized that, since the formula for An_1 was calculated without making any assumptions, the definition of An_1 is not restricted whatsoever. Hence, the numbers An_0 and An_1 differ on this point.

Appendix C

Calculation of the Effectiveness Factor for a First-order Reaction and the Geometry Factor for a Ring-shaped Catalyst

For a first-order reaction taking place in a ring-shaped catalyst pellet, the effectiveness factor was already calculated by Gunn [1]. That solution, however, is very complex because the mathematical techniques used are not the most suitable with which to solve the equations that arise. Therefore, the following solution derives another expression for the effectiveness factor.

The mass balance over a small volume element in the pellet (Figure 6.9) is given by

$$D_{e,A}\left\{\frac{1}{r}\frac{\partial}{\partial r}\left(r\frac{\partial C_A(r,h)}{\partial r}\right)+\frac{\partial^2 C_A(r,h)}{\partial h^2}\right\}=R(C_A) \tag{A.29}$$

For first-order kinetics, i.e. $R(C_A)=kC_A$ this can be rewritten as:

$$\frac{1}{r}\frac{\partial}{\partial r}\left(r\frac{\partial C_A(r,h)}{\partial r}\right)+\frac{\partial^2 C_A(r,h)}{\partial h^2}=\frac{k}{D_{e,A}}C_A(r,h) \tag{A.30}$$

This differential equation must be solved with the following boundary conditions:

$$h=0,\quad \forall r \;\rightarrow\; C_A=C_{A,s} \tag{A.31}$$

$$h=H,\quad \forall r \;\rightarrow\; C_A=C_{A,s} \tag{A.32}$$

$$r=R_i,\quad \forall h \;\rightarrow\; C_A=C_{A,s} \tag{A.33}$$

$$r=R_u,\quad \forall h \;\rightarrow\; C_A=C_{A,s} \tag{A.34}$$

It is convenient to introduce the following dimensionless groups:

$$\rho = \frac{r}{R_u} \tag{A.35}$$

$$\omega = \frac{h}{H} \tag{A.36}$$

$$\gamma(\rho,\omega) = \frac{C_A(\rho,\omega)}{C_{A,s}} \tag{A.37}$$

$$\iota = \frac{R_i}{R_u} \tag{A.38}$$

$$\lambda = \frac{H}{R_u} \tag{A.39}$$

$$\phi_{hc} = R_u \sqrt{\frac{k}{D_{e,A}}} \tag{A.40}$$

Using these dimensionless numbers the differential equation and boundary conditions become:

$$\frac{1}{\rho}\frac{\partial}{\partial \rho}\left(\frac{\partial\gamma(\rho,\omega)}{\partial\rho}\right) + \frac{1}{\lambda^2}\frac{\partial^2\gamma[\rho,\omega]}{\partial\omega^2} - \phi_{hc}{}^2\gamma(\rho,\omega) = 0 \tag{A.41}$$

$$\omega = 0,\ \forall\rho \rightarrow \gamma(\rho,\omega) = 1 \tag{A.42}$$

$$\omega = 1,\ \forall\rho \rightarrow \gamma(\rho,\omega) = 1 \tag{A.43}$$

$$\rho = \iota,\ \forall\omega \rightarrow \gamma(\rho,\omega) = 1 \tag{A.44}$$

$$\rho = 1,\ \forall\omega \rightarrow \gamma(\rho,\omega) = 1 \tag{A.45}$$

This differential equation can be solved by making a discrete Fourier sine expansion of $\gamma(\rho\ \omega)$ with respect to the variable ω. This yields

$$\gamma(\rho,\omega) = 1 - \frac{4\phi_{hc}^{\ 2}}{\pi}\sum_{j=0}^{\infty}\left[\frac{\sin\{(2j+1)\pi\omega\}}{(2j+1)a_j}\times\left\{1-\frac{P_0(a_j\iota,a_j\rho)+P_0(a_j\rho,a_j)}{P_0(a_j\iota,a_j)}\right\}\right] \tag{A.46}$$

with

$$a_j = \sqrt{\left\{\frac{(2j+1)\pi}{\lambda}\right\}^2 + \phi_{hc}^2} \tag{A.47}$$

and

$$P_0(a,b) = I_0(a)K_0(b) - I_0(b)K_0(a) \tag{A.48}$$

Since the generalized zeroth Aris number equals

$$An_0 = \left(\frac{V_P}{A_P}\right)^2\frac{k}{D_{e,A}} = \frac{1}{4}\times\left(\frac{1}{\lambda}+\frac{1}{1-\iota}\right)^{-2}\phi_{hc}^{\ 2} \tag{A.49}$$

Equation A.46 can be written as

$$\gamma(\rho,\omega) = 1 - An_0\frac{16}{\pi}\left(\frac{1}{\lambda}+\frac{1}{1-\iota}\right)^2\sum_{j=0}^{\infty}\left[\frac{\sin\{(2j+1)\pi\omega\}}{(2j+1)a_j}\times\left\{1-\frac{P_0(a_j\iota,a_j\rho)+P_0(a_j\rho,a_j)}{P_0(a_j\iota,a_j)}\right\}\right] \tag{A.50}$$

with

$$a_j = \sqrt{\left\{\frac{(2j+1)\pi}{\lambda}\right\}^2 + 4\left(\frac{1}{\lambda}+\frac{1}{1-\iota}\right)^2 An_0} \tag{A.51}$$

The effectiveness factor can be calculated by using either one of the following formulae:

$$\eta = \frac{2}{1-\iota^2}\int_0^1\int_\iota^1 \rho\gamma(\rho,\omega)\,d\rho d\omega \tag{A.52}$$

$$\eta = \frac{1}{2(1-\iota^2)An_0}\left(\frac{1}{\lambda}+\frac{1}{1-\iota}\right)^2\times\left\{\int_0^1\left(\frac{\partial\gamma(\rho,\omega)}{\partial\rho}\bigg|_{\rho=1} - \iota\frac{\partial\gamma(\rho,\omega)}{\partial\rho}\bigg|_{\rho=\iota}\right)d\omega - \frac{2}{\lambda^2}\int_\iota^1\rho\frac{\partial\gamma(\rho,\omega)}{\partial\omega}\bigg|_{\omega=0}d\rho\right\} \tag{A.53}$$

Both the Equations A.52 and A.53 yield

$$\eta = 1 - \frac{32}{\pi^2} \times \frac{An_0}{1-\iota^2} \times \left(\frac{1}{\lambda} + \frac{1}{1-\iota} \right)^2 \times \sum_{j=0}^{\infty} \left(\frac{1-\iota^2 - 2\chi(a_j \iota, a)}{(2j+1)^2 a_j^2} \right) \qquad \text{(A.54)}$$

with

$$\chi(a,b) = \frac{2 - bQ_0(a,b) - aQ_0(b,a)}{b^2 P_0(a,b)} \qquad \text{(A.55)}$$

Taking $P_0(a,b)$ as given by Equation 6.44 and with $Q_0(a,b)$ as the following cross-product of modified Bessel functions:

$$Q_0\,(a,b) = I_0\,(a)\,K_1\,(b) + I_1\,(b)\,K_0\,(a) \qquad \text{(A.56)}$$

Figure 6.17 has been constructed with the aid of Equation A.54.

The geometry factor Γ can be derived from Equations and A.54 as

$$\Gamma = -2 \left. \frac{\partial \eta_1 (An_0)}{\partial An_0} \right|_{An_0 = 0} = \frac{64}{\pi^4} \times \frac{1}{1-\iota^2} \left(1 + \frac{\lambda}{1-\iota} \right)^2 \sum_{j=0}^{\infty} \frac{1-\iota^2 - 2\chi(a_j \iota, a_j)}{(2j+1)^4} \qquad \text{(A.57)}$$

with $\chi(a,b)$ as defined in Equation A.55 and, since $An_0 = 0$, a_j equal to

$$a_j = \frac{(2j+1)\pi}{\lambda} \qquad \text{(A.58)}$$

References

1. Gunn, D.J. (1967) *Chem. Eng. Sci.*, **22**, 1439.

Appendix D

Intraparticle Pressure Gradients for Nondiluted Gases and Simple Reactions

In this appendix expressions are derived for the pressure gradient inside catalyst pellets for nondiluted gases with the aid of the dusty gas model, for general equations [1]. Assume a mixture of A, P and m inert components D_i ($i = 1, 2, \ldots, m$). A reacts to P according to:

$$A \rightarrow v_P P \tag{A.59}$$

Further restrictions are

- there is no surface diffusion inside the catalyst pores
- the catalyst pellet is isothermal, thus there will be no thermal diffusion;
- the external forces acting on the gas inside the pores are negligible.

Then according to the dusty-gas model we can write

- a force balance for component A;

$$\frac{\kappa_P J_A^* - \kappa_A J_P^*}{D_{AP}} + \sum_{i=1}^{m} \frac{\kappa_{D_i} J_A^* - \kappa_A J_{D_i}^*}{D_{AD_i}} + \frac{J_A^*}{D_{Ak}} = \frac{-\nabla(\kappa_A P)}{RT}(1+V_A) \tag{A.60}$$

- a force balance for component P

$$\frac{\kappa_A J_P^* - \kappa_P J_A^*}{D_{PA}} + \sum_{i=1}^{m} \frac{\kappa_{D_i} J_P^* - \kappa_P J_{D_i}^*}{D_{DP_i}} + \frac{J_P^*}{D_{PD_k}} = \frac{-\nabla(\kappa_P P)}{RT}(1+V_P) \tag{A.61}$$

- m force balance for components D_i

$$\frac{\kappa_A J_{D_i}^* - \kappa_{D_i} J_A^*}{D_{D_iA}} + \frac{\kappa_P J_{D_i}^* - \kappa_{D_i} J_P^*}{D_{D_iP}} \sum_{\substack{j=1 \\ j\neq i}}^{m} \frac{\kappa_{D_j} J_{D_i}^* - \kappa_{D_i} J_{D_j}^*}{D_{D_iD_j}} + \frac{J_{D_i}^*}{D_{D_ik}} \tag{A.62}$$

$$= \frac{-\nabla(\kappa_{D_i} P)}{RT}(1+V_{D_i}) \qquad (i=1, 2, \ldots, m)$$

In these equations $J_A{}^*$ is the mole flux of A in moles of A per second per square metre flowing through surface area of the catalyst pores. This is **not** the same as the mole flux in moles of A per second per square metre flowing through surface area of the catalyst pellet. This is elaborated in Appendix E. The term D_{AP} is the Maxwell diffusion coefficient of A in a binary mixture with P and D_{Ak} is the Knudsen diffusion coefficient of A inside the catalyst pores. κ_A is the mole fraction of A. $J_P{}^*$, D_{PDi}, D_{Pk}, κ_P etc. are similarly defined. V_A is a factor that accounts for viscous flow inside the pores. If V_A is much smaller than one, viscous flow can be neglected. We will neglect viscous flow for all components and substitute

$$V_A = 0 \tag{A.63}$$

$$V_P = 0 \tag{A.64}$$

$$V_{P_i} = 0 \tag{A.65}$$

The impact of this assumption is discussed at the end of this appendix.

The Maxwell diffusion coefficients D_{AP}, D_{ADi}, D_{PDi},... are independent of the gas composition, thus

$$D_{AP} = D_{PA} \tag{A.66}$$

$$D_{AD_i} = D_{D_iA} \qquad (i = 1, 2, ..., m) \tag{A.67}$$

$$D_{PD_i} = D_{D_iP} \qquad (i = 1, 2, ..., m) \tag{A.68}$$

$$D_{D_iD_j} = D_{D_jD_i} \qquad (i, j = 1, 2, ..., m) \tag{A.69}$$

A further simplification of Equations A.60-A.62 can be obtained by substituting

$$J_P^* = -\nu_P J_A^* \tag{A.70}$$

$$J_{D_i}^* = 0 \tag{A.71}$$

which follows from a mass balance.

With the aid of Equations A.63-A.71, Equations A.60-A.62 can be rewritten as

$$J_A^* \left(\frac{\kappa_{P+} \upsilon_P \kappa_A}{D_{AP}} + \sum_{i=1}^{m} \frac{\kappa_{D_i}}{D_{AD_i}} + \frac{1}{D_{Ak}} \right) = - \frac{\nabla(\kappa_A P)}{RT} \tag{A.72}$$

$$J_A^* \left(- \frac{\kappa_{P+} \upsilon_P \kappa_A}{D_{AP}} - \upsilon_P \sum_{i=1}^{m} \frac{\kappa_{D_i}}{D_{PD_i}} - \frac{\upsilon_P}{D_{Pk}} \right) = - \frac{\nabla(\kappa_P P)}{RT} \tag{A.73}$$

$$J_A^*\left(-\frac{\kappa_{D_i}}{D_{AD_i}}+\upsilon_P\frac{\kappa_{D_i}}{D_{PD_i}}\right)=-\frac{\nabla(\kappa_{D_i}P)}{RT} \qquad (i=1,2,\ldots,m) \tag{A.74}$$

Adding Equations A.73 and A.74 gives

$$J_A^*\left(\frac{1}{D_{Ak}}-\frac{\upsilon_P}{D_{Pk}}\right)=-\frac{1}{RT}\nabla\left\{(\kappa_A+\kappa_P+\sum_{i=1}^{m}\kappa_{D_i})P\right\}=-\frac{\nabla P}{RT} \tag{A.75}$$

For the Knudsen diffusion coefficients D_{Ak} and D_{Pk} we can write

$$D_{Ak}=\frac{2}{3}\times r_p\bar{v}_A=\frac{2}{3}\times r_p\sqrt{\frac{8RT}{\pi M_A}} \tag{A.76}$$

$$D_{Pk}=\frac{2}{3}\times r_p\bar{v}_P=\frac{2}{3}\times r_p\sqrt{\frac{8RT}{\pi M_P}} \tag{A.77}$$

where r_p is the pore radius, $\bar{v}$ the average velocity of the gas molecules and M the molar mass. Hence we may write:

$$\frac{D_{Pk}}{D_{Ak}}=\sqrt{\frac{M_A}{M_P}}=\sqrt{\upsilon_P} \tag{A.78}$$

Substitution of Equation A.78 into A.75 yields

$$\frac{J_A^*}{D_{Ak}}\times\left(\sqrt{\upsilon_P}-1\right)=\frac{\nabla P}{RT} \tag{A.79}$$

Comparison of Equations A.72 and A.79 gives

$$\frac{\nabla(\kappa_A P)}{\nabla P}=\frac{-1}{\sqrt{\upsilon_P}-1}\left\{1+(\kappa_P+\upsilon_P\kappa_A)\frac{D_{Ak}}{D_{Ap}}+\sum_{i=1}^{m}\left(\kappa_{D_i}\frac{D_{Ak}}{D_{AD_i}}\right)\right\} \tag{A.80}$$

For a certain given mole fraction κ_A the mole fractions κ_p and κ_{D_i} can be calculated from

$$\kappa_P=\{1+(\upsilon_P-1)\kappa_A\}\kappa_P\big|_{\kappa_A=0}-\upsilon_P\kappa_A \tag{A.81}$$

$$\kappa_{D_i}=\{1+(\upsilon_P-1)\kappa_A\}\kappa_{D_i}\big|_{\kappa_A=0} \qquad (i=1,2,\ldots,m) \tag{A.82}$$

where $\kappa_P|_{\kappa_A=0}$ and $\kappa_{D_i}|_{\kappa_A=0}$ are the mole fractions κ_P and κ_{D_i} that would have been obtained if all A were converted into P. It is convenient to substitute:

$$\kappa_P\big|_{\kappa_A=0} = 1-\sum_{i=1}^{m}\kappa_{D_i}\big|_{\kappa_A=0} \tag{A.83}$$

with Equation A.81 this becomes

$$\kappa_P = 1-\kappa_A-\{1+(\upsilon_P-1)\kappa_A\}\sum_{i=1}^{m}\kappa_{D_i}\big|_{\kappa_A=0} \tag{A.84}$$

Substitution of Equations A.82 and A.84 into A.80 yields

$$\frac{\nabla(\kappa_A P)}{\nabla P} = \frac{-1}{\sqrt{\upsilon_P}-1}\left\langle 1+\{1+(\upsilon_P-1)\kappa_A\}\times\frac{D_{Ak}}{D_{AP}}\times\left[1+\sum_{i=1}^{m}\left\{\kappa_{D_i}\big|_{\kappa_A=0}\left(\frac{D_{AP}}{D_{AD_i}}-1\right)\right\}\right]\right\rangle \tag{A.85}$$

Since the Maxwell diffusion coefficients are inversely proportional to the pressure, this can be written as

$$\frac{\nabla(\kappa_A P)}{\nabla P} = \frac{-1}{\sqrt{\upsilon_P}-1}\left\langle 1+\{1+(\upsilon_P-1)\kappa_A\}\times\frac{P}{P_s}\times\frac{D_{Ak}}{D_{AP}(P_s)}\times\left[1+\sum_{i=1}^{m}\left\{\kappa_{D_i}\big|_{\kappa_A=0}\left(\frac{D_{AP}(P_s)}{D_{AD_i}(P_s)}-1\right)\right\}\right]\right\rangle \tag{A.86}$$

Introduction of the modified Knudsen number Kn^*

$$Kn^* = \frac{D_{AP}(P_s)}{D_{Ak}}\times\left[1+\sum_{i=1}^{m}\left\{\kappa_{D_i}\big|_{\kappa_A=0}\left(\frac{D_{AP}(P_s)}{D_{AD_i}(P_s)}-1\right)\right\}\right]^{-1} \tag{A.87}$$

gives

$$\frac{\nabla(\kappa_A P)}{\nabla P} = \frac{-1}{\sqrt{\upsilon_P}-1}\left(1+\frac{1+(\upsilon_P-1)\kappa_A}{Kn^*}\times\frac{P}{P_s}\right) \tag{A.88}$$

This differential equation must be solved together with the boundary condition:

$$\kappa_A = \kappa_{A,s} \qquad \Leftrightarrow \qquad P = P_S \tag{A.89}$$

The result is

$$\left(\sqrt{\upsilon_P}-1\right)\kappa_{A,s}\times\frac{\kappa_A P}{\kappa_{A,s}P_S}$$
$$=\left\{\left(\sqrt{\upsilon_P}-1\right)\kappa_{A,s}+\frac{1}{2Kn^*}+1\right\}\times\exp\left\{-\frac{\left(\sqrt{\upsilon_P}+1\right)}{Kn^*}\left(\frac{P}{P_S}-1\right)\right\}-\frac{1}{2Kn^*}\left(\frac{P}{P_S}\right)^2-\frac{P}{P_S} \tag{A.90}$$

Remembering that

$$\frac{\kappa_A P}{\kappa_{A,s}P_s}=\frac{C_A}{C_{A,s}} \tag{A.91}$$

and defining

$$\sigma=\frac{P-P_s}{P_s} \tag{A.92}$$

enables Equation A.90 to be expressed as

$$\left(\sqrt{\upsilon_P}-1\right)\kappa_{A,s}\times\left(1-\frac{C_A}{C_{A,s}}\right)$$
$$=\frac{\sigma^2}{2Kn^*}+\sigma\left(1+\frac{1}{Kn^*}\right)+\left\{\left(\sqrt{\upsilon_P}-1\right)\kappa_{A,s}+1+\frac{1}{2Kn^*}\right\}\times\left\{1-\exp\left(-\frac{\sqrt{\upsilon_P}+1}{Kn^*}\times\sigma\right)\right\} \tag{A.93}$$

This is the equation given in the main text. This formula is rather complicated and hard to handle, so there is need for a simplified approximation. The first approximation to make is

$$\exp\left(-\frac{\sqrt{\upsilon_P}+1}{Kn^*}\times\sigma\right)\approx 1-\frac{\sqrt{\upsilon_P}+1}{Kn^*}\times\sigma \tag{A.94}$$

Equation A.93 then reduces to

$$\left(\sqrt{\upsilon_P}-1\right)\kappa_{A,s}\times\left(1-\frac{C_A}{C_{A,s}}\right)$$
$$=\frac{\sigma^2}{2Kn^*}+\frac{\left(Kn^*\right)^2+\left\{\left(\upsilon_P-1\right)\kappa_{A,s}+\sqrt{\upsilon_P}+2\right\}Kn^*+\frac{1}{2}\left(\sqrt{\upsilon_P}+1\right)}{\left(Kn^*\right)^2}\times\sigma \tag{A.95}$$

The second approximation is to neglect the second-order term in Equation A.95. This gives

$$\frac{\sigma^2}{2Kn^*}+\frac{\left(Kn^*\right)^2+\left\{(\upsilon_P-1)\kappa_{A,s}+\sqrt{\upsilon_P}+2\right\}Kn^*+\frac{1}{2}\left(\sqrt{\upsilon_P}+1\right)}{\left(Kn^*\right)^2}\times\sigma-\left(\sqrt{\upsilon_P}-1\right)\kappa_{A,s}\times\left(1-\frac{C_A}{C_{A,s}}\right)$$
$$=\frac{\left(Kn^*\right)^2+\left\{(\upsilon_P-1)\kappa_{A,s}+\sqrt{\upsilon_P}+2\right\}Kn^*+\frac{1}{2}\left(\sqrt{\upsilon_P}+1\right)}{\left(Kn^*\right)^2}\times\sigma-\left(\sqrt{\upsilon_P}-1\right)\kappa_{A,s}\times\left(1-\frac{C_A}{C_{A,s}}\right)\tag{A.96}$$

Then the dimensionless pressure σ can be calculated from

$$\sigma=\frac{\left(Kn^*\right)^2}{\left(Kn^*\right)^2+\left\{(\upsilon_P-1)\kappa_{A,s}+\sqrt{\upsilon_P}+2\right\}Kn^*+\frac{1}{2}\left(\sqrt{\upsilon_P}+1\right)}\times\left(\sqrt{\upsilon_P}-1\right)\kappa_{A,s}\times\left(1-\frac{C_A}{C_{A,s}}\right)\tag{A.97}$$

This is the approximation presented in the main text.

Now let us investigate when the approximation A.97 may be used instead of Equation A.93. The approximation

$$e^b\approx 1+b\tag{A.98}$$

is permissible if

$$b \ll 2\tag{A.99}$$

or

$$\frac{\sqrt{\upsilon_P}+1}{Kn^*}\times\sigma$$
$$\approx\frac{Kn^*}{\left(Kn^*\right)^2+\left\{(\upsilon_P-1)\kappa_{A,s}+\sqrt{\upsilon_P}+2\right\}Kn^*+\frac{1}{2}\left(\sqrt{\upsilon_P}+1\right)}\times(\upsilon_P-1)\kappa_{A,s}\times\left(1-\frac{C_A}{C_{A,s}}\right)\ll 2\tag{A.100}$$

The approximation

$$ax^2+bx+c\approx bx+c\tag{A.101}$$

is permissible around $ax^2+bx+c = 0$ if

$$\frac{ac}{b^2} << \frac{1}{2} \tag{A.102}$$

or

$$\frac{\left(Kn^*\right)^3}{\left\{\left(Kn^*\right)^2 + \left\{\left(\upsilon_P - 1\right)\kappa_{A,s} + \sqrt{\upsilon_P} + 2\right\}Kn^* + \frac{1}{2}\left(\sqrt{\upsilon_P} + 1\right)\right\}^2} \times \left(\sqrt{\upsilon_P} - 1\right)\kappa_{A,s} \times \left(1 - \frac{C_A}{C_{A,s}}\right) << 1 \tag{A.103}$$

The criteria A.100 and A.103 are always fulfilled for $Kn^* \downarrow 0$ or $Kn^* \to \infty$. Hence in those cases approximation A.97 will be exact. A maximum error in the approximation will be made for some value of $Kn^* \in [0,\infty[$. In the main text it is shown that even this maximum error is well below 10% for all practical cases (see Figure 7.6) so that the accuracy of approximation A.97 is sufficient.

The calculations presented in this appendix are valid only if viscous flow is negligible. This assumption still has to be proved to be justified. For that, viscous flow inside the catalyst pores for component A is examined; calculations for the other components are similar. Viscous flow is negligible if the number V_A is small enough, say smaller than 10%:

$$|V_A| < \frac{1}{10} \tag{A.104}$$

In the case for laminar Poiseuille flow in the catalyst pores, it can be shown [1] that

$$V_A = \frac{r_p^2}{8\eta_g} \times \frac{P}{D_{Ak}} \times \frac{\nabla P}{\nabla(\kappa_A P)} \tag{A.105}$$

where r_p is the pore radius and η_g the dynamic viscosity of the gas. For the Knudsen diffusion coefficient D_{Ak} we may write

$$D_{Ak} = \frac{2}{3} \times r_p \bar{v}_A = \frac{2}{3} \times r_p \sqrt{\frac{8RT}{\pi M_A}} \tag{A.106}$$

thus

$$r_p = \frac{3}{2} \times D_{Ak} \sqrt{\frac{\pi M_A}{8RT}} \tag{A.107}$$

If the average molar mass of the gas mixture equals , then for an ideal gas

$$<M> \times \frac{P}{RT} = \rho_g \tag{A.108}$$

Substitution of the Equations A.107 and A.108 into A.105 gives

$$V_A = \frac{9\pi}{256} \times \frac{M_A}{<M>} \times \frac{D_{Ak}}{\upsilon_g} \times \frac{\nabla P}{\nabla(\kappa_A P)} \tag{A.109}$$

According to Equation A.88, Equation A.109 can be expressed as

$$V_A = -\frac{9\pi}{256} \times \frac{M_A}{<M>} \times \left(\sqrt{\upsilon_P} - 1\right)\left(1 + \frac{1 + (\upsilon_P - 1)\kappa_A}{Kn^*} \times \frac{P}{P_S}\right)^{-1} \times \frac{D_{Ak}}{\upsilon_g} \tag{A.110}$$

Since

$$Kn^* = \frac{D_{AP}(P_S)}{D_{Ak}} \times \left[1 + \sum_{i=1}^{m} \left\{ \kappa_{D_i}\Big|_{\kappa_A = 0} \left(\frac{D_{AP}(P_S)}{D_{AD_i}(P_S)} - 1 \right) \right\} \right]^{-1} \tag{A.111}$$

Equation A.110 can be written as:

$$\begin{aligned} V_A = &-\frac{9\pi}{256} \times \frac{M_A}{<M>} \times \left(\sqrt{\upsilon_P} - 1\right) \times \left\{ Kn^* \times \frac{P_S}{P} + (\upsilon_P - 1)\kappa_A + 1 \right\}^{-1} \\ &\times \frac{P_S}{P} \times \frac{D_{AP}(P_S)}{\upsilon_g} \times \left[1 + \sum_{i=1}^{m} \left\{ \kappa_{D_i}\Big|_{\kappa_A = 0} \left(\frac{D_{AP}(P_S)}{D_{AD_i}(P_S)} - 1 \right) \right\} \right]^{-1} \end{aligned} \tag{A.112}$$

Since the kinematic viscosity ν_g is inversely proportional to the pressure, this can be written as

$$\begin{aligned} V_A = &-\frac{9\pi}{256} \times \frac{M_A}{<M>} \times \left(\sqrt{\upsilon_P} - 1\right) \times \left\{ Kn^* \times \frac{P_S}{P} + (\upsilon_P - 1)\kappa_A + 1 \right\}^{-1} \\ &\times \frac{D_{AP}(P_S)}{\upsilon_g(P_s)} \times \left[1 + \sum_{i=1}^{m} \left\{ \kappa_{D_i}\Big|_{\kappa_A = 0} \left(\frac{D_{AP}(P_S)}{D_{AD_i}(P_S)} - 1 \right) \right\} \right]^{-1} \end{aligned} \tag{A.113}$$

At this point it is convenient to introduce a modified Schmidt number

$$Sc^* = \frac{D_{AP}(P_S)}{\upsilon_g(P_s)} \times \left[1 + \sum_{i=1}^{m} \left\{ \kappa_{D_i}\Big|_{\kappa_A = 0} \left(\frac{D_{AP}(P_S)}{D_{AD_i}(P_S)} - 1 \right) \right\} \right]^{-1} \tag{A.114}$$

This can be seen as a Schmidt number generalized for multicomponent systems. Like the usual Schmidt number, its value will be close to one for gases and Sc* does not depend on the gas composition or pressure. Substitution of Equation A.114 in A.113 gives

$$V_A = -\frac{9\pi}{256} \times \frac{M_A}{<M>} \times \frac{\left(\sqrt{\upsilon_P}-1\right)Sc^*}{Kn^* \times \frac{P_S}{P} + \left(\upsilon_P - 1\right)\kappa_A + 1} \tag{A.115}$$

Hence, it is seen that the contribution of viscous flow is largest for $Kn^* = 0$, thus for very wide catalyst pores. For high values of Kn^*, viscous flow is overshadowed by Knudsen flow and V_A becomes zero. V_A will also become zero if the stoichiometric coefficient ν_P equals one (so no pressure gradient can build up) or if $\frac{M_A}{\langle M \rangle}$ or Sc^* equals zero (diffusion of the small molecules of component A is much faster than flow of the highly viscous gas mixture). For example, for hydrogenation reactions this often occurs for the hydrogen molecules, which are much faster moving than the other molecules. For a worst-case-analysis, assume $Kn^* = 0$:

$$V_A\Big|_{Kn^*\downarrow 0} = -\frac{9\pi}{256} \times \frac{M_A}{<M>} \times \frac{\left(\sqrt{\upsilon_P}-1\right)Sc^*}{\left(\upsilon_P - 1\right)\kappa_A + 1} \tag{A.116}$$

Furthermore, by substituting $\frac{9\pi}{256} \approx \frac{1}{10}$ and $Sc^* \approx 1$ (which will approximately be the case in practice) Equation A.116 becomes

$$V_A\Big|_{Kn^*\downarrow 0} = -\frac{1}{10} \times \frac{M_A}{<M>} \times \frac{\left(\sqrt{\upsilon_P}-1\right)}{\left(\upsilon_P - 1\right)\kappa_A + 1} \tag{A.117}$$

Substitution into Equation A.103 gives

$$\left| \frac{M_A}{<M>} \times \frac{\sqrt{\upsilon_P}-1}{1+\left(\upsilon_P - 1\right)\kappa_A} \right| < 1 \tag{A.118}$$

If this criterion is fulfilled, viscous flow can indeed be neglected. Figure A.3 shows a plot of ν_P versus κ_A for $\frac{M_A}{\langle M \rangle} = 1$. Within the shaded area viscous flow can be neglected with confidence. From the figure it is seen that viscous flow can be neglected in most practical cases, which is in agreement with published results.

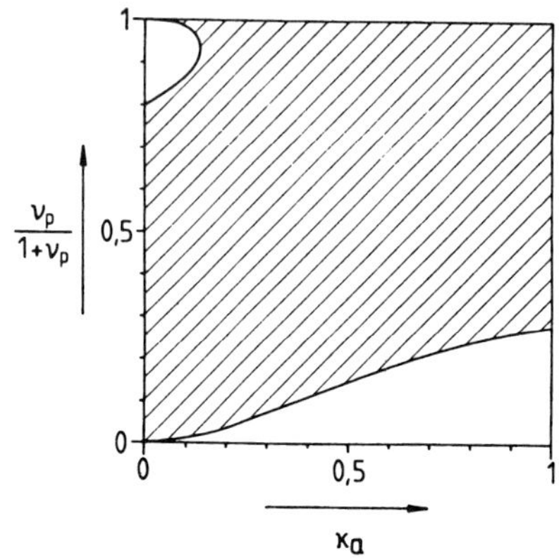

Figure A.3 Stoichiometric coefficient ν_p versus mole fraction κ_A for $M_A/\langle M \rangle = 1$. Within the shaded area viscous flow inside the catalyst pores may be neglected.

References

1. Mason, E.A., Malinauskas, E.A.P. (1983) Gas Transport in Porous Media: the Dusty-Gas Model, Amsterdam: Elsevier.

Appendix E

Effective Diffusion Coefficient as a Function of the Gas Composition for Nondiluted Gases and Simple Reactions

Consider a mixture of A, P and m inert components D_i ($i = 1, 2, \ldots, m$) in which A reacts to P according to:

$$A \rightarrow \upsilon_P P \tag{A.119}$$

Further, make the same assumptions as in Appendix D of no viscous flow, no surface diffusion, an isothermal catalyst pellet and no external forces. According to the dusty gas model a force balance for component A then yields:

$$\frac{\kappa_P J_A^* - \kappa_A J_P^*}{D_{AP}} + \sum_{i=1}^{m} \frac{\kappa_{D_i} J_A^* - \kappa_A J_{D_i}^*}{D_{AD_i}} + \frac{J_A^*}{D_{Ak}} = -\frac{\nabla(\kappa_A P)}{RT} = -\nabla C_A \tag{A.120}$$

where J_A* is the mole flux of A through the cross-sectional area of the catalyst pores. All the other symbols are as defined before. Since

$$J_P^* = -\upsilon_P J_A^* \tag{A.121}$$

and

$$J_{D_i}^* = 0 \tag{A.122}$$

we can write

$$J_A^* \left(\frac{\kappa_P + \upsilon_P \kappa_A}{D_{AP}} + \sum_{i=1}^{m} \frac{\kappa_{D_i}}{D_{AD_i}} + \frac{1}{D_{Ak}} \right) = -\nabla C_A \tag{A.123}$$

As discussed in Appendix D, κ_P and κ_{Di} can be expressed in κ_A via:

$$\kappa_P = 1-\kappa_A - \{1+(\upsilon_P-1)\kappa_A\}\sum_{i=1}^{m}\kappa_{D_i}\Big|_{\kappa_A=0} \tag{A.124}$$

$$\kappa_{D_i} = \{1+(\upsilon_P-1)\kappa_A\}\kappa_{D_i}\Big|_{\kappa_A=0} \qquad (i=1,2,\ldots,m) \tag{A.125}$$

Substitution of Equations A.124 and A.125 into A.123 yields

$$J_A^*\left\langle\frac{1}{D_{Ak}}+\frac{\{1+(\upsilon_P-1)\kappa_A\}}{D_{AP}}\times\left[1+\sum_{i=1}^{m}\left\{\kappa_{D_i}\Big|_{\kappa_A=0}\left(\frac{D_{AP}}{D_{AD_i}}-1\right)\right\}\right]\right\rangle=-\nabla C_A \tag{A.126}$$

Since the Maxwell diffusion coefficients are inversely proportional to the pressure

$$J_A^*\left\langle\frac{1}{D_{Ak}}+\frac{\{1+(\upsilon_P-1)\kappa_A\}}{D_{AP}(P_s)}\times\frac{P}{P_s}\times\left[1+\sum_{i=1}^{m}\left\{\kappa_{D_i}\Big|_{\kappa_A=0}\left(\frac{D_{AP}(P_s)}{D_{AD_i}(P_s)}-1\right)\right\}\right]\right\rangle=-\nabla C_A \tag{A.127}$$

Introduction of the modified Knudsen number

$$Kn^* = \frac{D_{AP}(P_s)}{D_{Ak}}\times\left[1+\sum_{i=1}^{m}\left\{\kappa_{D_i}\Big|_{\kappa_A=0}\left(\frac{D_{AP}(P_s)}{D_{AD_i}(P_s)}-1\right)\right\}\right]^{-1}$$

gives

$$J_A^*\times\frac{1}{D_{Ak}}\left\{1+\frac{1+(\upsilon_P-1)\kappa_A}{Kn^*}\times\frac{P}{P_S}\right\}=-\nabla C_A \tag{A.128}$$

By defining J_A as the mole flux of A in moles of A per second per square meter flowing through area of the catalyst pellet (Figure A.4), then

$$J_A = \frac{\varepsilon_P}{\gamma_P}J_A^* \tag{A.129}$$

where ε_p is the porosity of the catalyst pellet and γ_p the tortuosity of the catalyst pores. For example, if the porosity of the pellet is 25 %, the net flux through the cross-sectional area of the pellet is four times smaller than the flux through the cross-sectional area of the catalyst pores, because the cross-sectional area of the catalyst pellet is four times larger than the cross-sectional area of the pores. In addition, if the tortuosity of the catalyst pores is 2, the net distance covered inside the pellet is half the distance covered in-

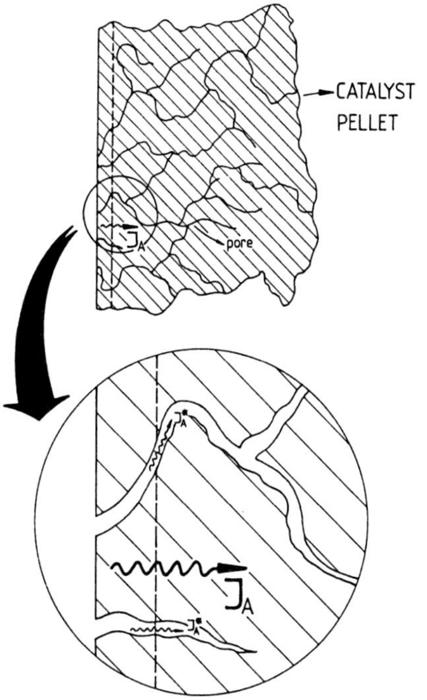

Figure A.4 Schematic drawing which illustrates the difference between the mole flux through the cross-sectional area of the pellet J_A and the mole flux through the cross-sectional area of the catalyst pores J_A*.

side the catalyst pores. This again decreases the net flux per cross-sectional pellet area twofold, so that in total J_A is eight times smaller than J_A*. Substitution of Equation A.128 into A.129 gives

$$J_A = -D_{Ak}\frac{\varepsilon_P}{\gamma_P}\left\{1+\frac{1+(\upsilon_P-1)\kappa_P}{Kn^*}\times\frac{P}{P_S}\right\}\times\nabla C_A \qquad \text{(A.130)}$$

For diluted gases this equation becomes

$$J_A = -D_{e,A}\times\nabla C_A \qquad \text{(A.131)}$$

Thus, the only effect of the gas being nondiluted is that the effective diffusion coefficient becomes a function of the gas composition and pressure. From Equations A.130 and A.131

$$D_{e,A} = D_{Ak}\frac{\varepsilon_P}{\gamma_P}\left\{1+\frac{1+(\upsilon_P-1)\kappa_A}{Kn^*}\times\frac{P}{P_S}\right\}^{-1} \qquad \text{(A.132)}$$

Appendix F

First Aris Number An_1 for Concentration-dependent Effective Diffusion Coefficients and Simple Reactions

If the effective diffusion coefficient depends on the concentration, the derivation for An_1 given in Appendix B does not hold. An expression for An_1, for concentration-dependent effective diffusion coefficients is therefore needed. The discussion is for arbitrary kinetics in an infinite slab; derivations for other catalyst geometries occur in a similar way.

If the effective diffusion coefficient is concentration dependent, the concentration profile inside an infinite slab is given by the following micro mass balance:

$$\frac{d}{dr}\left\{D_{e,A}(C_A)\frac{dC_A}{dr}\right\} = R(C_A) \tag{A.133}$$

Multiplying both sides by $D_{e,A}(C_A)\frac{dC_A}{dr}$ and integrating between $r = 0$ ($C_A = C_{A,m}$) and $r = r$ ($C_A = C_A$), we obtain:

$$\frac{dC_A}{dr} = \frac{\sqrt{2}}{D_{e,A}(C_A)}\sqrt{\int_{C_{A,m}}^{C_A} D_{e,A}(C_A)R(C_A)dC_A} \tag{A.134}$$

where we used the fact that $\frac{dC_A}{dr} = 0$ for $C_A = C_{A,m}$. For values of the effectiveness factor η close to one, both C_A and $C_{A,m}$ are close to $C_{A,s}$ and Equation A.134 can be linearized as follows:

$$\begin{aligned}\frac{1}{D_{e,A}(C_A)} &\approx \frac{1}{D_{e,A}(C_{A,s})}\left\{1 + \frac{C_A - C_{A,s}}{D_{e,A}(C_{A,s})}\left.\frac{\partial D_{e,A}(C_A)}{\partial C_A}\right|_{C_A=C_{A,s}}\right\}^{-1} \\ &\approx \frac{1}{D_{e,A}(C_{A,s})}\left\{1 - \frac{C_A - C_{A,s}}{D_{e,A}(C_{A,s})}\left.\frac{\partial D_{e,A}(C_A)}{\partial C_A}\right|_{C_A=C_{A,s}}\right\}\end{aligned} \tag{A.135}$$

$$\times \sqrt{\int_{C_{A,m}}^{C_A} D_{e,A}(C_A)\,R(C_A)\,dC_A}$$

$$\approx \left\langle \int_{C_{A,m}}^{C_A} \left[D_{e,A}(C_{A,s})\,R(C_{A,s}) + (C_A - C_{A,s}) \frac{\partial\{D_{e,A}(C_A)\,R(C_A)\}}{\partial C_A}\bigg|_{C_A=C_{A,s}} \right] dC_A \right\rangle^{\frac{1}{2}}$$

$$\approx \left[(C_A - C_{A,m})\,D_{e,A}(C_{A,s})\,R(C_{A,s}) + (C_A - C_{A,m})\left(\frac{1}{2}C_A + \frac{1}{2}C_{A,m} - C_{A,s}\right) \frac{\partial\{D_{e,A}(C_A)\,R(C_A)\}}{\partial C_A}\bigg|_{C_A=C_{A,s}} \right]^{\frac{1}{2}}$$

$$\approx \sqrt{C_A - C_{A,m}} \times \sqrt{D_{e,A}(C_{A,s})R(C_{A,s})} \times \left[1 + \frac{\frac{1}{4}C_A + \frac{1}{4}C_{A,m} - \frac{1}{2}C_{A,s}}{D_{e,A}(C_A)\,R(C_A)} \frac{\partial\{D_{e,A}(C_A)\,R(C_A)\}}{\partial C_A}\bigg|_{C_A=C_{A,s}} \right]$$

(A.136)

Substitution of Equations A.135 and A.136 into A134 yields

$$\frac{dC_A}{dr} \approx \sqrt{2} \times \sqrt{\frac{R(C_{A,s})}{D_{e,A}(C_{A,s})}} \times \sqrt{C_A - C_{A,m}}$$
$$\times \left\{ 1 + \frac{\frac{1}{4}C_A + \frac{1}{4}C_{A,m} - \frac{1}{2}C_{A,s}}{R(C_{A,s})} \frac{\partial R(C_A)}{\partial C_A}\bigg|_{C_A=C_{A,s}} + \frac{\frac{1}{2}C_{A,s} + \frac{1}{4}C_{A,m} - \frac{3}{4}C_A}{D_{e,A}(C_{A,s})} \frac{\partial D_{e,A}(C_A)}{\partial C_A}\bigg|_{C_A=C_{A,s}} \right\}$$

(A.137)

Integration of Equation A.137 between $r = 0$, equivalent to $C_A = C_{A,m}$, and $r = r$, equivalent to $C_A = C_A$, gives

$$r = Z_1 \int_{C_{A,m}}^{C_A} \frac{1 + \{C_{A,s} - C_{A,m}\}\left\{\frac{Z_2}{2} - \frac{Z_3}{2}\right\} - \{C_A - C_{A,m}\}\left\{\frac{Z_2}{4} - \frac{3Z_3}{4}\right\}}{\sqrt{C_A - C_{A,m}}} dC_A$$
$$= Z_1 \times 2\sqrt{C_A - C_{A,m}} \left[1 + \{C_{A,s} - C_{A,m}\}\left\{\frac{Z_2}{2} - \frac{Z_3}{2}\right\} - \frac{\{C_A - C_{A,m}\}}{3}\left\{\frac{Z_2}{4} - \frac{3Z_3}{4}\right\} \right]$$
$$= Z_1 \times 2\sqrt{C_A - C_{A,m}} \left[1 + \{C_{A,s} - C_{A,m}\}\left\{\frac{Z_2}{2} - \frac{Z_3}{2}\right\} - \{C_A - C_{A,m}\}\left\{\frac{Z_2}{12} - \frac{Z_3}{4}\right\} \right]$$

(A.138)

where Z_1, Z_2 and Z_3 are defined for convenience as

$$Z_1 = \sqrt{\frac{D_{e,A}(C_{A,s})}{2\,R(C_{A,s})}} \tag{A.139}$$

$$Z_2 = \frac{1}{R(C_{A,s})}\left.\frac{\partial R(C_A)}{\partial C_A}\right|_{C_A=C_{A,s}} \tag{A.140}$$

$$Z_3 = \frac{1}{D_{e,A}(C_{A,s})}\left.\frac{\partial D_{e,A}(C_A)}{\partial C_A}\right|_{C_A=C_{A,s}} \tag{A.141}$$

Substitution of $r = R$ and $C_A = C_{A,s}$ gives us:

$$R = Z_1 \times 2\sqrt{C_A - C_{A,m}}\left[1 + \{C_{A,s} - C_{A,m}\}\left\{\frac{5Z_2}{12} - \frac{Z_3}{4}\right\}\right] \tag{A.142}$$

The second-order approximation is

$$\sqrt{2(C_{A,s} - C_{A,m})} \approx R\sqrt{\frac{R(C_{A,s})}{D_{e,A}(C_{A,s})}} \times \left\{1 - (C_{A,s} - C_{A,m}) \times \left(\frac{5\,Z_2}{12} - \frac{Z_3}{4}\right)\right\} \tag{A.143}$$

From Equation A.137 the gradient on the outer surface of the slab can be calculated by substituting $C_A = C_{A,s}$. Combination with Equation A.143 then yields

$$\begin{aligned}
\left.\frac{dC_A}{dr}\right|_{r=R} &= \frac{\sqrt{C_{A,s} - C_{A,m}}}{Z_1}\left[1 - \{C_{A,s} - C_{A,m}\}\left\{\frac{Z_2}{4} + \frac{Z_3}{4}\right\}\right] \\
&= R \times \frac{R(C_{A,s})}{D_{e,A}(C_{A,s})} \times \left[1 + \{C_{A,s} - C_{A,m}\}\left\{\left(-\frac{1}{4} - \frac{5}{12}\right)Z_2 + \left(-\frac{1}{4} + \frac{1}{4}\right)Z_3\right\}\right. \\
&= R \times \frac{R(C_{A,s})}{D_{e,A}(C_{A,s})} \times \left\{1 - (C_{A,s} - C_{A,m}) \times \frac{2}{3} Z_2\right\} \\
&= R \times \frac{R(C_{A,s})}{D_{e,A}(C_{A,s})} \times \left\{1 - (C_{A,s} - C_{A,m}) \times \frac{2/3}{R(C_{A,s})}\left.\frac{\partial R(C_{A,s})}{\partial C_A}\right|_{C_A=C_{A,s}}\right\}
\end{aligned} \tag{A.144}$$

According to Equation A.143 in a first order approximation

$$C_{A,s} - C_{A,m} \approx \frac{R^2}{2}\frac{R(C_{A,s})}{D_{e,A}(C_{A,s})} \tag{A.145}$$

Substitution of Equation A.145 into A.144 yields

$$\left.\frac{dC_A}{dr}\right|_{r=R} \approx R \times \frac{R(C_{A,s})}{D_{e,A}(C_{A,s})} \times \left(1 - \frac{R^2}{3\,D_{e,A}(C_{A,s})} \left.\frac{\partial R(C_A)}{\partial C_A}\right|_{C_A=C_{A,s}}\right) \tag{A.146}$$

Since, for an infinite slab,

$$R = \frac{V_p}{A_p} \tag{A.147}$$

Equation A.146 can be written as

$$\left.\frac{dC_A}{dr}\right|_{r=R} \approx \frac{V_P}{A_P} \times \frac{R(C_{A,s})}{D_{e,A}(C_{A,s})} \times \left\{1 - \frac{1}{3} \times \left(\frac{V_P}{A_P}\right)^2 \times \frac{1}{D_{e,A}(C_{A,s})} \times \left.\frac{\partial R(C_A)}{\partial C_A}\right|_{C_A=C_{A,s}}\right. \tag{A.148}$$

The effectiveness factor can be calculated from

$$\eta \approx \frac{A_P\,D_{e,A}(C_{A,s})}{V_P\,R(C_{A,s})} \times \left.\frac{dC_A}{dr}\right|_{r=R} \approx 1 - \frac{1}{3} \times \left(\frac{V_P}{A_P}\right)^2 \times \frac{1}{D_{e,A}(C_{A,s})} \times \left.\frac{\partial R(C_A)}{\partial C_A}\right|_{C_A=C_{A,s}} \tag{A.149}$$

Then for values of η close to one,

$$1 - \eta^2 \approx \frac{2}{3} \times \left(\frac{V_P}{A_P}\right)^2 \frac{1}{D_{e,A}(C_{A,s})} \left.\frac{\partial R(C_A)}{\partial C_A}\right|_{C_A=C_{A,s}} \tag{A.150}$$

Hence, by definition, the first Aris number An_1 is given by

$$An_1 \approx \frac{2}{3} \times \left(\frac{V_P}{A_P}\right)^2 \frac{1}{D_{e,A}(C_{A,s})} \left.\frac{\partial R(C_A)}{\partial C_A}\right|_{C_A=C_{A,s}} \tag{A.151}$$

Or, since for an infinite slab the geometry factor Γ equals , we can write:

$$An_1 \approx \left(\frac{V_P}{A_P}\right)^2 \times \frac{\Gamma}{D_{e,A}(C_{A,s})} \left.\frac{\partial R(C_A)}{\partial C_A}\right|_{C_A=C_{A,s}} \tag{A.152}$$

which proves Equation 7.82 for arbitrary kinetics in an infinite slab. This formula for An_1 is in agreement with the formula for An_1 derived in Appendix B, despite the two entirely different routes used for their solution.

Subject Index

Author Index